Introduction to Sedimentology

To
The Memory of
My Mother

Introduction to Sedimentology

Second Edition

Supriya Mohan Sengupta D.Sc., F.N.A.
Ex-Professor & Head, Deptt. of Geology & Geophysics,
Indian Institute of Technology, Kharagpur

CBS

CBS Publishers & Distributors Pvt Ltd

New Delhi • Bengaluru • Chennai • Kochi • Kolkata • Lucknow • Mumbai
Hyderabad • Jharkhand • Nagpur • Patna • Pune • Uttarakhand

Introduction to Sedimentology

Second Edition

Copyright © Publisher

ISBN: 978-81-239-1491-6

Second Edition: 2007

Reprint: 2008, 2010, 2011, 2012, 2013, 2014, 2015, 2018, 2020, 2022

Published by Satish Kumar Jain and produced by Varun Jain for

CBS Publishers & Distributors Pvt Ltd

4819/XI Prahlad Street, 24 Ansari Road, Daryaganj, New Delhi 110 002, India

Ph: 011-23289259, 23266861, 23266867   Website: www.cbspd.com

Fax: 011-23243014   e-mail: delhi@cbspd.com;
cbspubs@airtelmail.in.

Corporate Office: 204 FIE, Industrial Area, Patparganj, Delhi 110 092, India

Ph: 011-4934 4934   Fax: 011-4934 4935   e-mail: publishing@cbspd.com;
publicity@cbspd.com

Branches

- **Bengaluru:** Seema House 2975, 17th Cross, KR Road, Banasankari 2nd Stage, Bengaluru 560 070, Karnataka, India
  Ph: +91-80-26771678/79   Fax: +91-80-26771680   e-mail: bangalore@cbspd.com
- **Chennai:** 7, Subbaraya Street, Shenoy Nagar, Chennai 600 030, Tamil Nadu, India
  Ph: +91-44-26680620, 26681266   Fax: +91-44-42032115   e-mail: chennai@cbspd.com
- **Kochi:** 42/1325, 1326, Power House Road, Opp KSEB, Power House, Ernakulam Kochi 682 018, Kerala, India
  Ph: +91-484-4059061-65,67   Fax: +91-484-4059065   e-mail: kochi@cbspd.com
- **Kolkata:** 147, Hind Ceramics Compound, 1st Floor, Nilgunj Road, Belghoria, Kolkata-700056, West Bengal, India
  Ph: +91-9096713055/7798394118, 9836841399   e-mail: kolkata@cbspd.com
- **Lucknow:** Basement, Khushnuma Complex, 7 Meerabai Marg (Behind Jawahar Bhawan),Lucknow-226001, UP, India
  Ph: +0522-4000032   e-mail: tiwari.lucknow@cbspd.com
- **Mumbai:** PWD Shed, Gala no 25/26, Ramchandra Bhatt Marg, Next to JJ Hospital Gate no. 2, Opp. Union
  Bank of India, Noorbaug, Mumbai-400009, Maharashtra, India
  Ph: 022-66661880/89   e-mail: mumbai@cbspd.com

Representatives

- Hyderabad   0-9885175004
- Patna   0-9334159340
- Jharkhand   0-9811541605
- Pune   0-9623451994
- Nagpur   0-9421945513
- Uttarakhand   0-9716462459

Printed at India Binding House, Noida, UP, India

PREFACE TO SECOND EDITION

It is more than ten years since the first edition of *Introduction to Sedimentology* appeared in print. It is gratifying to learn that students found the book useful. In the second edition a new chapter on Structures of Chemical and Biological Origin has been added. The discussion on carbonate sediments on shelf has been expanded and most of the other chapters have been revised and updated.

Prof. Asru Kumar Chaudhuri (Geological Studies Unit, Indian Statistical Institute, Kolkata) provided useful suggestions for the newly added chapter on Structures of Chemical and Biological Origin. Dr. Harendra Nath Bhattacharyay and Dr. Prabir Dasgupta (Department of Geology, Presidency College, Kolkata), Dr. Sarbani Patranabis Deb (Geological Studies Unit, Indian Statistical Institute, Kolkata) provided photographs for my use. Dr. Siddhartha Sankar Das (Department of Earth and Planetary Sciences, University of Allahabad), and Dr. Rajat Mazumder (Department of Geology, Asutosh College, Kolkata) provided useful suggestions. Dr. Saumitra Misra (IIT, Kharagpur) assisted in copy editing and proof reading. Mr. Chinmoy Mukherjee drew the revised and newly added text figures. Some well-known publishers permitted me to reproduce illustrations from their publications. Their names appear in appropriate places in the text.

Department of Ocean Engineering & Naval Architecture, IIT, Kharagpur provided facilities for this work. Indian National Science Academy, New Delhi, through the INSA Honorary Scientist Scheme, defrayed the cost of preparation of the manuscript. M/s CBS Publishers & Distributors, New Delhi took active interest in this publication. I am grateful to all the individuals and institutions mentioned above.

P-45/B, Hijli Cooperative Development
Society
Kharapgur (W.B.) 721 306

Supriya Mohan Sengupta

PREFACE TO FIRST EDITION

Sedimentology covers only a small part of the geology curriculum in most universities. Students of geology naturally do not find the time necessary to master comprehensive texts in sedimentology. While gathering piecemeal information from specialized texts they also often fail to appreciate the connections between the various aspects of the discipline.

The purpose of this book is to present a concise account of all the major branches of sedimentology (except chemical and biogenic structures) and to highlight the connecting links between them. The presentation is based on my conviction that in spite of the sophistications introduced in the current sedimentological literature, much of the subject can still be learnt by mastering a few basic principles. Once these have been grasped, it should not be difficult, following the references cited, to obtain specialized knowledge on any selected topic.

This book is expected to cater to the needs of students at various levels. The first four chapters are written in a style suitable for the undergraduate students. The first few sections of Chapters 7 (Tectonics and Sedimentation) and 8 (Stratigraphy and Sedimentation) should also be preferably taught at this level. The rest of the book is meant for those at the graduate level.

A book of this type could not have been prepared without borrowing heavily from published sources. I am grateful to the authors and publishers who have freely permitted me to reproduce information from their publications. Their names appear in appropriate places in the text. Dr. Indranil Banerjee of the Institute of Sedimentary and Petroleum Geology, Calgary, Canada, critically read an earlier version of the manuscript, provided valuable suggestions, and also made available to me a large volume of current sedimentological literature. I am indebted to him for all his help. I have also profited through discussions with Professor K. Naha of the Indian Institute of Technology (IIT), Kharagpur who reviewed some of the chapters.

Others who provided information during preparation of the manuscript and read earlier versions of the topics mentioned against their names include: Dr. Sibdas Bandyopadhyay (Indian Statistical Institute, Calcutta; Statistical Analysis), Dr. Abhijit Basu (Indiana University, Bloomington, USA; Provenance), Prof. Ajit K. Bhattacharyya (Jadavpur University, Calcutta; Depositional Environments), Prof. Amitabha Chakrabarti (IIT, Kharagpur; Sedimentary Structures, Tidal Flats), Prof. R. L. Folk (University of Texas, Austin, USA; Carbonate Sediments), Prof. G.M. Friedman

(North-eastern Science Foundation, Inc., USA; Sedimentary Textures). Dr. Sibdas Ghosh (Geological Survey of India, Calcutta), Prof. Mandakini Majumdar (IIT, Kharagpur; Hydraulics), Prof. G.V. Middleton (McMaster University, Hamilton, Canada; History of Sedimentology), Prof. Asoke Mookherjee (IIT, Kharagpur), Prof. H. Okada (Kyushu University, Japan; Siliciclastic Sedimentary Rocks), Prof. H. G. Reading (University of Oxford, Oxford, U.K.), Prof. K. K. Roy (IIT, Kharagpur; Wireline Logging) Prof. S. K. Sen (IIT, Kharagpur), Dr. R. N. Singh and Dr. R. Srinivasan (National Geophysical Research Institute, Hyderabad; Tectonics and Sedimentation) Prof. Å. Sundborg (Uppsala University, Uppsala, Sweden; Hydraulics and Sediment Transportation). Profs. A Chakrabarti, C. N. Rao and D. P. Sen of IIT, Kharagpur, and Dr. D. K. Saha of ISI, Calcutta allowed me to use thin sections and photographs from their unpublished collections. Dr. B. Mishra helped in photomicrography. Drs. Abha Chatterjee and B. P. Sandilya provided translations for the French and German terms respectively. I am grateful to all those mentioned above.

The contents of this book are based on the courses given by me at various levels at IIT, Kharagpur. I recall with appreciation the comments and suggestions received from my students from time to time. Some of my M. Tech. Students, particularly, Asesh Kumar Maji assisted in copy editing, indexing and computational work at different stages. My wife Ila (Sujata) took care of my mundane problems thereby leaving me free to work on the manuscript.

A large part of the cost of preparation of the manuscript was defrayed by the Curriculum Development Cell of IIT, Kharagpur under the Quality Improvement Programme (QIP). It is a pleasure to acknowledge the help and co-operation received from Prof. G. L. Datta and the QIP staff at various stages of preparation of the manuscript. Mr. Subrata Giri typed the final version of the manuscript. Mr. Tapan Sarkar drew most of the text figures. Mr. S. Ghosh provided photocopies of the halftone illustrations. But for the persuasion and interest of the staff of M/s Oxford & IBH Publishing Co. Pvt. Ltd., this work would not have been undertaken. I thank all of them.

Department of Geology &   **Supriya Sengupta**
Geophysics
IIT, Kharagpur

CONTENTS

ix

INTRODUCTION

SCOPE OF SEDIMENTOLOGY

Sedimentology deals with the processes and products of sedimentation. Sediments are produced either by disintegration and alteration of pre-existing rocks or by precipitation from solution. The particles ejected out of volcanoes and the dust particles of cosmic origin also add to the sediment mass of the earth in a limited way. Sediments are transported by running water, wind or moving ice to various depositional environments. The process of sediment transportation is often accompanied by the production of rhythmic bed forms or more complex structures which, when well preserved, provide clues to the palaeoenvironment and palaeocurrent.

Sediments produced out of mechanical or chemical processes are consolidated into sedimentary rocks by the pressure of overburden, recrystallization and cementation. Textures of sedimentary rocks bear the imprint of the nature of the changes (diagenesis) undergone during consolidation of sediments into sedimentary rocks. The whole process of rock decay, sediment transportation, deposition, precipitation and diagenesis takes place at or near the surface of the earth at normal pressure-temperature conditions (Fig. 1.1). This distinguishes the sedimentary processes from the igneous and metamorphic processes where a higher order of temperature and pressure is involved. Sedimentary rocks cover wide areas of the earth's surface although they account for only about 5% of the crustal volume. Thus they form only a thin veneer on the outermost part of the earth's surface. The average thickness of the sediment cover on the continental crust is about 2 km but in the ocean basin only about 1 km.

Sediments laid down in layers within depressions (basins) in the earth's crust are preserved to constitute a stratigraphic record. The ultimate aim of a sedimentological study is to unravel the chain of events responsible for production of particular stratigraphic sequences.

HISTORY OF DEVELOPMENT

The Early History

The pioneers responsible for laying the foundation of sedimentology were not geologists by profession. The basic principles of stratigraphy and sedimentation developed out of studies by naturalists belonging to different disciplines. In the mid-17th century, a Danish physician-cum-clergyman, Nicolaus Steno (1638 - 1687), noticed that layers of sediments are always laid down in water in a sequence in which the oldest one lies at the bottom and the youngest one on top. This observation led to the formulation of a very fundamental rule known as the *Law of Superposition*. Steno's second law, called the *Law of Original Horizontality*, states that the primary bedding of the layers of sediments laid down in water always parallels the surface of the earth. His third law, the *Law of Original Continuity*, states that all water-laid strata, must continue laterally. A corollary to this law, which came to be recognized by the eighteenth century, states that a truncation of an original sedimentary layer implies removal of the original sediments, either by erosion or due to faulting.

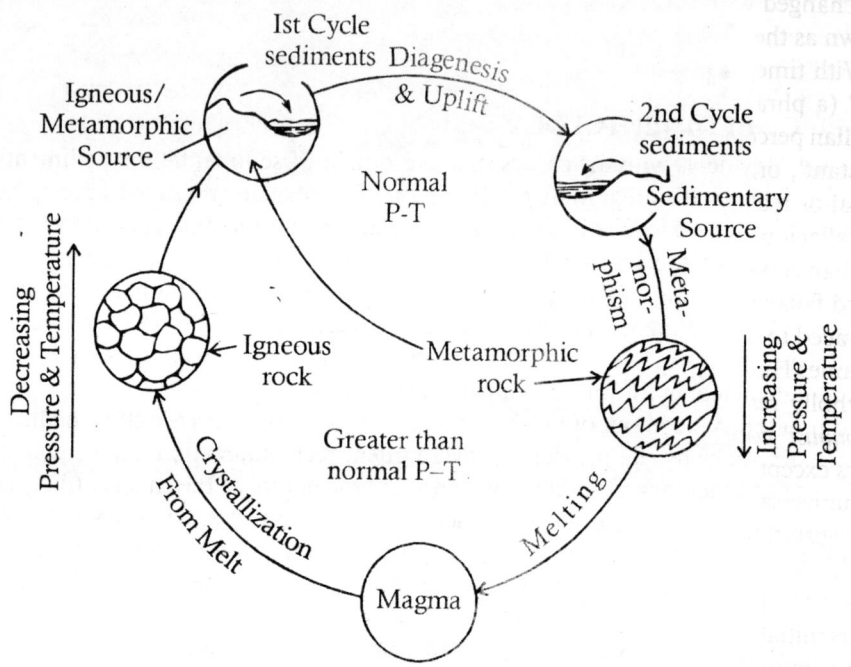

Fig. 1.1: Modern version of the fundamental rock cycle conceived by James Hutton nearly 200 years ago (based on Siever 1983).

These laws, which for centuries provided the basis for geological mapping and stratigraphic correlation, no doubt have their limitations. For example, ripple migration on a sediment bed may cause the original layers to be deposited at an angle to the horizontal plane. Similarly, variations in depositional condition at the time of sedimentation (facies change) can interrupt the lateral continuity of sediment layers.

Surprisingly, Steno's fundamental laws did not find practical application for nearly a century. Johann Gottlob Lehmann is believed to be one of the first to have applied the Law of Superposition to large-scale geological mapping in parts of Germany in the mid-18th Century. In 1815, a British civil engineer, William Smith (1769-1839), produced the first stratigraphic section from the records maintained at the construction site. His field studies also led to the production of the first geological map of Britain.

In 1785, James Hutton (1726-1797), a Scottish physician, recognized the cycle of weathering, erosion and transportation by running water (see Fig. 1.1). He also observed that after consolidation the sediments laid down in the sea produce stratified deposits. According to many, Hutton's appreciation of the immensity of geological time and his rejection of supernaturalism in providing explanations for geological phenomena marked the beginning of rational thinking in geology. Hutton's ideas were publicised during his lifetime by his friend, John Playfair, and nearly half a century later, by Charles Lyell (1797 - 1875). Lyell, who expanded and modified the Huttonian thesis, was quick to recognise that the changes that have taken place in the geological past can be explained by the processes that are taking place today. In other words, the laws of nature guiding the earth processes have remained invariant, although the earth itself has changed with time. This concept of uniformity in the earth processes came to be known as the *Principle of Uniformitarianism*.

With time, many corollaries of uniformitarianism like 'the present is a key to the past' (a phrase coined by Geikie in 1882), developed from the original Huttonian-Lyellian percept. Some of these, such as 'the rates of the earth processes have remained constant', or that uniformitarianism should be called 'actualism' because it refers to actual or real events, have been challenged by later workers on the ground that they are fallacious. In spite of these criticisms it must be remembered that the Huttonian-Lyellian concept has been the basis for most of our geological thinking. Many of the so called fallacious concepts developed from the beliefs of later workers and can hardly be traced to the original writings of James Hutton or Charles Lyell.

James Hutton was dedicated to the '*volcanist*' idea that the internal heat of the earth is wholly responsible for the earth processes. This was in marked contrast to the '*Neptunist*' concept, patronized by Abraham Gottlob Werner (1749 - 1817), that all rocks except recent volcanic lavas are produced from chemical or mechanical deposits in a universal ocean. So profound was the influence of the latter school on contemporary geologists that the idea of lateral change in depositional events (*facies change*), introduced by Gressly in 1833, did not find easy acceptance either in Europe or in North America for a long time. Geologists were still under the spell of the Neptunist belief that all rock layers initially laid down in a universal ocean, must also be continuous, without change.

An important event in mid-19th century was the observation by James Hall in the Appalachian mountains of North America that the thickest sediments accumulate in long, linear belts of subsidence in the earth's crust. This observation eventually led to the formulation of the concept of *geosyncline*, a model that dominated the sedimentological world for nearly a century.

Developments Since Late 19th Century

The late nineteenth and early twentieth centuries witnessed rapid developments. Studies by pioneers such as Henry Clifton Sorby (1826 - 1908) in England led to the foundation of sedimentology as we know it today. Sorby was not only interested in the study of sedimentary structures and their implications, but was also one of the first to make extensive use of the petrographic microscope in the study of sedimentary rocks. He produced two classical volumes on sedimentary petrography and also a monograph on cheritified limestones. A few years later, another important study on lithogenesis, by Johannes Walther, appeared in Germany. Walther also formulated the *Law of Facies Succession*, which still bears his name (see Chapter 9).

The first half of the twentieth century witnessed the initiation of specialised studies in many different branches of sedimentology. Systematic studies on the processes of sedimentation and their bearing on stratigraphy were initiated in North America by Grabau, Barrell and Schuchert. Studies by W.M. Davis (1850 - 1934) and G.K. Gilbert (1843 - 1918) and later by Twenhofel (1875 - 1957), established a close link between surface processes of the earth, sedimentation and stratigraphy. This marked the beginning of the use of the present as a key to the past. In 1944 sedimentologists' interests in modern sediments were revived with the publication of studies on the Mississippi delta by H.N. Fisk and his associates. Classical studies on modern eolian deposits (Bagnold 1941; McKee 1979), modern fluvial deposits (Hjulström 1935; Sundborg 1956; Leopold, Wolman and Miller 1964), and modern tidal sediments (Reineck since 1952, see Reineck and Singh, 1980; van Straaten 1954, 1954a; Ginsberg 1975) have been emulated by sedimentologists throughout the world for nearly half a century now (see Chapter 7).

Petrographic studies initiated by Sorby in England and Walther in Germany were continued, among others, by Grabau, Krynine and their students in North America (Chapter 3). The basic ideas on textural parameters of sediments were formulated during the first three decades of the present century (Udden 1914; Wentworth 1922; Wadell 1932, 1935; Zingg 1935; see Chapter 4). This was also the time when many of the procedures for sedimentological analyses were standardised.

The trend of laboratory simulation of sedimentary processes, initiated in France by Daubrée around 1870, was revived by G.K. Gilbert (1914) in North America. Since that time many of the hydraulic laboratories engaged in experimental studies on sedimentary processes have contributed greatly to our understanding of the basic processes involved in sedimentation. That such studies can be of importance to the perception of many fundamental geological problems was demonstrated in the Netherlands by Kuenen and Migliorini (1950). This was followed by many more contributions to experimental sedimentology throughout the world (see Chapter 5).

Vant Hoff's experiments on marine carbonates in the early years of this century and Correns' experimental studies around 1920 in Germany, marked the beginning of sedimentary geochemistry. This field of study is presently actively pursued, particularly in North America (see Krumbein and Garrels 1952; Garrels and MacKenzie 1971).

A major event of the present century was the revival of interest in palaeocurrent research, initiated in the nineteenth century in England by H.C. Sorby. Palaeocurrent studies by Brinkmann and also by Cloos in Germany around 1930 were rejuvenated in North America (see Pettijohn 1962), to be followed by development of many sophisticated techniques in palaeocurrent research (see Chapter 10).

The studies on tectonics and sedimentation, initiated in North America by James Hall around 1850, were pursued by Shvetsov in Russia, Bertrand in France and many others throughout the continent. Oceanographic expeditions, mainly by the Dutch sedimentologists and geophysicists in the East Indies provided valuable input for refinement of the geosynclinal model - ideas which were pursued and elaborated in North America (see Kay 1951). These ideas, together with the classical facies concept, which found acceptance in North America around 1930 and the petrographic studies initiated by Krynine 1948, provided the background for much of what was achieved during the first half of the twentieth century.

Recent Trends in Sedimentology

Developments until about the first half of the 20th century were mainly the outcome of individual efforts. The end of the Second World War saw a major change in this trend. Large-scale oceanographic expeditions by the major nations in the post-war period, development and deployment of submersibles, deep-toe equipment, side-scan sonar and undersea TV camera in subsea exploration, large-scale drilling operations at sea and refinement of core recovery techniques since 1950, have provided information, which was heretofore unobtainable. Much of what is known today about the silici-clastic and carbonate deposits of shallow seas has been revealed by these explorations.

Discovery of giant petroleum reserves in carbonate rocks provided impetus for the study of carbonate petrography and carbonate depositional environments. Intensive research on carbonates, initiated by the major oil companies, led to the discovery of what is known today about carbonate petrography and facies. An equally intensive investigation of the present day carbonate depositional environments in the Bahamas, Florida and the Persian Gulf also took place.

Exploration for petroleum opened up new dimensions in organic geochemistry also. Techniques were developed for working out the maturation and thermal history of sedimentary basins. Deployment of sophisticated equipment such as X-ray diffraction (XRD), X-ray fluorescence (XRF), scanning electron microscope (SEM), electron probe micro analysis (EPMA), cathodo-luminescence (CL) made investigation of fine-grained sediments and sedimentary rocks with a high degree of precision possible. Isotope studies (both stable and radio isotopes) provided clues to environmental interpretation. Application of mathematical and statistical techniques led to the investigation of subtle interdependence between lithological units and chemical constituents within stratigraphic sequences.

The ideas developed during the last few decades have been revolutionary in many ways. Information available from hundreds of thousands of boreholes drilled in search of hydrocarbon has diverted the attention of sedimentologists to the subsurface. Techniques have been developed for recognition of sedimentary facies from well logs, cores and well cuttings. Developments in the field of trace fossil study (*ichnology*) now allow interpretation of depositional environment from information which, until recently, was considered insufficient (see Chapter 6). The current emphasis is on facies modeling from both surface and subsurface data. Developments in the field of seismic exploration have introduced concepts such as *seismic and sequence stratigraphy* (see Chapter 9). New concepts such as *allostratigraphy and autostratigraphy* have also been introduced. Simultaneously, the traditional ideas on basin development are being modified from the point of view of plate tectonics.

Source

Albritton 1963 ; Dunbar and Rodgers 1957 ; Gilluly, Waters and Woodford 1975; Hubbert 1967; Mather and Mason 1939; McIntyre 1963; Miall 1978; Middleton 1973, 1978; Nelson 1985; Shea 1982; Zenger 1986.

LITERATURE ON SEDIMENTOLOGY

In the early part of the 20th century there were hardly any text and reference books on sedimentology available. Since 1950 the figure has been increasing almost exponentially.

Table 1.1 lists the more important text and reference books on sedimentology under three heads: 1) those dealing with general aspects of sedimentary processes and products, 2) those dealing with specialised sedimentological topics, and 3) those dealing with methodology for sedimentological research. All the texts mentioned in this table are written in English. Needless to say, many other well-written texts on sedimentology exist, particularly in the various European languages.

Table 1.1: Important text and reference works on sedimentology

Texts on specialized sedimentological topics	Texts on sedimentological techniques	Texts on general aspects of sedimentary processes and products	
			1910
		Hatch & Rastall (1913)	
			1920
	Milner (1922)	Twenhofel (1928)	
			1930
		Twenhofel (1932)	
	Krumbein & Pettijohn (1938)		1940
		Shrock (1948)	
		Pettijohn, 1st ed. (1949)	
Kuenen (1950)		Twenhofel (1950)	1950
		Krumbein & Sloss, 1st ed. (1951)	
		Dunbar & Rodgers (1957)	
		Pettijohn, 2nd ed. (1957)	
Carozzi (1960)			1960
	Milner, 2 vols. (1962)		
Potter & Pettijohn, 1st ed. (1963)		Krumbein & Sloss, 2nd ed. (1963)	
	Griffiths (1967)		
		Folk (1968)	
	Bouma (1969)		
Kukal (1970)			1970
Garrels & MacKenzie (1971)	Carver (1971)	Blatt, Middletion & Murray (1972)	
Pettijohn, Potter & Siever (1973),		Reineck & Singh 1st ed. (1973)	
Milliman (1974)		Pettijohn, 3rd ed. (1975)	
Bathurst (1975), Wilson (1975)		Selley (1976)	
Potter & Pettijohn, 2nd ed. (1977)		Friedman & Sanders (1978)	
Reading, 1st ed. (1978)		Reineck & Singh 2nd ed. (1980)	1980
Walker, 1st ed. (1978)			
Miall (1984)		Galloway & Hobday (1983)	
Walker, 2nd ed. (1984)			
Reading, 2nd ed. (1986)	Lindholm (1987)		
Shannon & Naylor (1989)			
Miall, 2nd ed. (1989)			1990
Tucker et al. (1990)		McLane (1995)	
Syvitski (1991)		Leeder (1999)	
Reading, 3rd ed. 1996		Nichols (1999)	

The results of current research on sedimentological topics are available in periodicals. Of these, the following three, published in the English language, are devoted wholly to sedimentology : Journal of Sedimentary Petrology (now called Journal of Sedimentary Research) published since 1931 by the Society of Economic Paleontologists and Mineralogists (renamed in 1989 as the SEPM Society for Sedimentary Geology), Tulsa, Oklahoma, USA ; Sedimentology, the Journal of the International Association of Sedimentologists (I.A.S.), published since 1962, and Sedimentary Geology (published since 1967 by Elsevier Publishing Company, Amsterdam). These three journals together publish more than 200 large research papers on sedimentological topics annually.

Sedimentological research papers are also regularly published by many other geological journals. Notable among them, in the English language, are Bulletin of the American Association of Petroleum Geologists (A.A.P.G.), Bulletin of the Geological Society of America, Marine Geology (Elsevier, Amsterdam), Journal of Geology (Chicago), Geological Journal (Wiley), Lithos (Oslo), Journal of Petroleum Science and Engineering (Elsevier), Palaeogeography, Palaeoclimatology, Palaeoecology (Elsevier), Basin Research (Blackwell), Canadian Journal of Earth Science (Ottawa) and Geology Today (Blackwell). Several memoirs, special publications, short course notes, reprint series of A.A.P.G., S.E.P.M., and I.A.S. and the Developments in Sedimentology Series of Elsevier are devoted wholly to sedimentological topics. The Lecture Notes Series (Springer) are also often of interest to sedimentologists.

Except for the publications of the Indian Association of Sedimentologists there is no Indian journal devoted wholly to sedimentology. All the Indian geological journals, however, frequently publish articles on sedimentology. Notable among them are: Indian Journal of Geology (formerly, Quarterly Journal of the Geological, Mining and Metallurgical Society of India) published from Kolkata (Calcutta) for over 70 years; Journal of the Geological Society of India (published from Bangalore since 1960) and Indian Journal of Earth Sciences (published since 1974 from Kolkata).

2

PROCESSES OF SEDIMENTATION

SURFACE PROCESSES AND ROCK WEATHERING

Sedimentary processes are largely controlled by just two agencies, water and wind. Thus the hydrological and wind cycles play very important roles in sedimentation. The hydrological cycle is schematically represented in Fig. 2.1.

About two-thirds of the precipitation that takes place on land finds its way back to the atmosphere through evaporation and transpiration. The rest flows back to the sea mostly as surface run-off through streams and partly as ground water. The actual amount of water involved in the different phases of movement varies from place to place, depending on local geography and climate but by and large a balance in the water budget is maintained.

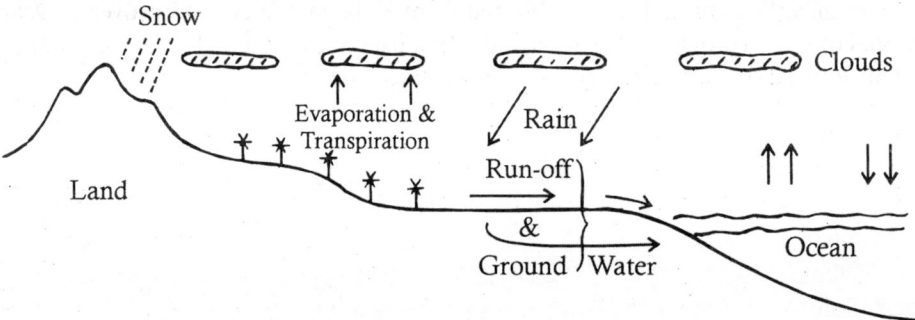

Fig. 2.1: Schematic representation of the hydrological cycle. The balance between evaporation and precipitation in an ideal tropical region is indicated by arrows.

Water is also responsible to a large extent for weathering and decomposition of pre-existing igneous, metamorphic, and sedimentary rocks. The actual amount of sediment yield depends not only on precipitation, but also on surface conditions, types of rocks exposed and density of vegetation cover. Rock weathering can take place by mechanical, chemical and biological actions. Freezing of water percolating into fractures and joints of rock bodies causes expansion and disintegration in cold climate (frost heaving), because water expands when it freezes. Joints within rock bodies allowing free access of water play important role in this context. Alternate heating and cooling during day and night, and growth of plant roots may also cause physical movement of the joint blocks.

Earthquakes, the heat of lightning and the impact between large bodies of displaced rock masses are some of the other causes of mechanical disaggregation. In drier climate, crystallization of soluble salts within cracks and fissures can disrupt rock bodies. Mechanical abrasion and deflation are important erosive processes in deserts.

Much of the actual disintegration of rocks takes place by chemical weathering. The process is particularly effective in tropical, humid climate where rainy seasons follow the summer months. Dissolved carbon dioxide ionizes rain-water into carbonic acid. Rain-water therefore, is generally acidic, with the pH ranging between 4 and 7. Lightning discharges and industrial contamination, particularly sulphurous gases, also produce some acids although in small quantities. The acidic water can take much of the natural carbonate into solution. Humus in the soil also makes natural limestones unstable. Oxidation of organic material in soils also releases CO_2, thereby producing an acidic microenvironment with a pH as low as 2.

Chemical weathering of silicate minerals is a much more complex process. In carbonated water, feldspars break down into clay minerals, silicic acid and carbonates (Holmes and Holmes 1978) :

$$6H_2O + CO_2 + 2KAISi_3O_8 = Al_2Si_2O_5(OH)_4 + 4SiO(OH)_2 + K_2CO_3$$
$$\text{orthoclase} \qquad \text{kaolinite} \quad \text{silicic 'acid'}$$

Similarly, albite also produces kaolinite:

$$2NaAlSi_3O_8 + 2CO_2 + 11H_2O = Al_2Si_2O_5(OH)_4 + 2Na + 2HCO_3 + 4H_4SiO_4$$
$$\text{albite} \qquad\qquad \text{kaolinite} \qquad\qquad \text{silicic acid}$$

Kaolinite, in fact, is often the first mineral to form as a product of chemical weathering. Depending on the chemical composition of the source rock being weathered, clay minerals such as smectite, illite, chlorite, and other oxides and hydrates may be produced. Pyrophyllite may be produced by breaking down of plagioclase.

Much of the soda generated by the process of chemical weathering is transported to the sea by running water, but potash remains in the soil to be absorbed later either by plant roots or by clay minerals. The average dissolved silica content of the world's rivers is about 13 ppm. Much of this silica possibly comes from feldspar rather than quartz. Solubility of silica increases with an increase in temperature and alkalinity of the medium. With a decrease in temperature and alkalinity, much of the dissolved silica precipitates in the intergranular space of sediments to provide bondage between the grains.

The trend of decomposition of the ferromagnesian minerals can be illustrated by the following example :

$$H_2O + CO_2 + Ca(Mg.Fe)(SiO_3)_2 \rightarrow 2SiO(OH)_2 + \text{soluble bicarbonates of Ca, Mg, Fe}$$
$$\text{diopside} \qquad \text{silicic 'acid'}$$

Similarly, chemical decomposition of biotite, hornblende and augite leads to the formation of clay and chloritic minerals. Bicarbonate of iron $Fe(HCO_3)_2$ is converted to limonite $Fe_2O_3.H_2O$ in the ground-water zone on coming into contact with oxygenated water (see Fig. 2.2A). Ferrous iron released from minerals during weathering is oxidized in soil to a ferric state. Common hydrated iron oxide minerals include limonite ($Fe_2O_3.H_2O$), goethite (FeO.OH), lepidocrocite (FeO.OH) and anhydrous dimorphs of hematite (n.Fe_2O_3). These minerals, very stable under oxygenated condition, continue to exist in the soil almost indefinitely (Fig.2.2B). Highly colourful, they impart yellow, orange or brownish red tones to the weathered soil surface. Some of these minerals also provide bondage between loose grains of sediments as ferruginous cement.

The product of weathering of rocks is called *regolith*. It is constituted of mineral particles, and is the main source of sediments. Regolith produced *in situ* is called residual regolith. It overlies the bed rock from which it is derived. Residual regolith, in turn, is overlain by surface soil. Regolith may also be transported from a distant source and deposited on the bed rock downslope, or on an older layer of regolith. The medium of transportation may be water, wind or ice. Transported regolith that accumulates at the foot of a slope is called *colluvium*. The regolith transported further downwards and deposited in the valley is called *alluvium*. The main agency for transportation of alluvium is the stream (Fig. 2.3).

Downslope movement of regolith, soil and hard rock under the influence of gravity is broadly designated as *mass wasting*. The process is given different names depending on the materials involved and their speed of movement. *Talus creep* involves rock bodies moving very slowly downslope. When the rocks move very fast they may cause *landslides*. Rock bodies moving still faster cause *rockfall*. Near the source regolith moves very slowly as soil creep and solifluction. When, aided by water, they move faster, they cause *mudflows*. Very fast moving regolith avalanches downslope.

Most of the sediments produced on the exposed surface of the earth find their way to the sea through rivers. Chemical analysis of sea-water, however, does not exactly match the composition of average river-water. This is because some of the chemicals arriving in the ocean are precipitated there. The time during which an element stays in the sea-water (called *residence time*) can be obtained by dividing the amount of an element present in the ocean by the amount of that element (tons/year) contributed by rivers to the sea. Sea-water contains large proportions of sodium chloride because residence time of Na^+ and Cl^- are of the order of 100 m.y. Since the residence times of Ca^{++} and Mg^{++} are much smaller (1 m.y. and 10 m.y. respectively), these elements are absorbed by marine organisms such as corals to build up their shells and tests. These shells eventually produce extensive coral colonies.

In the deeper parts of the ocean, where the temperature is low, pressure is high and the water is acidic due to dissolved gases, the carbonate is dissolved. This condition is likewise suitable for precipitation of dissolved silica, which has an extremely small residence time. Silica precipitation in deep ocean-waters in the form of chert is aided by microscopic marine organisms such as radiolaria and diatoms. Much of the deep ocean bottom is covered by siliceous ooze and also by extremely fine-grained, muddy sediments derived from land. It must be remembered that chemical precipitation is not confined to oceanic areas only. Natural springs, playas and caverns are some areas on land where chemical precipitation takes place.

MINERAL STABILITY

The process of weathering effects considerable change in source minerals of sediments. The degree to which a mineral can resist this change determines its stability.

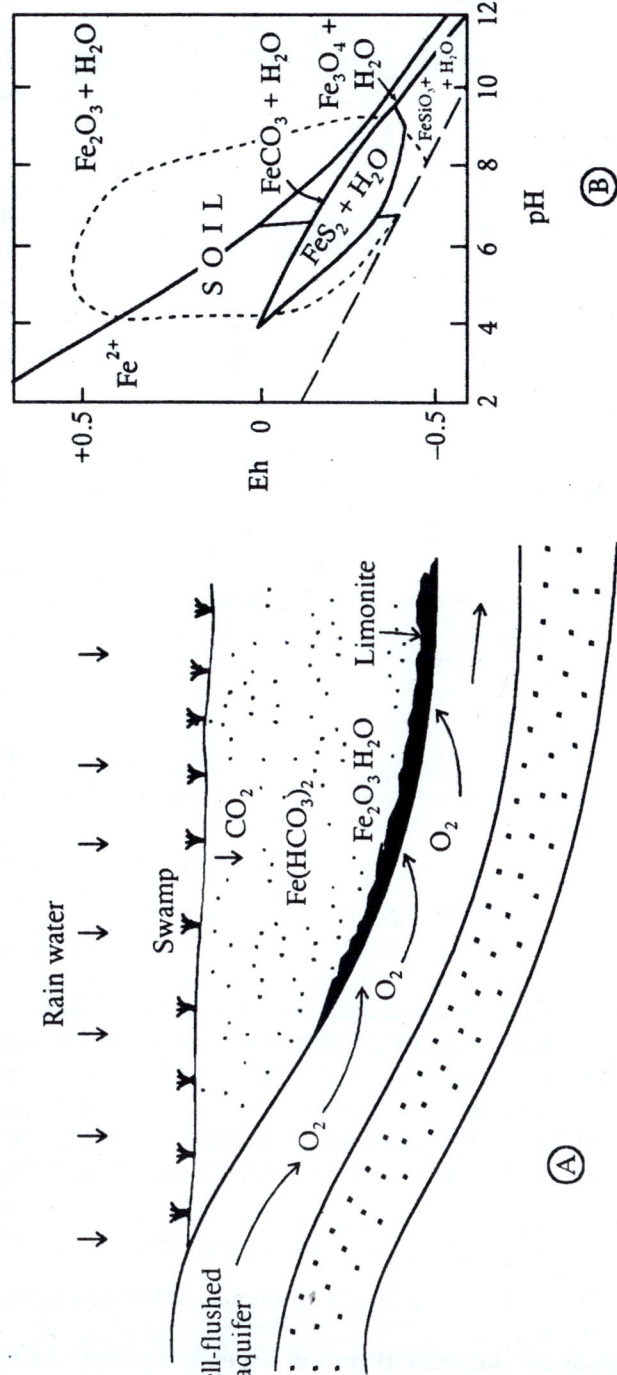

Fig. 2.2: (A) Diagammatic representation of the pocess of limonitization by oxidation of biocarbonate of iron in the ground-water zone (modified after Borchert 1960). (B) Stability fields of iron compounds under surface soil condition (partly redrawn from Garrels and Christ 1965).

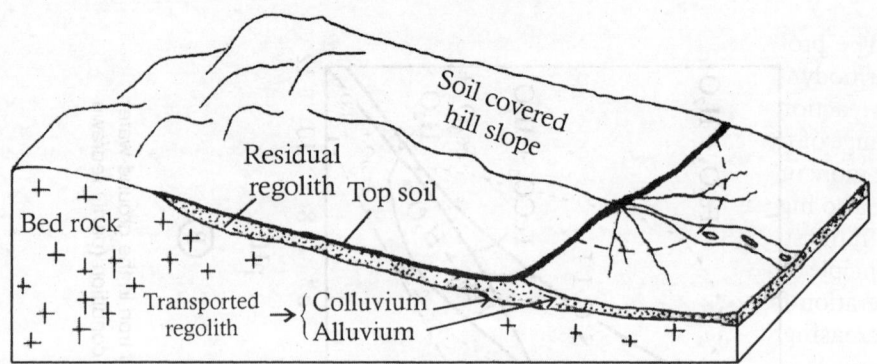

Fig. 2.3: Generation of regolith from weathering of bed rock (based on Strahler and Strahler 2002).

Interestingly, the stability series of rock forming silicates shows a trend almost opposite to that of crystallization as indicated by Bowen's reaction relation series (Table 2.1). For example, olivine, the earliest to crystallize from a melt, is also the first to be altered during chemical weathering. Quartz, the mineral last to crystallize, can survive chemical decay almost indefinitely (Goldich 1938).

Table 2.1: Relative stability of common rock forming silicates (modified after Goldich 1938)

Most stable	Quartz (200)	
	Muscovite (170)	
	Hornblende (160)	
		K-Feldspar
		Alkalic plagioclase (150-160)
	Biotite (150)	
	Augite (150)	Calcic plagioclase
Least stable	Olivine (133)	

N.B. Figures in parentheses indicate the proportion of anion to a cation content of 100 in each case.

The explanation of this lies in the anion to cation ratios of the minerals. The proportions of negative ions (anion) for the common rock-forming silicates are shown in parentheses in Table 2.1, assuming that the cation content is 100 in each case. The higher the anion proportion of a mineral, the lower its susceptibility to chemical weathering, which essentially is a process of oxidation or hydration (introduction of O^{2-} or OH^{-}). Thus olivine with a relative anion content of 133 will be the first to be affected by chemical weathering, whereas quartz, with a relative anion content of 200, will not change at all. Under this condition the process of weathering leads to the production of increasingly stable minerals. Two common examples of this process are:

1. olivine (133) → chlorite (180) or serpentine (180) → limonite (200).
2. feldspar (160) → hydromica (200) → clay minerals (225) → bauxite (250 - 300).

(figures in parentheses indicate proportions of anions relative to a cation content of 100 in each case).

The actual amount of chemical alteration depends not only on the chemical composition and crystal structure of a mineral, but also on local climate, mainly temperature and humidity, as well as the chemistry of the pore fluid.

The problem of mineral stability can also be studied from the point of view of thermodynamics. In a chemical reaction the sum of free energies of formation of all the reaction products minus the free energies of the reactants measures the free energy change of the reaction. The more negative the value of free energy change of a mineral, the more unstable it is. Hence its chances of reaction to produce a more stable product are also higher.

In the stability series obtained by Curtis (1976), according to the above mentioned principle olivine (most negative) will alter first and muscovite (least negative) will resist alteration for a long time. The common rock-forming silicates, arranged in order of decreasingly negative free energy, are as follows: olivine, pyroxene, amphibole, plagioclase, microcline, muscovite. The trend is the same as that obtained by Goldich (1938, see Table 2.1).

While free energy indicates the susceptibility of a mineral to chemical reaction, the physical stability of a mineral can be expressed in terms of its resistance to abrasion. The 'durability' of a mineral has been experimentally shown to be related to its hardness. The scale of resistance to abrasion, as worked out by Thiel (1945), for example, is as follows (starting with the mineral having least resistance) : barite, siderite, fluorite, goethite, enstatite, kyanite, hematite, augite, apatite, hypersthene, rutile, hornblende, zircon, epidote, garnet, staurolite, microcline, tourmaline, quartz.

Many minerals in this list are *'heavy minerals'*, that is, they have a specific gravity higher than that of quartz ($\rho = 2.65$) or feldspar ($\rho = 2.54 - 2.76$). For practical purposes, all minerals having a specific gravity higher than that of bromoform ($\rho = 2.87$ at 20° C) are listed as heavy minerals. According to stability the heavy minerals may be grouped as follows (modified by Pettijohn 1975, after Sindowski 1949) :

Most stable: zircon, rutile, tourmaline.

Intermediate : staurolite, kyanite, hornblende.

Least stable: (garnet), augite, apatite, olivine.

Because of their stability, zircon, rutile and tourmaline survive prolonged weathering and concentrate in the older soil profiles. Garnet is generally considered to be more stable than that shown here.

SOURCE OF SEDIMENTS

The sediments derived directly from crystalline igneous and metamorphic rocks are the first-cycle sediments. Sediments produced out of lithified first-cycle sediments are termed reworked (polycyclic) sediments (see Fig. 1.1). Reworking can be detected from textural evidence. During the process of cementation of a first-cycle quartz sand precipitation of silica on detrital quartz grains may produce overgrowths. If the edges and corners of the overgrowth on a quartz grain of first-cycle origin are found to be well rounded, we know that the grain of quartz is of second-cycle origin.

The source area and the source rock of a sediment and also, according to some authors, the climate, relief and the environment prevailing at the source, are collectively called its *provenance* (French: *provenier* meaning 'to originate or to come from'). The mineralogical constituents of terrigenous sediments bear the imprint of the source rock and also, to some extent, its climate and relief. Quartz, feldspar and heavy minerals are particularly suitable for deciphering provenance.

Quartz and Feldspar as Provenance Indicators

Quartz, the most abundant derital constituent in sedimentary rocks, is often used as an indicator of provenance, but opinion differs about the specific characters of a quartz grain which are indicative of its origin. Undulatory extinction ('strain shadows'), traditionally taken to be indicative of a metamorphic source, has been found in quartz grains of plutonic origin also (Blatt and Christie 1963). Acicular inclusions, long believed to be a characteristic of quartz of granites, have also been detected in metamorphic rocks. No significant difference between shapes of quartz grains of igneous and metamorphic origin have been statistically established, although this was, at one time, believed to be an important discriminating criterion. A group of geologists at Indiana University, after a thorough review of the problem, came to the conclusion that strongly undulose and finely polycrystalline quartz grains are indicative of metamorphic source rocks, but the quartz grains of plutonic and volcanic origin are characteristically strain free (Basu 1985). The Indiana group developed a method of discriminating sedimentary rocks of plutonic, low- and high-rank metamorphic parentage by plotting the proportions of the following four parameters in a *'diamond diagram'* : undulatory quartz, non-undulatory quartz, polycrystalline quartz, and the number of crystal units per polycrystalline grains (Fig. 2.4).

As it is very stable, quartz survives prolonged weathering and erosion. The ratio of quartz to feldspar (plus unstable rock fragments) can therefore be used as an index of mineralogical maturity of a sediment of mechanical origin. Comparative studies of sand populations transported under various climatic and environmental regimes have shown that a unique combination of extreme climate, relief, transportation, and rate of sedimentation is necessary for elimination of all the unstable fractions leading to a highly matured, first-cycle quartz sand. This imprint of climate, although preserved for the first 75 km of transportation in high-gradient streams, is rapidly destroyed as soon as a high-energy marine environment is reached. For this reason, palaeoclimatic interpretation from mineralogical composition is feasible only for terrigenous, first-cycle sand (Suttner *et al.*, 1981).

Feldspar, which ranks only next to quartz in abundance, may also act as an indicator of provenance. The chemical composition of feldspar may be useful for this purpose. Homogeneous alkali feldspars more sodic than Ab_{50} are generally of volcanic origin. Alkalic feldspars more potassic than Or_{88} are derived from plutonic and metamorphic sources (Trevena and Nash 1981). Identification of the structural states of detrital alkali feldspars may also lead to recognition of the parent rock. Detrital feldspars from metamorphic sources are mostly microcline, while those from volcanic sources are generally sanidine. Alkali feldspars with relatively higher Al/Si structural states are likely to be associated with rapidly cooling outer shell of a pluton rather than its core. This change in the degree of ordering of feldspar is reflected in the stratigraphic sequence produced out of gradual unroofing of batholiths that have served as a source of sediments (Basu 1977). A quick method for the determination of the structural states of alkali feldspars for this purpose was developed by Basu and Suttner (1975).

In the modern big rivers of the world the proportion of quartz to feldspar increases when the climate is tropical to subtropical, suggesting a lesser chance of survival of

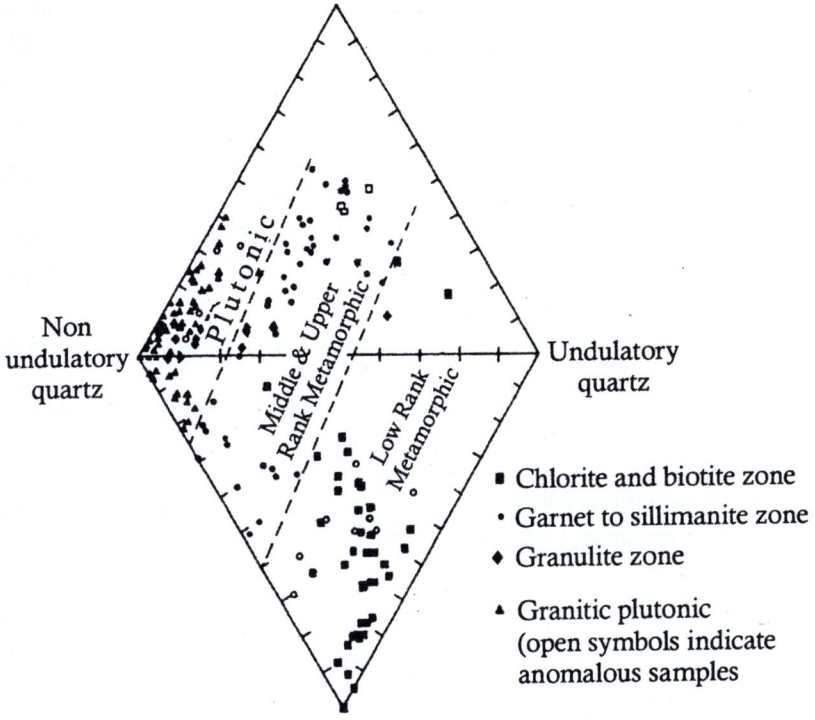

Polycrystalline quartz
(2-3 crystal units per grain; ≥ 75%)
of total polycrystalline quartz

Non undulatory quartz

Undulatory quartz

Plutonic

Middle & Upper Rank Metamorphic

Low Rank Metamorphic

■ Chlorite and biotite zone

• Garnet to sillimanite zone

♦ Granulite zone

▲ Granitic plutonic
(open symbols indicate
anomalous samples

Polycrystalline quartz
(>3 crystal units per grain;> 25%
of total polycrystalline quartz)

Fig. 2.4: Diamond diagram showing distribution of quartz grains of different varieties
with respect to plutonic igneous and metamorphic sources (after Basu *et al.*,
1975). With permission of SEPM (Society for Sedimentary Geology).

feldspar in humid climate (Potter 1978, 1978a). The ratio of quartz to feldspar in the
first cycle fluvial sand is > 1 in the humid Appalachians and < 1 in the relatively arid
Rocky mountains of the United States (Basu 1976). The foregoing suggests the lesser
chance of survival of feldspar in humid climate but it may not be altogether impossible
for feldspar to survive prolonged transportation in humid climate although its size
decreases by cleaving. The latter conclusion is reached from the finding that irrespective
of climate, the proportion of feldspar increases with a decrease in grain-size. A certain
amount of climatic rigour, either arid or cold, was long believed essential for the
preservation of fresh feldspar grains in sediments, but a later discovery of fresh feldspar
under humid, tropical, South American climate led to the belief that short-distance
transportation followed by quick burial is the key to the preservation of fresh feldspars
in nature (Pettijohn 1975). This implies sudden uplift of the source area and quick

subsidence of the basin of accumulation, but the effects of climate and tectonism need not be mutually exclusive in the preservation of fresh feldspars in sediments.

According to current thinking, the proportion of quartz and the relatively unstable elements such as feldspar and rock fragments in a sediment population is dependent not only on the source rock and climate, but also to some extent on the degree of crustal stability (Dickinson and Suczek 1979). These ideas will be discussed later, under tectonics and sedimentation (Chapter 8).

Heavy Minerals as Provenance Indicator

While minerals such as quartz and feldspar can be derived from a variety of rocks, the heavy minerals, or more particularly, assemblages of certain suites of heavy minerals, are restricted to some distinctive source rocks only. Heavy mineral assemblages identified with major source rocks, igneous, sedimentary, and metamorphic, are shown in Table 2.2.

Table 2.2: Heavy mineral assemblages of the major source rocks (condensed from Feo-Codecido 1956)

Heavy mineral assemblage	Source rock
Apatite, biotite, hornblende, monazite, muscovite, rutile, pink tourmaline, zircon	Acid igneous rocks
Fluorite, garnet, muscovite, topaz, blue tourmaline, monazite	Granite pegmatites
Augite, diopside, hypersthene, magnetite, olivine, chromite	Basic igneous rocks
Andalusite, garnet, staurolite, zoisite, epidote kyanite, sillimanite	Metamorphic rocks
Barite, iron ores, rutile, tourmaline, zircon (rounded grains common)	Reworked sediments

In vertical stratigraphic columns heavy minerals often occur in a reverse order of stability. For example, tourmaline, zircon and rutile, the most stable of the heavy minerals, occur in the older beds, whereas much less stable amphibole, pyroxene and olivine are often found in the younger horizons. As an explanation of this mode of occurrence it has been suggested that the older beds have lost the less stable minerals due to the action of the circulating ground-water or interstratal solutions which have altered or removed these minerals from the rock. If this is actually the case, they can hardly be used as indicators of provenance (Pettijohn 1975, p. 499).

3

SEDIMENTARY PETROLOGY

After transportation and deposition loose sediments are consolidated into sedimentary rocks by a process called *diagenesis*. The sedimentary rocks thus formed may be broadly divided into two groups: *exogenetic* and *endogenetic*. Exogenetic rocks are the products of fragmentation of the source rock outside the basin of sedimentation. Endogenetic rocks are produced out of precipitation from solution within the basin. The exogenetic and endogenetic rocks can be distinguished by their textures. While the endogenetic rocks show an interlocking or crystalline fabric ('*non-clastic*' texture), the exogenetic rocks show discrete framework grains separated by voids. The latter texture is called '*clastic*' from the Greek *clastos*, meaning broken.

Classification of sedimentary rocks into clastic and non-clastic, though popular among sedimentologists, may at times lead to ambiguity. A chemically precipitated sediment for example, may also be transported from its place of origin after fragmentation and deposited elsewhere to form a sedimentary rock. Such rocks, although essentially endogenetic, may show a clastic texture. To avoid confusion in such cases, Folk (1968) introduced the following terms: *terrigenous, orthochemical*, and *allochemical*.

Terrigenous rocks are produced from sediments derived from land areas located outside the basin. The process of derivation is purely mechanical but not all terrigenous sediments are the products of weathering and erosion. Fragments of volcanic origin and the products of crushing by glaciers also add to the terrigenous sediments. The orthochemical or truly chemical rocks are produced by chemical or biochemical precipitation within the basin. Evaporites such as halite, anhydrite and gypsum and non-evaporites such as limestone, dolomite and ironstone are examples of such precipitates. Intermediate between the terrigenous and orthochemical rocks are the allochemical rocks (*allo* means false). Produced initially within a basin by chemical precipitation, these rocks are fragmented, transported and redeposited elsewhere. Both terrigenous and allochemical rocks show a '*clastic*' texture. Only the orthochemical rocks show a non-clastic texture. Sediments of chemical and terrigenous origin may combine to form rocks of a hybrid nature. Calcareous and carbonaceous shales are examples of such hybrid rocks. While accepting the rationality of Folk's scheme of classification, it must be appreciated that the terms clastic and non-clastic have thoroughly pervaded sedimentological literature and cannot be totally ignored at present. These terms will be used in the present text whenever indispensable.

Of all the major groups of sedimentary rocks found on the earth, only three, sandstone, shale and limestone, together account for about 95% of the total. Theoretical calculations, based on the idea that the average chemical composition of the sedimentary rocks must match the average composition of the igneous rocks from which they are derived, lead to the following figures of abundance of the major rock types : shale (81%), sandstone (11%) and limestone (8%) (Garrels and Mackenzie 1971). Actual measurements of the stratigraphic record, however, show a smaller proportion of shale and a higher proportion of sandstone and limestone (Pettijohn 1975).

ROCKS OF MECHANICAL ORIGIN

Sedimentary particles of mechanical (terrigenous) origin are classified according to their size. Particles larger than 2 mm constitute gravel (Latin-*rudite*), those between 2 and 1/16 mm constitute sand (Latin-*arenite*), while those finer than 1/16 mm constitute silt and clay (Latin-*lutite*). Accordingly, sedimentary rocks of mechanical origin are classified broadly into three groups: rudaceous, arenaceous and lutaceous (also called argillaceous). This scheme of classification is followed here for describing rocks of mechanical origin.

Rudaceous Sedimentary Rocks (Conglomerates and Breccias)

Unconsolidated sediments coarser than 2 mm are grouped under the term gravel. Indurated gravels are called *conglomerates*. Since gravel is classified as granule, pebble, cobble and boulder (see Table 4.1), these terms are often used as prefixes to indicate the size of the dominant fragments in a conglomerate (e.g., pebble conglomerate, boulder conglomerate). While the particles constituting a conglomerate are essentially rounded, a breccia consists of gravel-size clasts which are angular. The term *rubble* is applied to an assemblage of loose angular fragments of gravel size. Breccias need not be of sedimentary origin only. For example, fault breccias are produced by tectonic activity. Large, well-rounded particles of volcanic origin (volcanic bombs) constitute a rock called an *agglomerate*.

Like other sedimentary rocks of mechanical origin the space between the framework grains of a conglomerate is filled by finer particles. These fine particles constitute the *matrix* of the rock. Gravel and sand fragments can combine in various proportions to produce a conglomerate. The resultant rock is named accordingly.

Conglomerates may be classified either in a purely descriptive way (example : chert conglomerate, limestone conglomerate) or according to the mode of their origin (fluvial conglomerate, glacial conglomerate).

In the comprehensive genetic classification proposed by Pettijohn (1975), conglomerates are grouped into four major categories which are geologically meaningful: *epiclastic, pyroclastic, cataclastic* and *meteoric* (see Table 3.1).

Epiclastic conglomerates are the commonest of all conglomerates. These consist of gravels of terrigenous origin. The conglomerates of fluvial origin as also those generated on beaches by wave action are of this type. The intraformational conglomerates, which are also epiclastic, are composed of clasts derived from within the basin. Shale-pebbles, produced out of mud cracks, are good examples of intraformational conglomerates, but the process may be more complicated.

Extraformational epiclastic conglomerates include *oligomicts, petromicts (polymicts)*, and *diamicts*. Oligomictic conglomerates consist of pebbles of homogeneous composition. Petromictic (polymictic) conglomerates consist of pebbles of varied rock types. Diamictic conglomerates are characterized by a high proportion of matrix. These may be of glacial (*tills* or *tillites*) or non-glacial (*tilloid*) origin. Pebbly mudstones (*argillites*) are also included in this group. Conglomerates found in alluvial fan deposits are good examples of tilloids.

The clast-supported conglomerates belonging to the basal Bababudhan Group of South India may be cited as an Indian example of oligomictic conglomerate of Archaean age. Vein quartz pebbles predominate within these conglomerates but some quartzite pebbles (metacherts?) are also present. These conglomerates are believed to have been derived from a granitic or gneissic source and deposited on a braided stream plain (Srinivasan and Ojakangas 1986). The Talchir (Lower Permian) diamictities of India are some of the well-known Indian glacial conglomerates (Fig. 3.1). The pebbles within these conglomerates show striations and grooves on the surface and flattening of faces parallel to the longest axis (Fig. 3.2). Talchir diamictites of similar character have also been reported, among others, by Casshyap and Tewari (1982), Sen (1991).

Other important types of conglomerates include *pyroclastic conglomerates of* volcanic origin, *cataclastic conglomerates* produced by earth movements and *meteoric conglomerates* produced by meteoric impact. Pyroclastic conglomerates, produced by agglomeration of pieces of lava which solidified in flight, having been hurled into the air during volcanic eruption, are termed *agglomerates*.

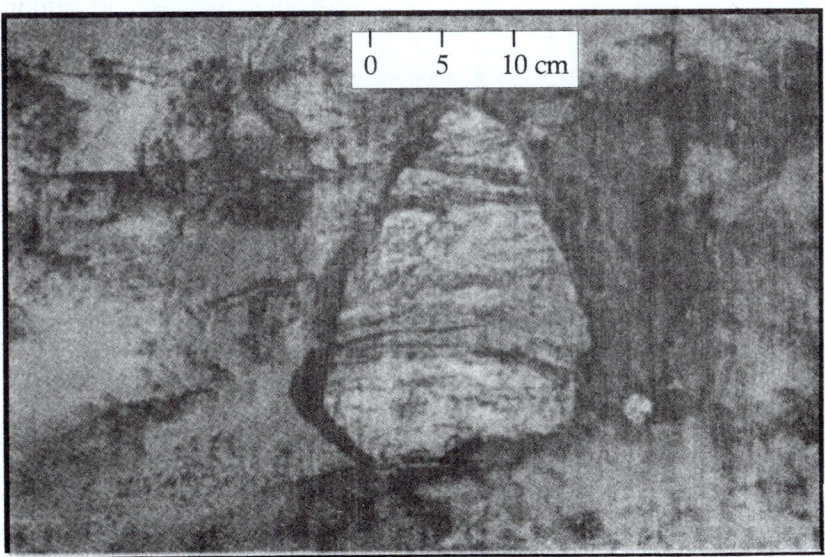

Fig. 3.1: A pentagonal granite gneiss boulder within Talchir diamictite. Saharjuri coalfield, India.

Arenaceous Sedimentary Rocks (Sandstones)

Terrigenous sedimentary rocks composed mainly (~70%) of sand-size particles are called *sandstones*. Sand may be deposited in a variety of environments ranging from

Table 3.1: Classification of conglomerates and breccias (modified after Pettijohn 1975)

Type of conglomerate	Name of conglomerate	Description	Origin and examples
Epiclastic	Oligomictic conglomerate	Consists of pebbles of same composition. Matrix proportion low.	Derived from vein quartz or a monomineralic rock (most fluvial, wave or beach conglomerate.).
	Polymictic (Petromictic of Pettijohn) Conglomerate	Consists of pebbles of varied types. Matrix proportion low.	Derived from rocks of varying composition (fluvial, wave or beach conglomerate).
	Conglomeratic argillite	Laminated argillites with ice rafted pebbles.	e.g. varves.
	Tillites (Diamictite)	High matrix conglomerates with striated, faceted, pentagonal boulders.	Glacial.
	Tilloid (Diamictite)	Chaotic assemblage of non-glacial boulders in muddy matrix.	Non-glacial (example: conglamerates originating from subaqueous gravity flow as in alluvial fans.
	Intraformational conglomerate	Mud, or shale-pebble intraclasts.	Temporary phases of subaerial erosion.
Pyroclastic	Volcanic Breccia and agglomerate	Blocks of previously deposited vlocanic material. Volcanic bombs solidified in flight.	Volcanic
Cataclastic	Breccia	Angular fragment, crushed pieces, fault gouge, slickenside blocks.	Slumping, folding, faulting etc.
Meteoritic	Impact Breccia	Effects of shock recognizable.	Meteoritic impact.

marine to non-marine, through mixed (see Chapter 7), sand-size fragments of quartz, feldsprs, micas, heavy minerals and rock fragments constitute the bulk of framework grains of sandstone. Rocks composed of carbonate or pyroclastic grains are not treated as sandstones. The voids between the framework grains of sandstones are filled either by finer grains of terrigenous origin or by chemically precipitated cement. Under a microscope the cement shows a crystalline texture but the matrix is composed of discrete, detrital particles. Matrix is a relative term. Particulate materials (irrespective of their actual sizes) distinctly smaller than the framework grains constitute the matrix of a terrigenous rock. In the case of a conglomerate, sand-size grains constitute the matrix, whereas in sandstones the matrix is composed of fine, mud-size grains only.

Classification of Sandstones

Sandstones are classified by observable petrographic criteria. Proportions of quartz, feldspar, rock fragments, and mica are commonly used for this purpose. Being the most stable of all detrital grains, quartz survives prolonged weathering and abrasion at the source and also during transportation. A sandstone composed wholly of quartz grains is considered the most matured. For this reason the ratio of quartz to the relatively unstable components, feldspar and rock fragments, is used as an indicator of the mineralogical maturity of a sandstone. The proportion of feldspars and rock fragments is indicative of the nature of the source rock.

When sand is transported in nature by running water, the finer fractions are carried in suspension while the coarser ones are moved along the bottom. The very fine grained clay is carried as colloidal suspension. The clay particles, being electrically charged, adhare to each other to increase the viscosity of the transporting fluid. Grain-to-grain interaction of even the non-electrically charged fine particles might also increase the resisting stress within the fluid, thereby increasing its apparent viscosity (Bagnold 1956, see also Leeder 1999, p. 126). The finer fractions eventually contribute to the matrix of the sandstone. The fluidity of a running body of water being inversely proportional to its suspended load, it is argued that a large amount of matrix in a sandstone is indicative of low fluidity of the transporting agency. The matrix content of a sandstone is thus important for its interpretative value (Pettijohn 1975). This idea, however, has been challenged by some authors on the ground that a part or whole of the matrix may be of secondary origin. Increase of matrix content of many sandstones with increase of age or with decrease of framework grain-size led to such a conclusion (Dickinson 1970; Okada 1971).

Many different schemes of classification of sandstones have been proposed from time to time by various authors using the four criteria - quartz, feldspar, rock fragments, and matrix. The difficulty involved in plotting four variables on two-dimensional diagrams prompted some authors to use two separate triangles—one devoted to high-matrix sandstones (called *wackes*) and the other to low-matrix sandstones (called *arenites*). In each case the end members of the triangles are quartz, feldspar and rock fragments (Gilbert 1954). Arkosic and lithic are the adjectives used for sandstones having high (> 25%) feldspathic and lithic contents respectively.

While the schemes using two separate triangles for wackes and arenites have practical advantages, they divide sandstones into two rigid compartments. In reality, however, a sandstone may have any amount of matrix ranging from zero to about 30%. The idea of a continuous change in the matrix content of sandstones was introduced by Dott (1964) who used a three-dimensional prism instead of two fixed triangles. Sandstones of various matrix content were represented by triangular sections of this prism. The boundary between arenites and wackes was arbitrarily placed by Dott at a matrix content of 15%. The end members of each triangle were quartz, feldspar, and rock fragments. This technique of representation is a closer simulation of the real situation than that proposed by other authors.

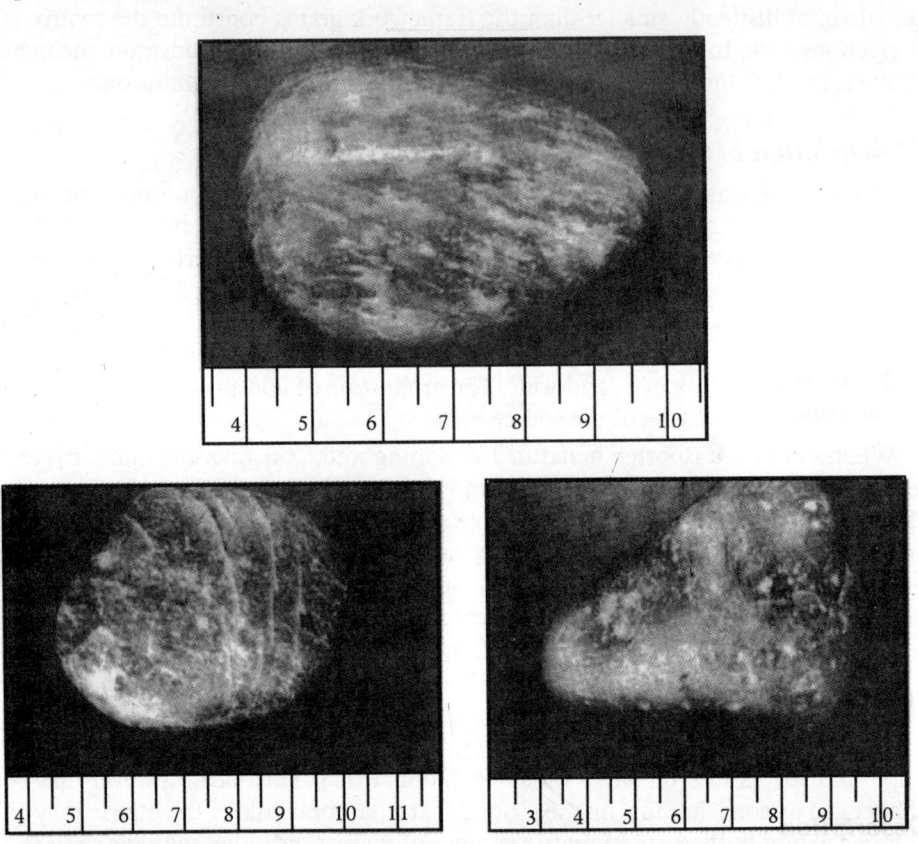

Fig. 3.2: Typical glaciated pebbles from the Talchir Boulder Bed, Giridih, India. Note pentagonal shapes and striated, pitted surfaces. Scale in cm.

Figure 3.3 gives a modified version of Dott's scheme of classification. The demarcation between quartz arenite and quartz wacke here has been put at 12.5% (or 1/8th of the field of view of the microscope, which is easier to estimate than 15%). The boundary between sandstone and mudstone has been put at 30% because a sandstone, by definition, must have about 70% sand-size grains. Quartz and metaquartzite fragments have been grouped together because they behave similarly from the point of view of mineralogical maturity. Chert grains have been included in rock fragments because chert is a distinct indicator of provenance (Okada 1971). For reasons to be discussed later, graywackes have been plotted in a separate triangle at the high matrix end of the prism. The daughter triangle of Fig. 3.3 gives a detailed classification of lithic arenites as proposed by Folk (1968).

Sandstones with appreciable quantities of slate, phyllite, mica schist and other phylloid fragments of low-grade metamorphism are termed phyllarenites. Those containing large proportions of detrital limestone fragments are called *calclithites*. Sandstones with chert or volcanic fragments are named chert and volcanic arenites respectively.

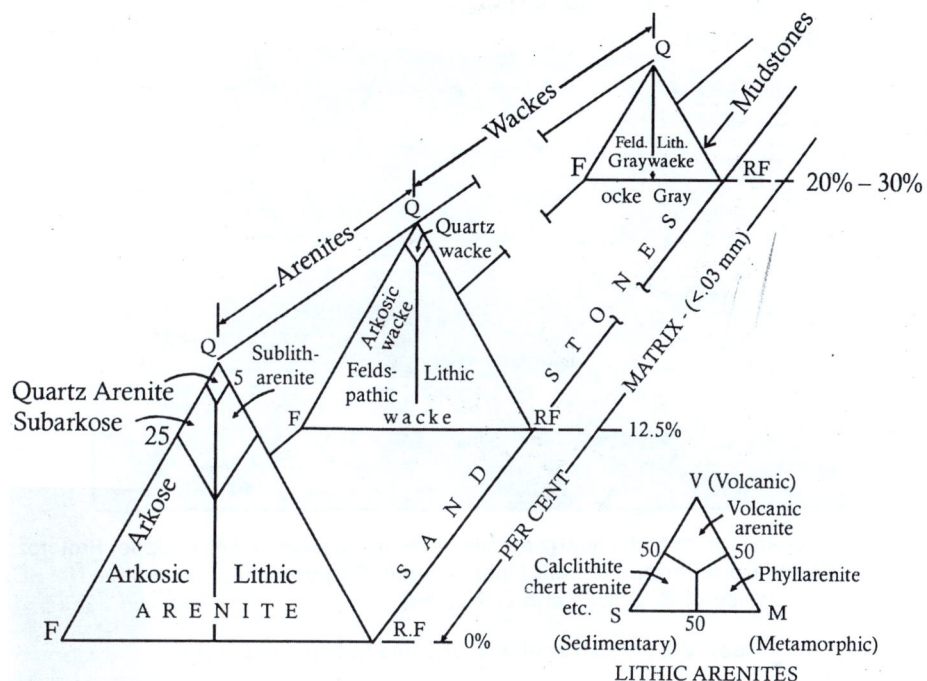

Fig. 3.3: A revised version of Dott's (1964) scheme of classification of sandstones. Q = quartz + metaquartzite, F = feldspars, R.F = rock fragments. The daughter triangle shows subdivision of lithic arenites after Folk (1968).

Description of Sandstones

Quartz Arenite (also called Orthoquartzite)

Quartz arenites consist almost entirely of quartz grains cemented together by silica or welded together by predissolved quartz (see diagenesis). Chemical precipitates other than silica may also serve as cement. The quartz grains constituting the framework are often well rounded and well sorted (Fig. 3.4). A small amount of heavy minerals of stable variety (tourmaline, zircon, rutile, ilmenite) may be present in quartz arenites. The Rewa sandstones of the Vindhyan Supergroup may be cited as an Indian example of quartz arenite. Modal analyses of representative samples of Rewa sandstones are presented in table 3.2.

The detrital grains of quartz in arenites may be produced by erosion of vein quartz. They may also be derived from one of the following source rocks through prolonged weathering, erosion and transportation under high energy conditions (Folk 1968).

1. Plutonic (granitic) source → arkose → subarkose → quartz arenite.
2. Metamorphic source → graywacke → subgraywacke → quartz arenite.
3. Sedimentary source → submature orthoquartzite → mature orthoquartzite (quartz arenite).

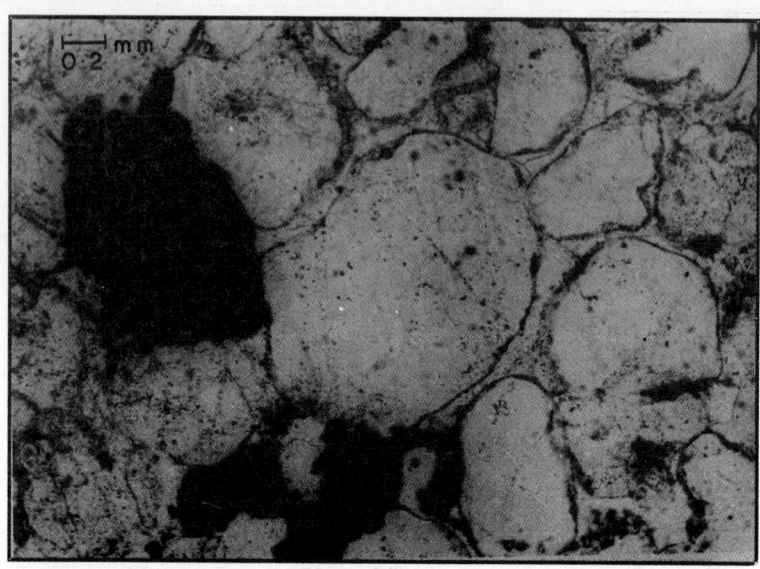

Fig. 3.4: Photomicrograph of a quartz arenite under plane polarized light. Upper Proterozoic Chandarpur Formation near Mahasamund, Raipur, India (thin section No. SCH-3, courtesy of S. S. Das and C. N. Rao).

Table 3.2 : Modal analyses of some Indian sandstones

Sample No.	Quartz & Quartzite	Feldspars	Micas	Rock frag- ments	Cement (calca- reous)	Cement (sili- ceous)	Cement (ferru- ginous)	Matrix	Total
Quartz Arenites									
20.4*	93.90	-	-	0.80	-	3.45	1.72	-	99.87
20.1*	88.53	-	-	0.18	-	5.83	5.46	-	100.00
20.10*	86.16	-	-	0.21	-	3.04	10.38	-	99.79
Quartz Wackes									
AM 18W+	66.00	3.00	8.00	-	2.00	-	3.00	18.00	100.00
AM 13+	72.00	2.00	3.00	-	-	-	7.00	16.00	100.00
AM 8+	75.00	2.00	2.00	-	-	-	5.00	16.00	100.00
Arkosic Wacke & Calcareous Arkose									
M 30 ..	52.57	25.11	3.14	-	-	-	4.18	14.99	99.99
M 46 ..	45.57	20.05	-	-	30.86	-	-	3.52	100.00
M 5 ..	46.54	20.66	13.90	-	-	-	0.21	18.69	100.00

* Rewa sandstones, Vindhyan Supergroup (Precambrian), Mihar, India (Basumallick 1962).
+ Sandstones from Ironstone Shale Formation (Permian), Raniganj Coalfield, India (A. Mandal 1985, Unpub. M.Sc thesis, IIT, Kharagpur).
.. Sandstones from Raniganj Formation (Permian), Jharia Coalfield, India (S. Nayak 1985 Unpub. M.Sc thesis, IIT, Kharagpur).

Chemically, quartz arenites consist of a very large proportion (> 95%) of SiO_2 and small amounts of CaO, CO_2, Al_2O_3, FeO + Fe_2O_3. The chemical composition is controlled mainly by the type and proportion of heavy minerals and cementing materials in the rock (see Table 3.3).

Quartz Wacke

Sandstones with a fairly large proportion (say > 12.5%) of fine-grained matrix are termed quartz wackes. The presence of a large proportion of matrix, a wide range of grain-size variation, lack of rounding, and sorting of the framework grains are indicative of textural immaturity. The presence of feldspars and rock fragments of various types indicates mineralogical immaturity. A fine-grained matrix is the dominant binding material in these rocks. A large amount of matrix in wackes leaves hardly any space for cement (Fig. 3.5). In some quartz wackes replacement of the matrix by chemically precipitated cement may create the illusion of an arenite, but the textural immaturity of these sandstones is clearly indicated by lack of rounding and sorting of the framework grains. A high proportion of trapped detrital clay in the interstices of framework grains (*protomatrix*), reflects an abundance of suspension load in the transporting medium, as in a turbid river during flood, but a certain amount of matrix may also be produced at a later stage by diagenetic alteration of sand-size grains (*epimatrix*), or deformation and squashing of pelitic fragments (*pseudomatrix*). Recrystallized material occurring between framework grains constitutes *orthomatrix*. The sandstones belonging to the Barakar, Ironstone Shale and Raniganj Formations of the Indian Gondwana are good examples of quartz wackes. Mineralogical compositions of some of the typical quartz wackes are given in table 3.2.

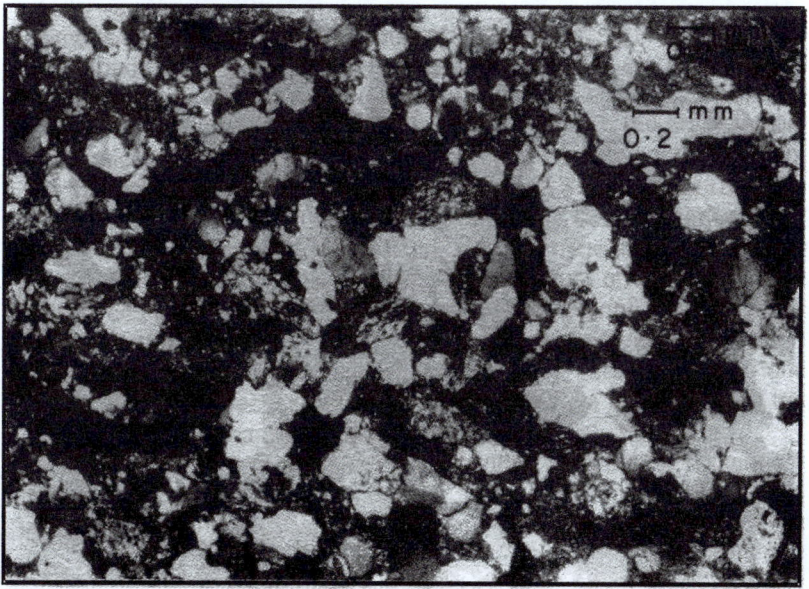

Fig. 3.5: Photomicrograph of a quartz wacke under cross-polarizers. Nahan Group, Koshailia Nala, Kalka, India (thin section No. DP 25/68, courtesy of D. P. Sen).

Feldspathic Sandstone and Arkose

A sandstone with a high (> 10%) proportion of feldspar is designated as a feldspathic sandstone. When the matrix content of such a rock is also high, the rock is called a feldspathic wacke. Low matrix feldspathic sandstones are classified as arkose or subarkose depending on whether the feldspar content is greater or less than 25% respectively. Arkose therefore is the name restricted to arkosic arenites with a high (> 25%) proportion of feldspar and a very low amount (< 10%) of matrix (Fig. 3.6). The term arkose should be used with caution. Many feldspar-bearing sandstones which seem to be arkose at first glance, turn out on detailed scrutiny to be subarkose or arkosic wacke.

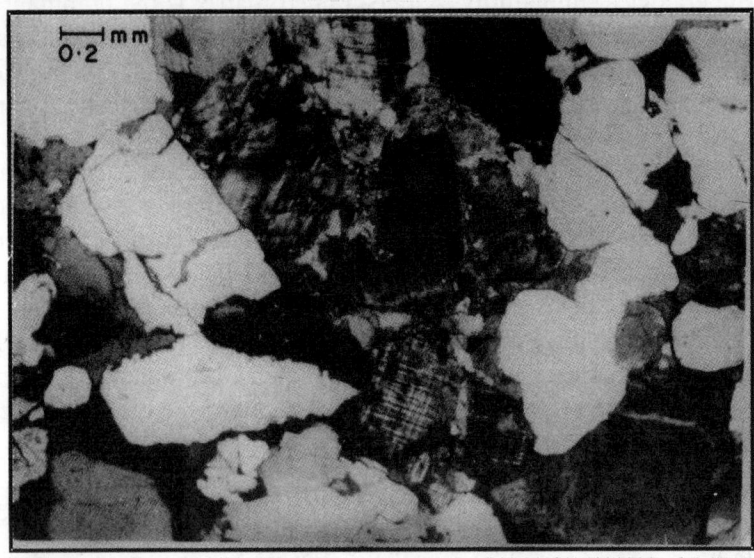

Fig. 3.6: Photomicrograph of an arkose under cross-polarizers. Note large grains of feldspars. Barakar Formation, Central Godavari Valley, India (thin section No. SSG-1, author's collection).

Next to quartz and feldspar, the mineral which is frequently present in arkose is mica, both of the muscovite and biotite varieties (Table 3.2). Calcite is a common cementing material. Since arkoses are commonly derived from granitic sources, the proportion of K-feldspar always exceeds plagioclase. In chemical analysis this is reflected by a higher proportion of K_2O and a lower proportion of Na_2O (see Table 3.3).

The presence of large proportions of fresh grains of an unstable mineral such as feldspar in arkose is puzzling. For a long time sedimentologists maintained that lack of moisture in the environment is essential for preservation of freshness of the feldspar grains. A warm-arid or cold-dry condition was suggested as the essential prerequisite. A later discovery of fresh feldspars in regoliths produced under humid, tropical conditions led to the conclusion that the key factor in the preservation of feldspar is tectonics, not climate. Quick burial without much transportation, due to sudden upheaval of the source and rapid subsidence of the area of deposition, may lead to preservation of fresh feldspar grains. It should be remembered, however, that both climate and tectonism could have played important roles in the preservation of fresh feldspars in some sandstones.

Table 3.3: Average chemical composition of major sandstone types (data from Pettijohn 1975, Table 7.3; averaged by Potter, Maynard and Pryor 1980)

	Orthoquartzite (wt.%)	Graywacke (wt.%)	Arkose (wt.%)
SiO_2	95.4	66.7	77.1
Al_2O_3	1.1	13.5	8.7
Fe_2O_3	0.4	1.6	1.5
FeO	0.2	3.5	0.7
MgO	0.1	2.1	0.5
CaO	1.6	2.5	2.7
Na_2O	0.1	2.9	1.5
K_2O	0.2	2.0	2.8
CO_2	1.1	1.2	2.8
C	–	0.1	–
H_2O	0.3	3.0	0.9

Graywacke

Graywackes, which look very much like volcanic igneous rocks in hand specimens, belong to an exotic group of sandstones. Typical graywackes, as described by Krynine (1948), contain a medium- to fine-grained, angular framework of quartz, chert and rock fragments such as slate, schist, phyllite, set in large proportion of fine-grained matrix. Silt-size grains of illite, chlorite, micas, and various alteration products of basic igneous rocks constitute the matrix. Feldspars, abundant in graywackes derived from high-rank metamorphic sources, are usually of the sodic plagioclase variety.

Graywackes, often found in association with a huge thickness of shales and radiolarian chert, are believed to be the product of sand carried into the deep sea by *density (turbidity)* currents although not all turbidites are graywackes (for example, see Okada 1966). Typical graywackes are commonly associated with altered basic intrusives ('*greenstones*') and often grade into volcanic wackes. Most of these rocks have been reported from orogenic belts of Precambrian to Cenozoic age.

The term greywacke was originally assigned by the 18th century miners in the Harz mountain in Germany to a rock with an exceptionally high matrix content (hence called grau + wacke, the German for gray, pasty matter; 'gray grit' or 'dark, dirty sand'; Krynine 1948). Since that time the term has been used for different types of rocks or even assemblages of rocks. The definition of graywackes became so ambiguous that any dark-coloured rock not positively known to be igneous came to be jestfully referred to as a greywacke in the Canadian Shield. For these reasons questions have been raised about the desirability of retaining the term greywacke. Tectonic or turbidite origin is implied in the classical usage of the term, but rocks similar to graywackes have been reported from fluvio-estuarine sediments also, while clean, non-graywacke sand has been found interbedded with argillaceous sediments of deep-water origin. Compositions of the type greywacke of the Harz mountain of Germany, and those of some arkose of France are known to overlap (Okada 1971). Considering all these, some authors recommend replacement of the term greywacke by a more general term, such as 'wacke' (Okada 1971; Friedman and Sanders 1978; Fritz and Moore 1988). In his three-component scheme of classification Okada (1971) recommend use of only two groups of sandstones : arenites and wackes. The feldspathic and lithic wackes of this scheme encompass all sandstones, including those that have been traditionally designated as graywackes.

The term greywacke and arkose have long pervaded sedimentological literature and discontinuation of these terms would merely add to the existing confusion. Moreover, graywackes, with their abundant matrix, and angular framework grains, create a distinctive textural pattern (see Figs. 3.7 and 3.8). It is proposed here to include these sandstones in a new triangular field at the high matrix end of the revised Dott diagram (Fig. 3.3). Consequently, the term greywacke has been removed from its usual place in the original Dott diagram of 1964.

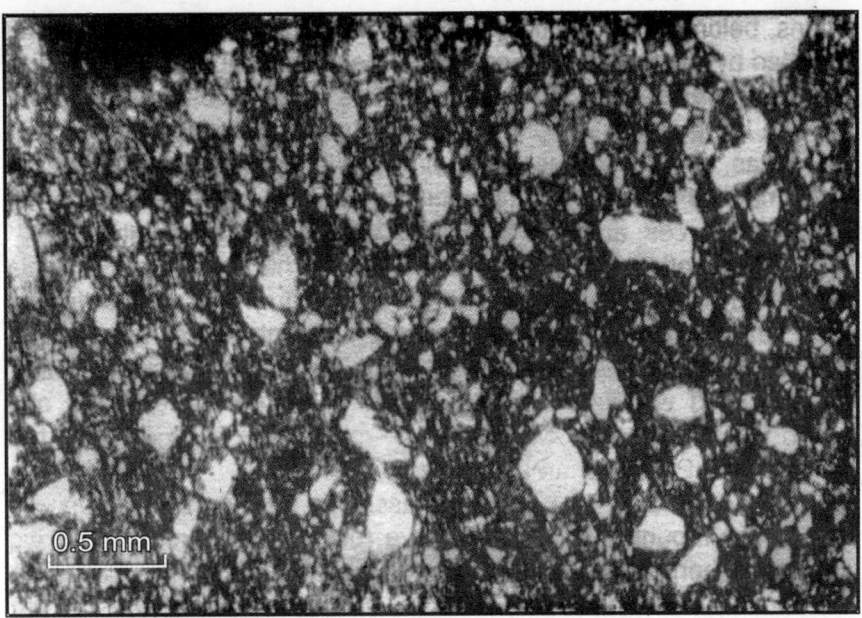

Fig. 3.7: Photomicrograph of a graywacke, Chandpur Formation (Cambro-Ordovician?), north-west of Nainital, India. Plane polarized light. Note high proportion of matrix (thin section No. N-1, author's collection).

The problem of the occurrence of a large proportion of matrix in graywackes was studied experimentally (Kuenen 1966, Kuenen and Sengupta 1970). More recent experimens (Ghosh, Mazumdar, Saha and Sengupta 1986) indicate that, contrary to common belief, a wide range of grain-sizes, from coarse to fine, may be simultaneously deposited from a suspension when the flow velocity of a stream carrying a large amount of suspension load is decreased. The fine-grained particles deposited in the intergranular pores of the coarser framework grains might, in such cases, produce a texture similar to that of a greywacke.

Chemical Composition of Sandstones

Average chemical compositions of orthoquartzites, arkoses and graywackes are shown in Table 3.3. As expected, an orthoquartzite consists almost wholly of silica, whereas the proportion of alumina is quite high both in arkose and greywacke. The presence of aluminosilicates, like feldspars, accounts for the alumina in these rocks. In arkose potash exceeds soda because of the presence of potash feldspars but the proportion is often reversed in graywackes due to the presence of sodic plagioclase.

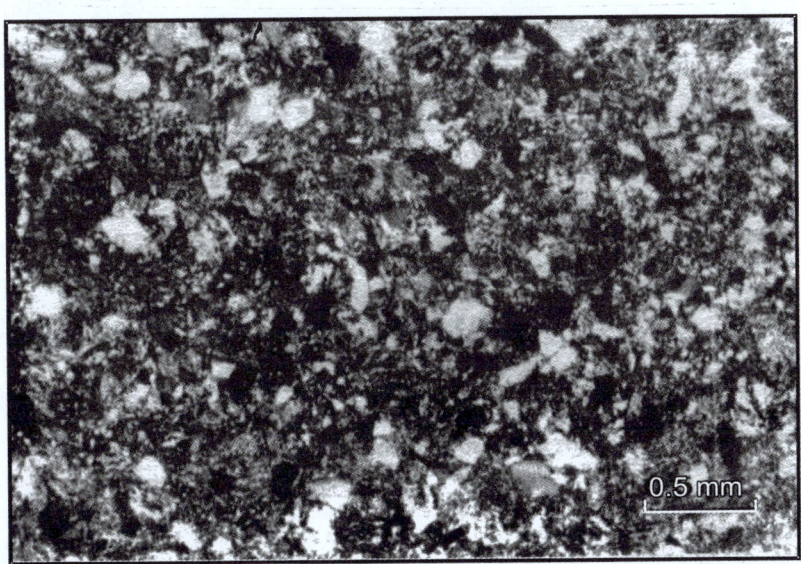

0.5 mm

Fig. 3.8: Photomicrograph of a graywacke from Mt. Marion Formation (Middle Devonian), north-east of Monroe, USA. Note angular framework grains 'floating' in a high proportion of matrix (Plane polarized light). Author's collection.

Table 3.4: Average bulk chemical composition of Gondwana sandstones from Raniganj basin, India (after Suttner and Dutta 1986)

	North America		Gondwana, India				
	LA	*AR*	*I*	*II*	*IV*	*V*	*VI*
(Blatt et al. 1972)			*(n = 5)*	*(n = 7)*	*(n = 11)*	*(n = 2)*	*(n = 4)*
SiO$_2$	66.1	77.1	70.30	79.60	72.70	93.50	92.15
Al$_2$O$_3$	8.1	8.7	10.50	7.53	9.31	2.46	3.12
K$_2$O	1.3	2.8	1.99	1.71	2.25	0.78	0.35
Na$_2$O	0.9	1.5	1.29	0.41	1.20	0.04	0.02
CaO	6.2	2.7	2.34	1.23	3.68	0.08	0.05
FeO	1.4	0.7	3.42	1.01	2.90	0.60	1.31
MgO	2.4	0.5	0.68	0.49	0.77	0.07	0.06
MnO	-	-	0.58	0.03	0.07	0.01	0.02
TiO$_2$	-	-	0.27	0.29	0.47	0.11	0.48
P$_2$O$_5$	-	-	0.36	0.25	0.26	0.50	0.35
Ignition loss	-	-	5.50	4.51	4.18	-	-
SiO$_2$/Al$_2$O$_3$	8.3	8.7	6.7	10.9	8.2	38.4	29.5

LA –	average lithic arenite; AR-average arkose.
VI –	sandstones from upper part of Panchet Formation.
IV –	sandstones from upper part of Barren Measures ('Ironstone Shale') Formation + Lower Panchet Formation.
II –	sandstones from Barakar and lower part of the Barren Measures ('Ironstone Shale') Formation.

Chemical analyses of Indian sedimentary rocks are rare. The average chemical composition of a few Gondwana sandstones from the Raniganj coalfield, West Bengal, are reproduced in Table 3.4 from Suttner and Dutta (1986). The average chemical composition of lithic arenites (LA) and arkoses (AR) is also included in this table. In each of these analyses potash exceeds soda, indicating the presence of K-feldspars. The calcium found in these rocks is apparently derived from the calcium carbonate cement.

Average modal analyses of the Gondwana sandstones from the Jharia and Raniganj basins have already been shown in Table 3.2. The feldspars of the sandstones belonging to the Ironstone Shale Formation of Raniganj coalfield account for high Al_2O_3 content of these rocks. The presence of appreciable quantities of FeO and CaO is obviously due to the ferruginous and carbonate cements.

The chemical and mineralogical composition of graywackes may vary considerably, depending on their occurrence in relation to particular geotectonic settings. These are discussed in detail in Chapter 8.

Lutaceous (Argillaceous) Sedimentary Rocks

Lutaceous sedimentary rocks are composed mainly of grains smaller than sand size. Sediments finer than sand but coarser than clay are classified as silt. Silt and clay together constitute mud. Indurated silt is called siltstone. Indurated, non-laminated mud forms mudstone. When mudstone is laminated and fissile, the rock is designated as shale. About 1/3rd to 2/3rd of such rocks are composed of clay-size particles, the rest being silt (Table 3.5). The fissility of shales is attributed to parallel arrangement of the constituent flaky minerals. Black shales, rich in organic matter, are particularly fissile while calcareous shales have no fissility at all.

Table 3.5: Classification of indurated argillaceous rocks (more than 50% of grains are finer than sand). (Modified after Potter, Maynard and Pryor 1980)

Proportion of clay-size constituents (rest mostly silt)	0-32%	33-65%	66-100%
Beds > 10 mm thick	Bedded Siltstone	Mudstone	Claystone
Laminae < 10 mm thick	Laminated Siltstone	Mudstone (Shale)	Clay-shale

Shale constitute more than half of the stratigraphic record and hence are important. A study of shales is not as easy as that of sandstones, however, because the grain-size involved is very small. A study of the clay minerals which constitute a substantial part of all shales is possible only with advanced techniques. Shale chemistry is considered more important than shale petrology. SiO_2 and Al_2O_3 together constitute the dominant part of common shales. H_2O and CO_2 constitute about 10%. Five other oxides, Fe_2O_3, FeO, K_2O, MgO and CaO account for about 15-20% of common shales. Potash generally exceeds soda. In Table 3.6 the chemical compositions of some Indian shales of Precambrian and Permian age have been compared with that of the average shale composition.

Table 3.6: Chemical composition of shales

	1	2	3	4	5	6
SiO_2	58.10	58.93	62.15	39.05	48.06	64.57
Al_2O_3	15.40	18.17	11.38	29.07	19.31	11.93
Fe_2O_3	4.02	9.55	6.09	0.51	1.56	13.47
FeO	2.45	–	–	24.97	26.05	–
MgO	2.44	2.90	5.06	0.34	2.10	1.31
CaO	3.11	0.13	4.01	2.22	4.12	0.87
Na_2O	1.30	0.40	nd	1.36	nd	0.84
K_2O	3.24	5.78	4.07	0.95	nd	–
TiO_2	–	0.92	0.68	0.03	Tr	0.80
MnO	–	0.03	0.06	0.36	0.12	0.02
P_2O_5	–	0.12	0.09	0.06	nd	0.20
S	–	0.005	nd	nd	nd	2.27
Cr_2O_3	–	0.02	0.01	–	–	–
NiO	–	0.004	0.001	–	–	–
BaO	–	0.05	0.04	–	–	–
CO_2	2.63					
C LOI	0.80	3.18	7.52	*	*	4.12
H_2O	5.00					

1. Average shale (Pettijohn 1975, Table 8.7, averaged by Potter *et al.* 1980).
2. & 3. Precambrian shales from Varikunta-Zangamrajupalle, Cuddapah, India (Geological Survey of India 1986, source: V. Subrahmanyam).
4. Naturally fused Permian shale ('para lava'), Bokaro Gondwana Basin, India (Fermor 1918).
5. Naturally fused Permian shale ('para lava'), Jharia Gondwana Basin, India (Sengupta 1957).
   * The volatiles have escaped during spontaneous combustion and fusion.
6. Bijaigarh Shale, Precambrian, Vindhyan 'System' (Awasthi 1964, cited by Singh 1980).
   nd = not determined.

Composition and Classification of Lutaceous Rocks

Besides clay minerals, which constitute the bulk, some grains of quartz, feldspar, carbonates and organic matter are found in all shales. The non-terrigenous constituents include glauconite, volcanic glass, biogenic silica and some phosphatic compounds. Shales may be classified by their colours, which reflect chemical composition. With increasing Fe^{3+}/Fe^{2+} ratio, the colour of a shale changes from grey to red, through green and yellow (McBride 1974). Red shales, found near the surface, indicate the presence of hydrated ferric oxide minerals, which are formed under oxidizing conditions. Since an arid or semi-arid condition helps oxidation, red shales may be taken to be indicative of an arid palaeo-environment. A red colour is also common in deep-marine clays. This is partly due to oxidation of the ferruginous minerals carried into the sea by streams and partly due to the volcanic dusts that settle into the deep sea from the atmosphere. Clay-size ferruginous matter, settling through the oxygen-rich sea-water at an extremely slow rate, may become oxidized over the years.

Similarity in the chemical composition of some deep-sea clays and an average igneous rock suggests their volcanic origin (Dunbar and Rodgers 1957, p. 57).

The proportion of organic carbon is another important constituent influencing the colour of shales. With an increasing amount of organic carbon, grey shales tend to be black. The presence of pyrite and siderite together with trace elements such as vanadium, nickel and uranium in black shales suggests that these shales were deposited under a reducing environment (see discussions under euxinic basins).

Clay Minerals in Shales

Clay minerals (hydrous aluminosilicates) such as kaolinite, illite and smectite are important constituents of shales. Gibbsite (aluminium hydroxide) is also common. The type of clay mineral present in a particular shale depends on the nature of the source rock, depositional environment, climate and also, to some extent, the degree of diagenesis undergone. Kaolinite is generally the product of warm, humid, tropical conditions of weathering. In areas of low rainfall the proportion of smectite increases compared to kaolinite. Illite, very common in shales, may be derived either from pre-existing shales in the source area or by diagenesis of clay minerals such as smectite. Chlorite can also be produced in this way. In fact, illite and chlorite are common in older shales while kaolinite and smectite are characteristic of younger shales.

Depositional Environment of Shales

A certain amount of low-energy condition is essential for deposition of all fine-grained sediments. The deposits of the deep-sea include blue-grey mud of terrigenous origin, green mud of glauconitic origin, and red mud produced either out of volcanic dusts or from clay-size particles of ferruginous sediments washed into the sea. Planktonic fauna may be associated with these deposits when they are laid down at relatively shallow depths. At greater depths siliceous fauna predominates. Ocean-floor clay deposits may extend for hundreds of kilometres. Elongate bodies of silt may be deposited on the inner and outer shelf areas, if the energy condition there is sufficiently low.

Deposition of silt and clay is in no way restricted to a marine environment only. Finely suspended silt and clay spilling over river channels during high flood may produce extensive sheets of mud in the flood basins. Settling by flocculation on overbank areas also causes deposition of mud. In large lakes the coarser particles are trapped near the shore but the finer mud travels into the lake centre, to be dropped there from suspension. These form extensive, ovate muddy deposits at the lake bottom. Invertebrate shells of various types, spores and algae are common in these deposits. In closed basins, whether marine or non-marine, restricted circulation may lead to the deposition of black shales. Nearshore areas of fluctuating marine and continental conditions can also trap a large amount of mud. Deltas, lagoons and tidal flats are examples of mixed environments which can have large deposits of silt and clay.

Many shale sections in stratigraphic record demonstrate cyclicity. These may be either varves or non-varves, such as alteration of shale and limestone. Some of these features are discussed later (Chapter 7).

ROCKS OF CHEMICAL AND BIOCHEMICAL ORIGIN

Carbonates

Sedimentary rocks of chemical and biochemical origin are divided into two groups : evaporites and non-evaporites. Non-evaporites constitute the bulk of chemical and biochemical rocks. Among non-evaporites the most important are the carbonate rocks. The carbonate minerals forming these rocks include calcite (rhombohedral $CaCO_3$), aragonite (orthorhombic $CaCO_3$) and dolomite (rhombohedral $CaCO_3.MgCO_3$). Rhombohedral iron carbonate, $FeCO_3$ (siderite) also constitutes a small proportion of the naturally occurring carbonates. The term limestone is generally restricted to the calcitic and aragonitic rocks. Limestones composed essentially of the mineral dolomite are called *dolostones*. Approximately one quarter of the stratigraphic record is constituted of limestones and dolostones. Limestones occur in all ages, from the Precambrian to the Recent, but they are most common in the Late Precambrian (Proterozoic). Limestones are also abundant in the early Palaeozoic.

Chemistry of Carbonate Precipitation

Precipitation of calcium carbonate in nature is dependent on a cycle of natural processes termed the carbonate cycle. These processes involve formation of carbonic acid from water and carbon dioxide ($CO_2 + H_2O \Leftrightarrow H_2CO_3$) followed by dissolution of carbonic acid in two stages (McLane 1995, p. 228).

$$H_2CO_3 \Leftrightarrow H^+ + HCO_3^-$$

$$HCO_3^- \Leftrightarrow H^+ + CO_3^{2-}$$

Precipitation of calcium carbonate takes place according to the following reaction:

$$Ca^2 + CO_3^{2-} \rightarrow CaCO_3$$

The natural processes aiding in the above mentioned chemical reactions are: increase of temperature, evaporation, upwelling of deep marine waters to zones of lower pressure and higher temperature, photosynthesis, and bacterial decay.

All limestones contain some quantities of Si, Al, and Fe. These are the acid insoluble residues (Table 3.7). In natural calcites the amount of $MgCO_3$ molecule is less than about 4 mole percent. These are the low-Mg calcites. High-Mg calcites contain much higher proportion (10-18 mole per cent) of $MgCO_3$. The chemical composition of some Indian limestones of different geological ages is compared to the composition of the well-known limestones from Europe and USA in Table 4.7. An average carbonate rock, like the one tabulated in the first column of this table, consists of some quantity of MgO together with large proportions of CaO and CO_2. In fact, calcite ($CaCO_3$) and magnesite ($MgCO_3$) form an isomorphous series within which two varieties of calcite occur. One of the varieties designated as high magnesian calcite, contains more than 10 mole per cent of $MgCO_3$. The other, designated as low-magnesian calcite, contains less than 8 mole per cent $MgCO_3$. This division is arbitrary. A calcite containing 9 mole per cent $MgCO_3$ may be designated either high-or low-magnesian calcite (Friedman and Sanders, 1978).

Carbonate sediments, it must be remembered, form close to the area of deposition either by inorganic precipitation or by organic activity. The chemical conditions prevailing at the site of deposition therefore, play important roles in carbonate sedimentation. Oxygen isotopes in limestones have been used extensively for interpreting *palaeotemperature* and *palaeosalinity*. The problem in palaeotemperature interpretation is that the oxygen isotope is affected, not only by temperature variation but also by physiology and life habitat of organisms. Interpretation of palaeosalinity from oxygen isotopes gives more dependable and consistent results. Isotope studies can also be used in the interpretation of diagenetic and cementation history of limestones.

Table 3.7: Chemical composition of limestones and dolomites (percentage of major oxides)

	1	2	3	4	5	6	7	8
SiO_2	5.19	1.15	0.09	8.774	15.88	10.51	11.44	4.12
Al_2O_3	0.81	0.45		0.762	6.79	3.28	2.85	1.24
Fe_2O_3	0.54	–	} 0.11	0.720	} 8.69	} 3.36	1.15	0.92
FeO		0.26		0.745			0.78	0.16
MgO	7.90	0.56	0.35	15.363	1.08	15.49	17.50	2.03
CaO	42.61	53.80	55.37	30.319	34.06	25.79	24.24	50.42
Na_2O	0.05	} 0.07	–	0.994	nd	nd	1.26	0.84
K_2O	0.33		0.04	0.784	1.43	1.74	0.86	0.67
TiO_2	0.06	–	–	0.141	0.26	0.14	0.33	0.22
MnO	0.05	–	–	0.105	1.71	0.62	0.39	0.30
P_2O_5	0.04	–	–	0.109	0.03	0.02	0.45	0.36
S	–	–	–	–	0.005	nd	–	–
Cr_2O_3	–	–	–	–	0.01	nd	–	–
BaO	–	–	–	–	0.19	0.03	–	–
H_2O^+	0.56	0.69	} 0.32	LOI 41.624	29.86	38.94	0.55	0.77
H_2O^-	0.21	0.23						
CO_2	41.58	42.69	43.11				38.26	38.65

1. Composite analyses of 345 limestones (Clarke 1924)*
2. Lithographic limestone, Solenhofen, Bavaria (Clarke 1924)*
3. Travertine Hotsprings, Yellowstone, Wyoming (Clarke 1924)*
4. Average composition of 38 limestone samples, Krol Formation, Simla Himalayas (Bagati 1979).
5. Limestone, 6. Dolomite, both from Varikunta-Zangamrajupalle, Cuddapah, India (Geological Survey of India 1986, source : V. Subrahmanyam).
7. Fawn dolomitic limestone (microdolsparite), Son Valley, India (Mehrotra *et al.* 1975).
8. Rhotas limestone (micrite), Son Valley, India (Mehrotra *et al.* 1975).
Note : nd = not determined * Cited by Pettijohn 1975, Table 10.4.

Trace Elements in Limestones

A number of *trace elements*, particularly, Fe^{2+}, Mn^{2+}, Sr^{2+}, constituting 10^{-2} to 10^{-3} weight per cent of the rock, are also found in limestones (see Table 3.8). Ions of these trace elements replace the host ions having similar charge and ionic radius. Later they find their way into the interstitial space between crystal boundaries. These trace elements attach themselves to the silicates, oxides or carbonates, or may occur as substitutes of Intensive studies on these trace elements have been carried out with the hope of obtaining clues to the microenvironment of deposition and diagenesis of carbonates. For example, the proportion of iron and manganese in carbonates is determined largely by the state of oxidation (redox potential) during carbonate precipitation which in turn is controlled by organic respiration, bacterial decay, and photosynthesis. However, such environmental interpretations may not always be correct. Sr, commonly believed to be characteristic of lagoons and central reefs, is also found in evaporites and inland lake deposits (Flügel 1982). Sr in aragonite, once thought to be phylogenetically controlled, is now believed to be dependent also on environmental temperature. Non-reefal carbonate sediments are characterized by trace elements like Ni, Co, V, Ba, and occasionally, also Sr. The presence of sodium in carbonates is considered by some to be indicative of precipitation in saline water. High concentration of P in the carbonates from the Krol Formation of the Simla Himalayas has been taken to be indicative of near shore lagoonal environment (Table 3.8). Similarly, high concentration of Mn and P are believed to be indicative of warm, humid climate (Bagati 1979). (See Tucker *et al.* 1990, McLane 1995 for further details on trace elements in limestones).

Table 3.8: Average concentration of trace elements (in ppm) in relation to carbonate microfacies, Krol Formation, Simla Himalaya (Bagati 1979)

Trace elements	Dolomicrite #1	Chert-bearing dolomicrite	Dolomicrite #2	Dolosparite	Siliceous oolite-bearing oodolomicrudite	Intradolosparudite	Micrite	Sandy micrite	Total no. of samples
	4 samples	4 samples	2 samples	5 samples	1 samples	4 samples	2 samples	1 samples	23
Pb	60	50	25	50	50	70	58	50	
Ga	68	145	-	43	55	113	97	88	
V	59	-	75	84	140	14	-	-	
Bi	27	187	30	92	15	129	162	62	
Sn	30	60	-	37	-	57	28	20	
Cu	69	44	30	43	45	42	19	50	
Ni	50	45	25	50	50	50	50	50	
Cr	-	35	-	-	-	21	-	25	
Ba	19	200	15	26	75	99	82	90	
Sr	91	160	135	126	140	131	175	290	
Zn	112	25	13	25	25	112	25	40	

- Sought but not detected.

Note: For definition of rock names such as dolomicrite see text under dolomite.

Limestone Forming Environments

Limestone is found today in almost all environments, both marine and non-marine (Fig. 3.9). Seawater being mildly alkaline (pH ~ 8) allows precipitation of limestone. Carbonate saturation in water is highest in tropical to warm, temperate seas. A small increase in temperature or alkalinity in these areas may cause the carbonate to precipitate. The type of carbonate sediments deposited on marine shelves is discussed in detail in Chapter 7, under facies associations.

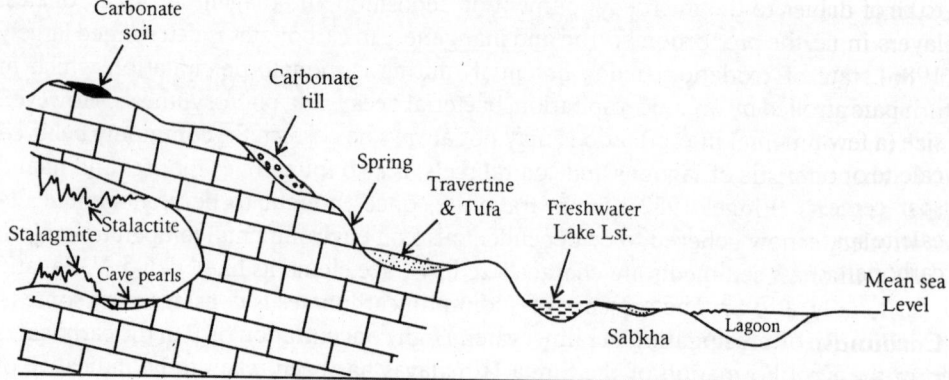

Fig. 3.9: Depositional environment of non-marine carbonates. (based on Wolf 1978).

Limestones of non-marine origin include *algal mounds* formed in fresh water lakes. These are mixtures of carbonates and argillaceous sediments. The corresponding rock is called *marlstone.* Travertines in hot springs, spongy, porous carbonate deposits (*tufa*) associated with natural springs; dense, elongate, vertical bodies (*stalactites* and *stalagmites*) formed within caves, are other examples of freshwater limestone deposits (Fig. 3.9). All these are mostly calcitic limestones. In the arid, desert environment, and also in some supratidal areas, dolomite occurs together with gypsum and anhydrite. The environment of deposition of this association is discussed under *sabkhas* (p. 154 footnote).

In dry and semi-arid regions thin layers of carbonate occur on the surface soil together with some silica, alumina and iron oxide. These are produced by evaporation of lime-rich water drawn to the surface from the vadose zone by capillary action. These deposits, recognizable in ancient stratigraphic rocks also, are named *caliche.* Such features have been described from many localities including the Permian Capitan reef complex of southern New Mexico and west Texas (Dunham 1969). The indurated counterpart of caliche is the *calcrete,* also called *kankar* in India, *nari* in the Middle East, and *kalkkruste* in Germany. In Chile and Peru the term *caliche* is used for deposits of impure nitrate of soda together with sulphates, iodates and phosphates. *Karst* (named after the Karst region of Yugoslavia) implies a particular type of topography, well known for its scenic beauty. Carbonate dissolution and precipitation within caverns by percolating ground water and underground streams are responsible for generation of karsts.

Pisolites consist of pea-sized (2-10 mm in diameter) grains commonly made of calcium carbonate. Pisolites of other minerals like bauxite, limonite, hematite are also known. In cross-section pisolites show concentric laminae similar to those found in oolites. Such features, often found in caves and mines, are also called *cave pearls.* They are common in geyser and hot spring deposits. Generation of growth-laminae by chemical precipitation followed by their abrasion due to contact with neighbouring grains in agitated water is the process commonly suggested for the development of pisolites (Donahue 1969). Some pisolites of the Carlsbad facies, previously regarded as algal onkoids produced in shallow lagoon behind the reef rock, have been shown to be of diagenetic origin (Dunham 1969). However, *'pisoids'* containing radial fibrous-layers in the Miocene reef talus of Egypt are of algal onkoid origin (Aissaoui *et al.* 1986).

Spars are equant, blocky, anhedral chemically precipitated carbonates of fairly large size (a few micrometres to several millimetres). They are composed of low-magnesium calcite or ordered dolomite, but never of high-magnesium calcite. Calcite and dolomite spar cement occurs very early in the diagenetic history of a carbonate rock. Some calcite spar (*pseudospar*) can also be the product of neomorphism of micrite or of fibrous carbonate cement (McLane 1995 p. 239).

Carbonate Petrography and Classification

Limestones can form by precipitation within a basin or may be redeposited elsewhere, after erosion and transportation from their place of origin by water or air. The *in situ* limestones have been termed *autochthonous* by Pettijohn (Folk prefers the term *orthochemical* for such rocks) while the term allochthonous is used for the transported variety (*allochemical* of Folk). Their texture and structure can distinguish the two types of limestones. While the grains of allochthonous limestones can be well sorted as to size and have their interstices filled with clear calcite cement, the grains of autochthonous limestones are usually unsorted. The intergranular spaces of such limestones are filled with lime mud. The allochthonous limestones show evidence of transportation, such as cross-bedded structures. The autochthonous limestones have articulated organic structures like undisturbed coral colonies. Carbonate rocks made up of materials eroded from older, lithified carbonate sediments have been termed *calclithite* by Folk (1959) to distinguish them from *calcarenites*, a term coined by Grabau (1904) for rocks composed of carbonate sand produced by chemical or biochemical precipitation.

The transported and redeposited allochemical components of carbonates àre divided into the following classes by Folk (1959).

Intraclasts: Reworked fragments of penecontemporaneous carbonate sediments.

Fossils (skeletals): Broken or unbroken skeletons and tests of carbonate secreting organisms.

Oolites (ooids): Spherical carbonate bodies, 0.1 to 1.0 mm in diameter having concentric or radial internal structure.

Pellets (peloids): Extremely small (0.03-0.15 mm) carbonate bodies lacking internal structures.

Intraclasts are made up of microcrystalline or fine-grained carbonates. Ripped-apart fragments derived from the underlying desiccated beds by strong currents often constitute the intraclasts. These often show 'curl-up' desiccation structures. The fossils may range from invertebrates like foraminifer, corals, echinoderms, bryozoans, and brachiopods to broken skeletal remains of vertebrates or even shark teeth. The fossil assemblages found in limestones often characterize a particular geological age or environment. Oolites generally occur in clusters. Recent oolites are composed of concentric layers of aragonite around a nucleus, which may or may not be a carbonate. Ancient oolites are generally calcitic. Oolites having radial internal structures often alternate with those having tangential structures. Opinion varies as to whether the radial or tangential structures are of primary origin. Most of the pellets represent excretra of animals ('fecal pellets') but pellets of non-organic, chemical origin are also known. ·

Tucker *et al.* (1990) used the term *peloid* to describe sand size, spherical grains composed of microcrystalline carbonate. These have been described as a group of polygenetic grains originating out of any of the following constituents: faecal pellets, calcareous algae, micritised grains, or simply mud clasts.

The orthochemical components are divided into two groups: micrite and sparite.

Micrite: Microcrystalline calcitic deposits of a size comparable to terrigenous clay (0.001 to 0.004 mm). The constituent grains can hardly be resolved, except under high magnification. White or grey (with waxy to dull lustre) to nearly black in hand specimens, micrite is believed to be deposited as biochemical ooze or produced by attrition of organic shells. Some of the ancient micrites are possibly of diagenetic origin. Micrites may consolidate to form carbonate rocks.

Sparite ('spar' or sparry calcite): Calcite crystals deposited within veins, pore spaces or fissures are called sparite. Individual spar crystals generally range from 0.01 to 0.10 mm in size. They may exceed 1.0 mm in diameter in exceptional cases. Some of the ancient spars are believed to be the products of recrystallization of micrite. Sparry calcite is deposited only as a secondary vein or pore-filling material.

Folk's scheme of classification of limestones uses various combinations of the allochemical and orthochemical components (see Fig. 3.10). Each term used in Folk's scheme is divided into two parts: a prefix and a stem (Table 3.9). The prefix indicates the nature of the allochemical component (intraclast, fossil, oolite or pellet). The stem indicates the nature of the orthochemical component (micrite or sparite). The speciality of Folk's scheme is that the nomenclature gives a complete petrographic description of the limestone. Thus Folk's scheme of limestone classification is so popular among petrographers.

A quick method for on-the-spot classification of limestones was developed by Dunham (1962) with a view to helping well-site and field geologists. This classification scheme is based on easily identifiable characters—whether the limestone is grain-supported or mud-supported. This scheme has been popular, particularly among petroleum geologists, because it provides a quick method of identification of rock properties useful for reservoir evaluation. A comparative study of the schemes proposed by Folk (1959), Dunham (1962) and Pettijohn (1975) is given in Table 3.10. The major groups of limestones identified by Dunham (1962) are described below (see Fig. 3.14).

Grainstones of Dunham (1962) are rocks having a grain-supported framework. The intergranular space is filled with clear calcite cement. The carbonate detritus may consist of fragments of pre-existing limestones (intrasparite of Folk), fossil fragments (biosparite of Folk) or oolites (oosparite of Folk). Depending on the grain-size of the constituting carbonate grains, a grainstone may be grouped either as *calcarenite* (sand-size grains) or *calcirudite* (larger than sand-size) of Grabau (1904).

The sandbanks of the Bahamas are similar to the ancient calcarenites. This has led many researchers to presume that the ancient calcarenites were deposited either in shallow seas under the influence of strong tidal action, or as subaerial carbonate dunes. Calcarenites, in fact, can be produced in marine conditions when oolites and skeletal debris of sand size are bound together by calcareous mud. Sparry calcite may also serve as the cementing material for calcarenites (Pettijohn 1975, p. 353).

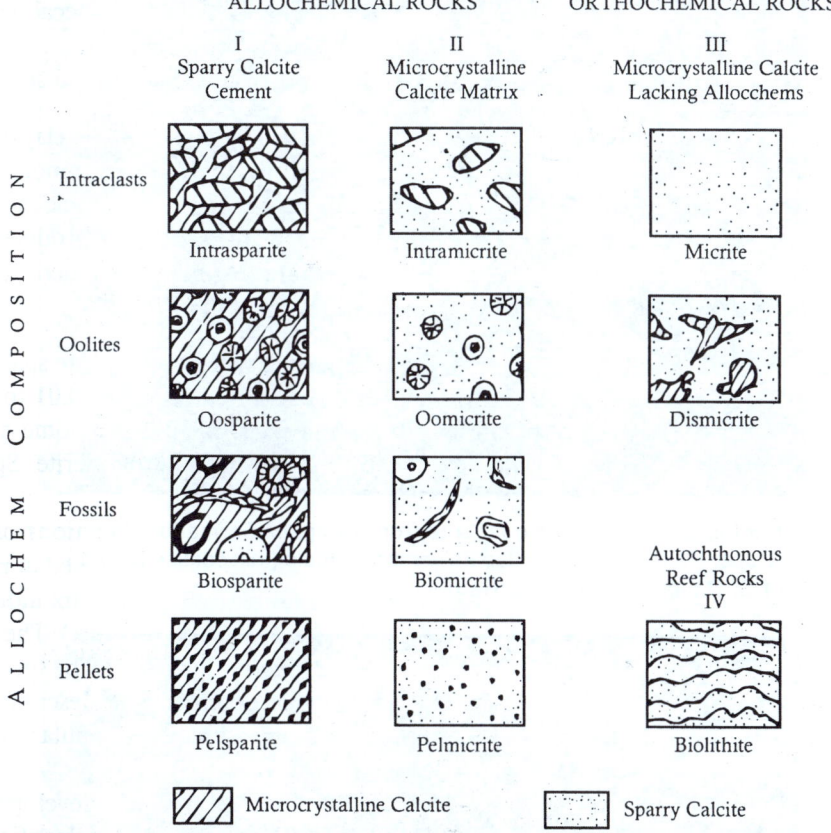

Fig. 3.10: Graphic representation of Folk's scheme for classification of limestones (redrawn from Folk 1959). With permission of the author and American Association of Petroleum Geologists.

Boundstone is the same as reefoidal carbonate. This represents a loose network of carbonate with large voids. The framework is constituted of organisms such as corals, algae and bryozoa.

Wackestones are micrites having more than 10% coarse grains. These may be composed of skeletons of organisms or other materials.

Packstone has a grain-supported framework but contains a matrix of mud.

Mudstone consists of extremely fine-grained carbonate, comparable to the calcilutite of Grabau (1904) or the micrite of Folk (1959). The fine-grained, perfectly homogeneous variety of mudstone used for lithographic purposes is called a *lithographic stone*.

Mudstones may be produced by attrition, or may be chemically or biochemically precipitated. The choice of the term mudstone to designate a particular type of limestone is not a happy one because argillaceous rocks of terrigenous origin and composed of silt-clay are also called mudstones (see Table 3.5).

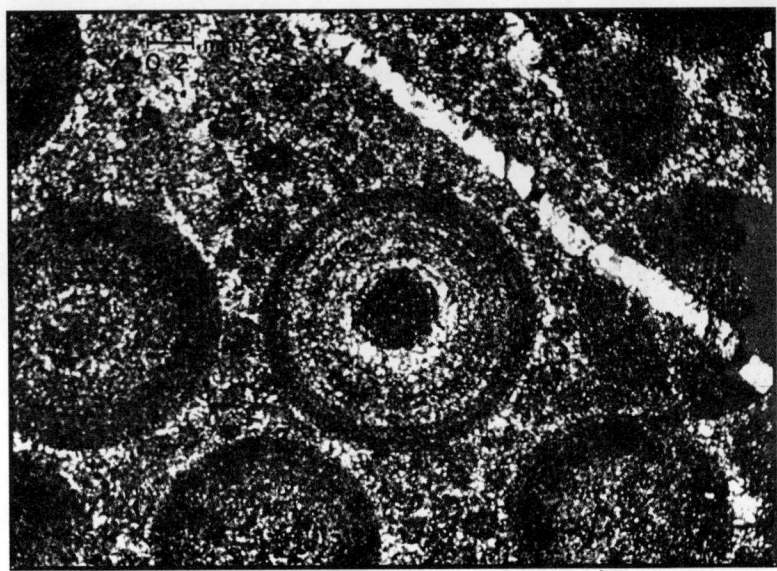

Fig. 3.11: Photomicrograph of an oosparite from Vempalle Formation, Papaghani Group (Upper Proterozoic), Cuddapah, Vempalle, India (thin section No. A11/3, courtesy of D. Dasgupta and C. N. Rao).

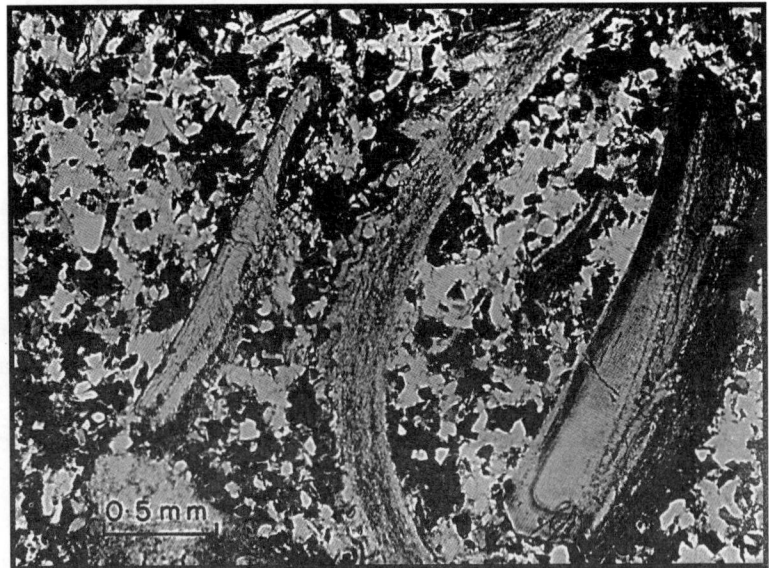

Fig. 3.12: Photomicrograph of a biosparite from 'Shell Limestone'. Trichinopoly (thin section No. 6, author's collection). Cross polarized light.

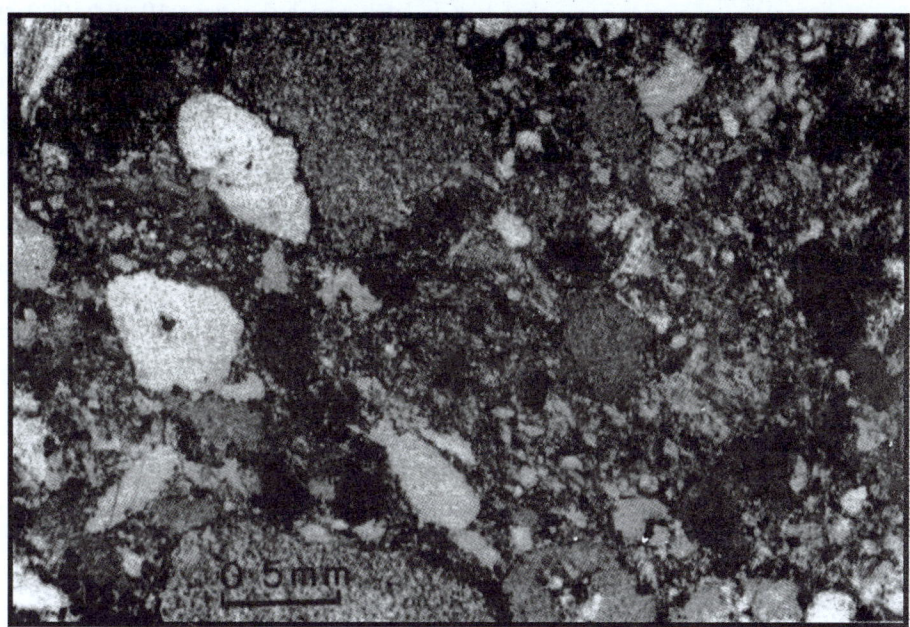

Fig. 3.13: Photomicrograph of an intrasparite. Locality unknown (thin section supplied by Hindustan Minerals & N.H.S.S. Co., Kolkata.

Depositional texture recognizable					Texture not recognizable
Original components not bound together during deposition				Original Components bound together during deposition	
Contains mud (particles less than 20 microns)		Lacks mud			
Mud-supported		Grain-supported			
grains <10%	grains >10%	mud >10%	mud <10%		
Mudstone	Wackestone	Packstone	Grainstone	Boundstone	Crystalline Carbonate

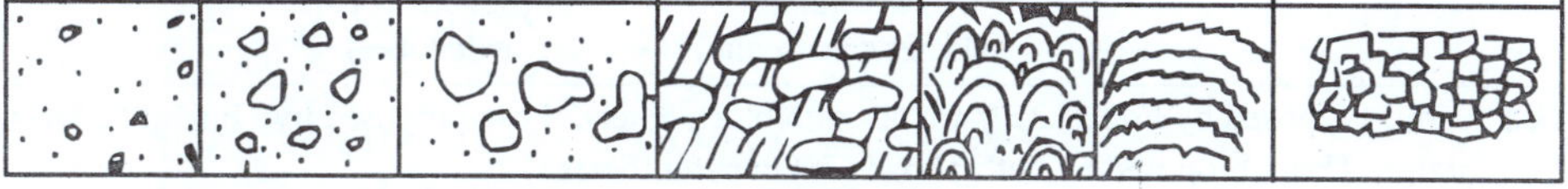

Fig. 3.14: Dunham's (1962) scheme of classification of carbonate rocks according to depositional texture (modified).

Dolomite and Dolomitization

When more than 50% of a carbonate rock is composed of the mineral dolomite ($CaCO_3.MgCO_3$), the rock is called a *dolostone* (Shrock 1948) or simply a *dolomite*. Pettijohn would name a rock dolomite only if the proportion of dolomite in it exceeds

90%. When calcite occupies more that 10% in such a rock (between 10 and 50%), the resultant rock is named a '*calcitic dolomite.*' A dolomite with an appreciable proportion of ferrous carbonate is called a ferrodolomite ('ferrandolomite', Pettiijohn 1975 pp. 359-360). Ferrous carbonate entering into a solid solution with dolomite imparts a buff colour to the rock upon oxidation. In hand specimens dolomite is often identified by this colour, which is uncommon in calcite limestones. The other common impurities in natural dolomites are anhydrite and gypsum.

Table 3.9: Folk's scheme of classification of limestones (1959). Terms proposed by Grabau (1904) and Pettijohn (1975) have been inserted for comparison

Dominant Allochemical Component	Grain size	Orthochemical Component		
		Allochems > 10%		Allochems < 10%
		sparite > micrite	micrite > sparite	Microcrystalline rocks
Intraclasts	> sand	INTRASPARRUDITE *Calcirudite* (Intraformational conglomerate)	INTRAMICRUDITE	
	sand	INTRASPARITE *calcarenite* (lithic calcarenite)	INTRAMICRITE	
Oolites	> sand	OOSPARRUDITE *Calcirudite* (pisolite)	OOMICRUDITE	
	sand	OOSPARITE *Calcarenite* (oolitic calcarenite)	OOMICRITE	
Fossils	> sand	BIOSPARRUDITE *Calcirudite* (coquina)	BIOMICRUDITE (coquinoid limestones)	
	sand	BIOSPARITE *Calcarenite* (biocalcarenite)	BIOMICRITE (fossiliferous calcilutite)	
Pellets	< sand	PELSPARITE *Calcilutite* (Pellet calcarenite)	PELMICRITE (Pelletiferous calcilutite)	

Spanning right column: Micrite and Dolomicrite (Calcilutite)

Note: Terms proposed by Folk (1959) are given in block letters, those proposed by Grabau (1904) in italics. Those proposed by Pettijohn (1975) in parentheses.

    In ancient geological records dolomites are more common than calcite limestones although, strangely enough, dolomite-forming environments are extremely rare in the world today. Only some of the sabkha flats of the Persian Gulf (see under evaporates), the playa lakes of California, the hypersaline lagoons of South Australia, the supratidal areas of Florida and the Bahamas are at present known to be congenial to the precipitation of dolomite.

Table 3.10: A comparative study of the major schemes of classification of limestones

Criterion	Authority		Classification	Name
Grain-size	Grabau (1904)	1.	> Sand size	Calcirudite
		2.	Sand size	Calcarenite
		3.	< Sand size	Calcilutite
Mode of deposition	Pettijohn (1975)	1.	Precipitated *in situ*	Autochthonous
		2.	Transported from outside	Allochthonous
Mode of deposition	Folk (1959)	1.	Precipitated *in situ*	Orthochemical Sparite Micrite
		2.	Transported from outside or locally deposited	Allochemical Intraclasts Oolites Fossils Pellets
Textural features Nature of binding of original components	Dunham (1962)	1.	Components bound together during deposition	Boundstone
		2.	Components not bound together	
			2 (A) Rock is mudsupported grains < 10% grains > 10%	Mudstone Wackestone
			2 (B) Rock is grain-supported	Packstone Grainstone

Seawater is an abundant source of magnesium ions, but dolomite does not widely precipitate in seawater. This is because presence of SO_4^{2-} in seawater inhibits precipitation of dolomite. A little SO_4^{2-} is enough to inhibit dolomitization of calcite, but aragonite can still be dolomitized under this condition (Tucker *et al.* 1990, p. 369). Sulphate in solution, while inhibiting dolomitization, does not totally stop precipitation of dolomite. In fact, in some of the coastal sabkhas dolomite and gypsum are known to precipitate together.

Folk and Land (1975) studied the influence of salinity and Mg/Ca ratio on precipitation of calcite, dolomite and aragonite in different natural environments (Fig. 3.15). It emerged from their study that salinity and Mg/Ca content of the medium are the two most important factors controlling precipitation of calcite and dolomite. When salinity is low, dolomite can precipitate at a Mg/Ca ratio of 1:1. With an increase in salinity, a much higher concentration of Mg is needed for precipitation of dolomite.

Two types of dolomites are recognized: (*i*) primary and (*ii*) secondary. The primary dolomites are very fine grained, smaller than 0.020 mm generally. These, named dolomicrite by Folk, contain unmistakable signs of primary deposition like laminations, ripple marks, and mud cracks. Microbeddings of calcite and dolomite, calcite intraclasts within dolostones, and dolostones laterally passing into clacitic limestones also indicate primary nature of dolomites. The secondary dolomites are the coarsely crystalline, saccharoidal rocks. These bear evidence of replacement like faint outlines ('ghosts') of former grains in rocks composed wholly of dolomite. Preservation of original shapes of dolomitized fossils, presence of grains of detritus particles within dolomites, dolomite veins cutting into calcite limestones, dolomite rhombs cutting the boundary of calcitic fossils are also some of the sure indications of the secondary nature of dolomite. Replacement dolomites are of a secondary nature. Several theories have been proposed from time to time to explain the process of replacement of calcite by dolomite. The more important of these are briefly discussed below.

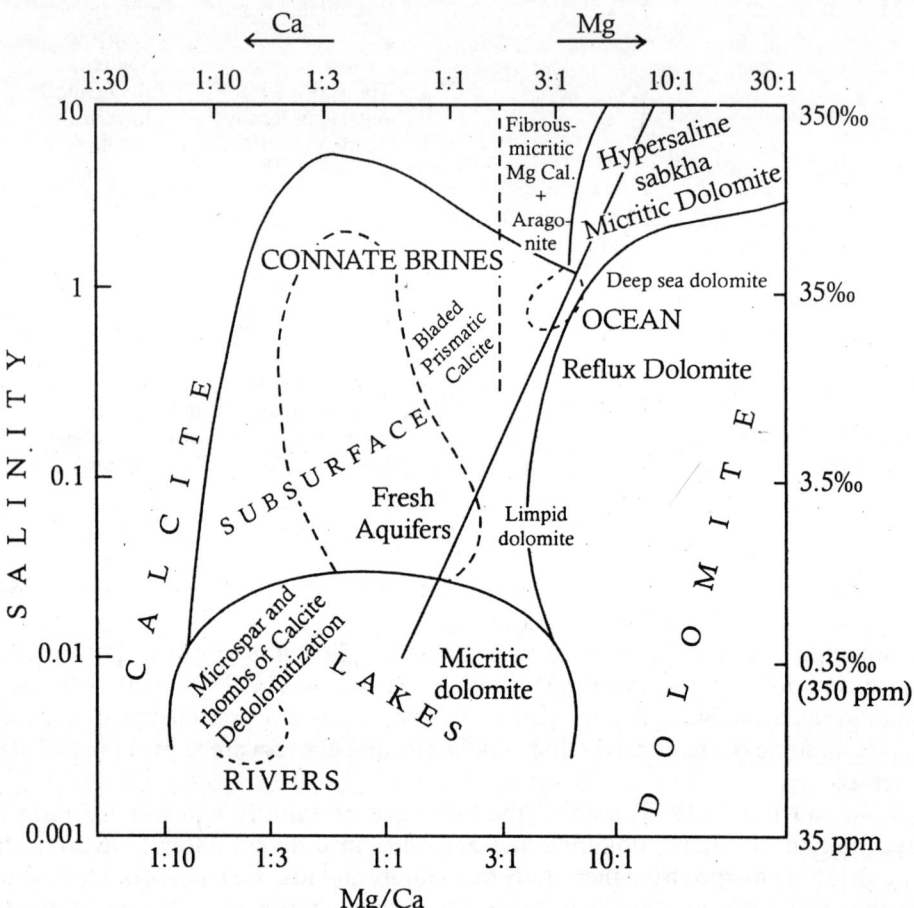

Fig. 3.15: Influence of salinity and Mg/Ca ratio on precipitation of calcite, dolomite and aragonite in different natural environments (compiled from diagrams by Folk and Land 1975).

Seawater entering restricted lagoons evaporates fast, leaving a hot, dense, alkaline and saline fluid (Fig.3.16). This water loses a part of its calcium to the carbonate secreting organisms. Further loss of calcium due to precipitation of calcium bearing evaporates like gypsum and anhydrite on landward shoals leads to hot, hypersaline brine of high Mg/Ca content. Because of its high density this brine invades the surrounding carbonates, replacing calcite by dolomite. This *seepage reflux hypothesis* of Adams and Rhodes was very popular in the 1960s. However, Hsü and Siegenthaler (1969) doubted the efficiency of the seepage mechanism alone to dolomitize large masses of calcite of low permeability. In their model of *evaporative pumping* they suggested that evaporation of the pore water in sediments draw phreatic seawater into the surface enriching its dissolved ion content and aiding in dolomitization.

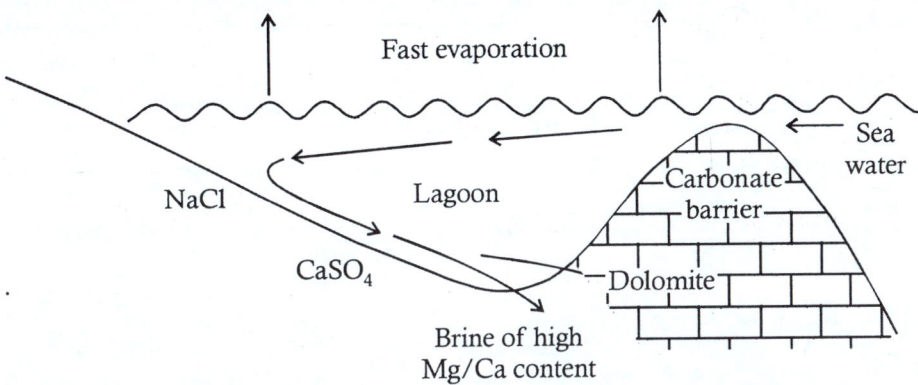

Fig. 3.16: Schematic representation of the seepage-reflux mechanism leading to dolomitization (based on Adams and Rhodes 1960).

Localised replacement of calcite by dolomite along post-depositional structures like faults and fissures might have produced what has been termed as epigenetic dolostones. Syngenetic dolostones could have developed penecontemporaneously by the reaction of lime carbonate with magnesium bearing seawater (Friedman and Sanders 1978, p.179).

Observations on the coastal sabkha of the Persian Gulf led Patterson and Kinsman (1982) to suggest that evaporation of seawater from the supratidal surface of coastal sabkha lead to precipitation of gypsum and accentuation of the Mg/Ca·ratio. This saline water of high magnesium concentration dolomitizes the surrounding calcareous sediment as it seeps into it.

Mixing of the meteoric freshwater with seawater might result to a fluid of low salinity and high Mg/Ca ratio—a condition ideal for the deposition of magnesium. As the zone of mixing of freshwater and marine phreatic water moves across a body of carbonate, it dolomitizes the sediment. (McLane 1995, pp. 246-247).

Dolomitization involves large-scale recrystallization. The process often leads to the generation of a mosaic of euhedral dolomite crystals, producing a texture called *idiotopic fabric*. When the crystal faces are not so well developed a mosaic of anhedral dolomite crystal results (Fig. 3.17). This is called the *xenotopic fabric* (Friedman 1965). Intermediate textures are called *hypidiotopic*.

As early as 1836 it was suggested that replacement of calcite limestone by dolomite results in a volume shrinkage of 12-13%. This idea, subsequently discarded, became popular again, following the studies by Holt in 1948. It is now believed that shrinkage of c-axis of crystals of dolomite creates voids during replacement. Because of the secondary porosity acquired in this way dolomite bodies may act as effective reservoirs of petroleum (Levorsen 1958, p.121). However, modern workers have pointed out that volume shrinkage can take place only when replacement occurs within a closed system, a situation rarely found in nature (Longman 1981).

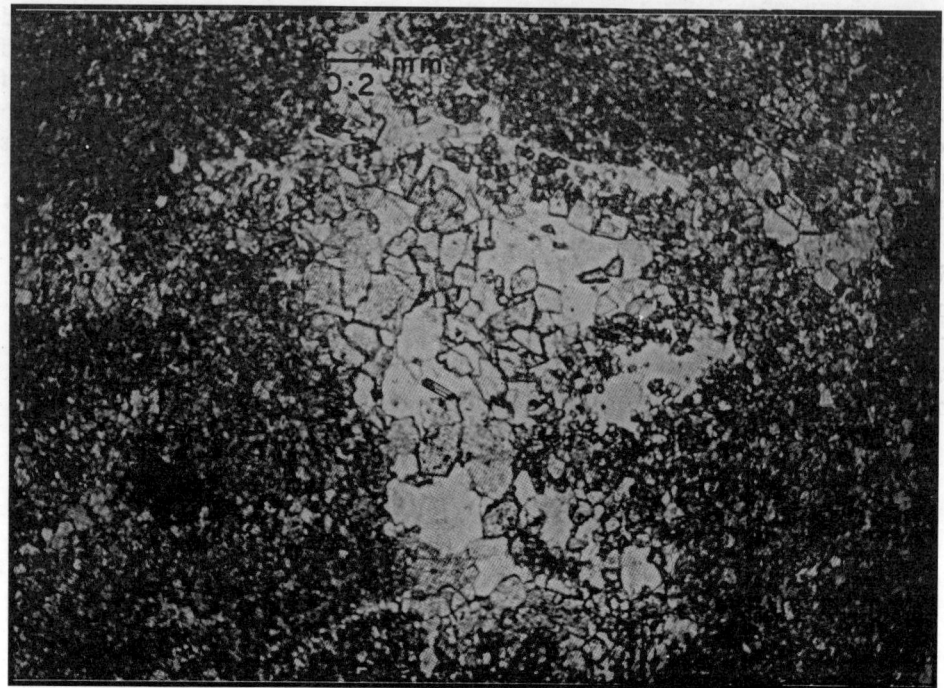

Fig. 3.17: Sparry dolomite crystals filling a cavity in a limestone. Note progressive increase in crystal size towards the centre. Location: Pulivendla, India (thin section no. H 13(11), courtesy of D.K. Dasgupta and C.N. Rao).

Chemical Sediments Other Than Carbonates

The chemical sediments other than carbonates belong to two groups of precipitates, primary and secondary. The primary precipitates include evaporites such as chlorides, nitrates and sulphates. The secondary precipitates include non-evaporites such as the siliceous, phosphatic and ferruginous deposits.

Evaporites

Evaporites are precipitated from solution and concentrated by evaporation. Sodium chloride (halite), and calcium sulphates (gypsum and anhydrite) are common examples of evaporates. Potassium chloride (sylvite), hydrated magnesium chlorides (bischovite and epsomite) are relatively rare varieties of evaporite. All these rock-forming salts are deposited by the process of evaporation of sea-water. The sequence of deposition followed by the common rock-forming evaporites is as follows (beginning with the least soluble mineral, which is always the first to be deposited): gypsum or anhydrite, sylvite and halite.

Areas of extreme aridity and local areas of internal drainage with a high rate of evaporation, as in continental or coastal sabkhas, are ideal for precipitation of evaporites. Another ideal situation for evaporite deposition is a barred (silled) basin formed as a bay or cut-off arm of a sea (see discussions on euxinic or silled basins). Circulation in such a closed basin is maintained by the influx of salt-laden sea-water flowing above the sill and return (reflux) of this water into the sea as a subsurface current. The amount of evaporite deposited into the closed basin is governed by the amount of salt flowing into the basin and the amount carried back to the sea during reflux. In this process the least soluble salts (anhydrite and gypsum) are deposited near the sill and the highly soluble salts (halite and sylvite) are deposited away from the sill, in the distal part of the basin (see Chapter 7, Fig. 7.20).

Due to precipitation of the calcium ions in the proximal part of the basin in the form of gypsum and anhydrite, the sea-water becomes enriched in magnesium. The magnesium-rich sea-water flowing back into the sea causes dolomitization of the Ca-rich carbonates precipitated on the sill. This is what is commonly called the 'reflux theory' of dolomitization. The physical environments of evaporite deposition and the main facies found in these environments are summarized in Fig. 3.18.

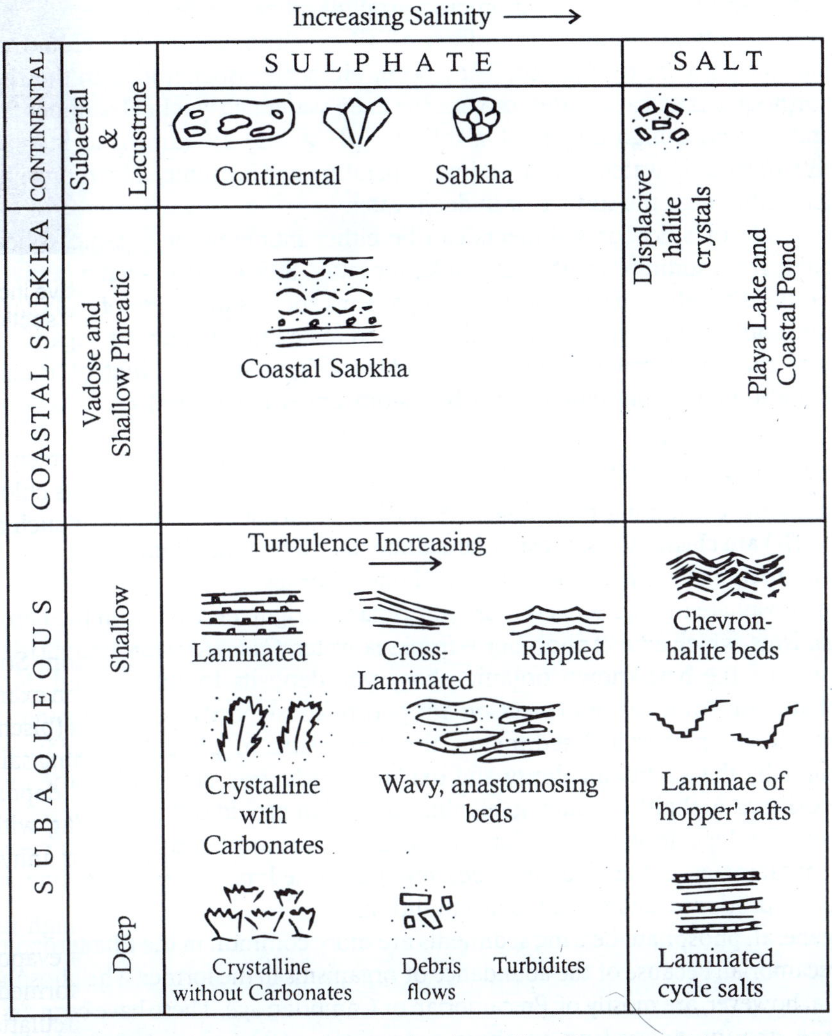

Fig. 3.18: Physical environments of evaporite deposition (based on Schreiber *et al.* 1976 and Kendall 1984).

Evaporites are found in all geological ages, from the Precambrian to the Recent, but they are possibly most common in the Permian. The best examples are the Permian salt deposits of Texas, Mexico and New Mexico. The well-known evaporite deposits of north-west Europe, Germany, and the North Sea are also of Permian age. Also well known are the salt deposits of the Persian Gulf belonging to a wide range of geological ages, from the Permian to the Tertiary. The Rann of Kutch of Western India, the coastal sabkha of the Persian Gulf, the Great Bitter Lake of Suez, the Death Valley of the USA and the Dead Sea are some of the well-known evaporite depositional areas of the present day.

Siliceous Sediments

The common siliceous minerals of sedimentary origin include chert, flint, opal, chalcedony and a variety of other microcrystalline silica. Under the microscope they show the aggregate polarization effect of silica although the individual crystal boundaries cannot be recognized. They occur either in association with the shallow-water orthoquartzite-carbonates, or with the deep-water geosynclinal sediments. High temperature and a high degree of alkalinity (pH > 8.5) are known to increase the solubility of silica in water. A decrease in temperature and alkalinity allows precipitation of silica, either near the surface or in deep water.

The origin of siliceous sediments can be either inorganic or organic. Silica may precipitate from sodium silicate gels in alkaline lakes. Silica of deep-water geosynclinal association is ascribed to silica-secreting organisms such as the oozes of radiolaria and diatoms. Some of the bedded cherts are believed to be lithified ooze of organic origin. Some of the chert can also be or secondary origin—the product of diagenetic differentiation and replacement (see discussions under diagenesis).

Phosphatic Sediments

All sediments containing more than 7.8% P_2O_5 (equivalent to an apatite content of about 20%) are classified as phosphatic sediments. Those containing more than 19.5% of P_2O_5 (apatite content of 50%) are called phosphorites.

The phosphate in sediments is mostly of organic origin, derived from the hard parts of organisms, which extract phosphorus from sea-water. *Guano* (organic excreta) deposits are some of the best-known organic phosphate deposits found on land. Nodular phosphatic conglomerates and metasomatic phosphatic deposits are of inorganic origin.

Like all other chemical sediments, the deposition of phosphatic sediments is controlled by the acidity-alkalinity and oxidation-reduction potential of the medium of deposition. A slightly reducing, alkaline condition provides the environment most congenial for deposition of phosphate (see Fig. 3.19). Phosphates can occur either as a platform facies, as in the glauconite-bearing marine sediments, or as a basinal (deep-sea) facies produced under anaerobic conditions.

In general, phosphate-bearing sediments are more common in the Phanerozoic than the Precambrian because of the abundance of organisms in the former. The phosphorites in India, however, are mostly of Precambrian or Cambrian age. These have been assigned to three groups according to their association (Banerjee 1986): 1) bedded cryptocrystalline phosphorites associated with dolomitic stromatolites; 2) Intraclastic and microcrystalline phosphorites associated with carbonates, black shales and dark cherts; 3) phosphorites associated with ferruginous sediments and residual weathering products.

All the above-mentioned phosphorites are of marine origin. Silled (restricted) basin conditions, controlled by basement topography, were responsible for the high concentration (10-35%) of P_2O_5 in most cases. The specific conditions of deposition of these phosphates have been discussed in detail by Banerjee (1986). In Rajasthan, a restricted condition was provided by the presence of a shallow carbonate embankment along a Precambrian basement ridge. Similar conditions also existed in the Himalayas where the phosphorites were deposited over the Upper Proterozoic Krol Formation. In Peninsular India, dolomites of Proterozoic age provided the silled basin conditions for the Vindhyan phosphorites. Association with stromatolites in many cases indicates that micro-organisms played important roles in concentrating phosphorus from sea-water. In some cases, as in the Bijawar rocks of Central India, processes such as diagenetic mobilization, weathering, leaching, precipitation in voids and fissures also helped in concentrating phosphorites. Phosphorites found in association with ironstones in the Raniganj Coalfield, India are of shallow-water origin.

Iron-bearing Sediments

Iron-bearing sediments include *banded (bedded) iron formations* (BIF), ironstones, ferruginous ooids and fossil fragments. The bedded iron formations consist of alternating thin layers of silica and ferruginous minerals of chemical origin. The iron-bearing minerals are mostly oxides (magnetite and hematite). Silica occurs as quartz, chert or jasper. The bedded iron rocks, found in all the Precambrian cratons of the world, are often modified by diagenesis and metamorphism. The ironstone formations which are of younger age, are composed mostly of iron carbonate (siderite) with some iron silicates (chamosite). Ooids and fossil fragments, often replaced by iron oxide minerals, are also known to occur within ironstones. The siderites are often altered to limonite near the surface.

As in the case of other chemical sediments, acidity-alkalinity (pH) and oxidation-reduction potential (Eh) play important roles in the precipitation of iron-bearing minerals. As indicated by the Krumbein-Garrels diagram (Fig. 3.19), an oxidizing and alkaline condition is ideal for precipitation of the iron oxide minerals. The production of near-surface carbonate or bicarbonate of iron could be wholly inorganic (James 1966). The inorganic iron may be transported together with clay minerals as colloidal iron oxide. Organic matters can also play an important role in the transportation of iron because large proportions of ferric oxide can be kept in colloidal suspension by small quantities of organic matter. Following a more modern approach to the problem, Berner (1981) concludes that hematite and goethite are precipitated under oxic ($C_{O_2} \geq 10^{-6}$) conditions, while pyrite and marcasite are precipitated in anoxic ($C_{O_2} < 10^{-6}$) conditions. Siderite is formed under anoxic conditions which are nonsulphidic $\left(C_{H_2S} < 10^{-6} \right)$. These conclusions are based on the measurement of dissolved oxygen and dissolved sulphide in modern sediments at the time of authigenic mineral formation. C in the notations refers to concentration of oxygen or total sulphide in terms of moles per litre.

The conditions leading to simultaneous deposition of silica and iron oxides are not fully understood but a marine or restricted marine environment of sedimentation, somewhat isolated from the open sea, is believed to be responsible for deposition of the banded iron formations of Precambrian age. A Precambrian ocean with an upper oxic layer and a bottom anoxic layer containing Fe and SiO_2 is the commonly assumed model. Large-scale precipitation of BIF followed the upwelling of deep, Fe–SiO_2–rich waters to the newly developed continental margins and shelves.

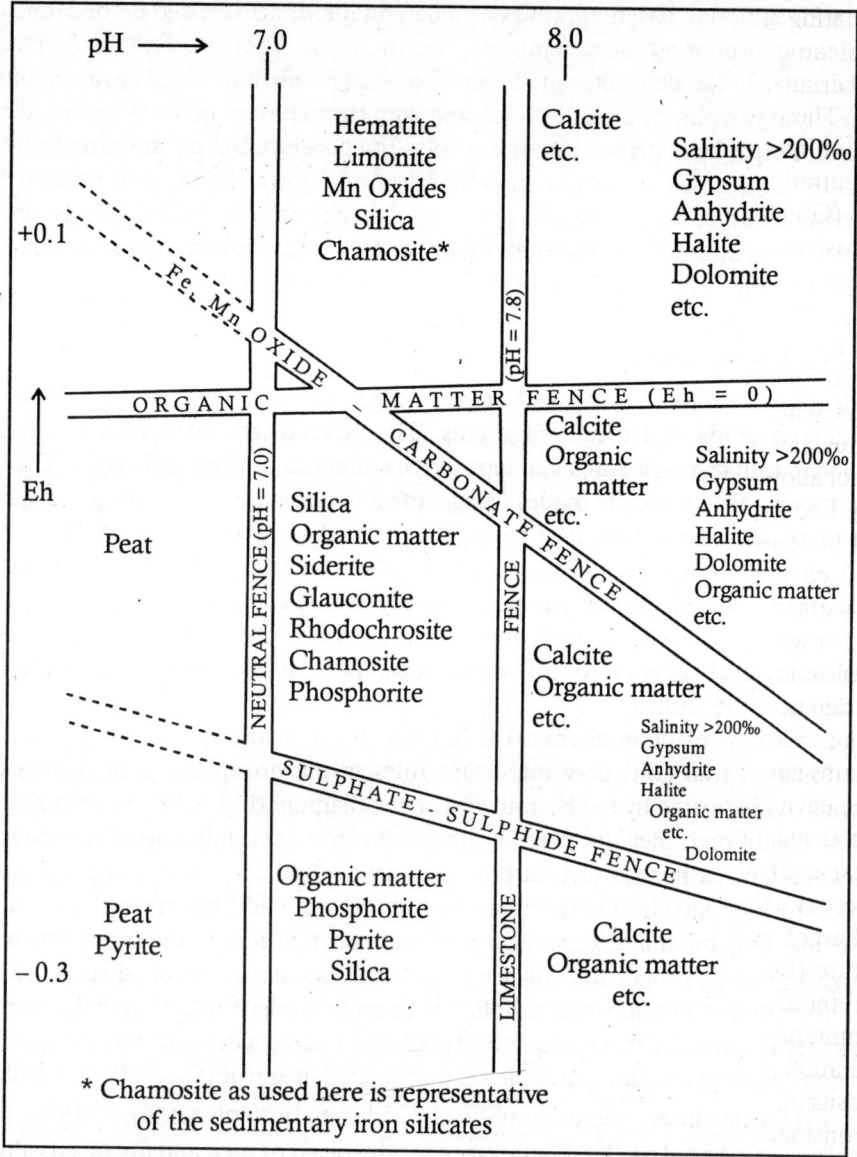

Fig. 3.19: Fence diagram illustrating influence of Eh and pH on generation of mineral assemblages in sedimentary rocks (after Krumbein and Garrels 1952).

In some cases, as in the Precambrian rocks of the Lake Superior District, silica and iron, produced from weathering of landmasses, were deposited in alternate layers under freshwater lacustrine conditions. The model proposed for such a deposition by Hough (1958) envisages a lacustrine condition wherein settling of the dead, oxygen-consuming organisms from the surface creates a reducing condition in the lake bottom which is cold as well as acidic. The condition is ideal for precipitation of silica, while iron brought in suspension by streams during the summer remains in solution to be precipitated as iron oxide only when the lake 'turns over' (see Chapter 7) during the autumn to produce an oxidizing environment. Repetition of the process leads to the production of alternating bands of iron oxide and silica.

Volcanic activity, sea-bottom reactions on clastic materials and deep-seated weathering processes on land are some of the commonly suggested sources of iron and silica. These conditions, it has been suggested, were repeated periodically but occurred on a unique scale in the early Proterozoic (2.0–2.5 b.y.), when an extraordinary concentration of BIF (92% of the total) was formed in almost all continents round the globe (Goodwin 1982).

Erosion under climatic conditions similar to what prevails today will not allow transportation of large quantities of iron in suspension without a substantial amount of accompanying detritus, but surprisingly most of the thick sequences of the Precambrian iron deposits are devoid of clastics. As an explanation, it has been proposed that the Precambrian atmosphere had a large content of CO_2, resulting in acidic surface waters, which allowed easy leaching and transportation of iron with no accompanying detritus. It is also possible that near-absence of oxygen in the Precambrian atmosphere did not allow oxides of iron, the chief inhibitor in the process of iron transportation, to form. Local concentrations of iron oxides in this model were explained as the result of local biological photosynthesis. As an alternative to these suggestions, it has been proposed that the sluggish surface run-off over a senile, deeply weathered Precambrian land surface did not allow transportation of particles coarser than particulate iron oxide together with fine clay. Many workers are of the opinion, however, that the contributions of volcanic, hydrothermal or biological processes in the generation of Precambrian iron deposits cannot be ruled out altogether (Stanton 1972).

The Precambrian *Banded Hematite Jasper* (BHJ) Formation of Noamundi, India contains numerous syndepositional structures suggesting their deposition in a shallow-water environment. The possible biological origin for at least a part of the iron and silica is suggested by the presence of microstructures resembling organic forms (Rai, Sarkar and Paul 1980; Sarkar 1984). According to one view, the banded iron formations represent exhalative phases of volcanism along a fracture zone. Iron and silica, discharged into an inner sea, were precipitated out in alternating bands through this fracture (Banerji 1977). A source of pyroclastics and exhalatives (volcanic and/or hydrothermal) adjacent to the existing BHJ occurrence together with a rapid rate of precipitation of the detrital-pyroclastic/chemical sediments close to their place of generation has been suggested as an alternative hypothesis for the iron and manganese deposits of the Noamundi Basin (Dasgupta and Batabyal 1988).

Ironstones are also known from the Ironstone Shale Formation of the Raniganj Coalfield in West Bengal, and its stratigraphic equivalent in other Gondwana basins of India. Apart from iron and silica, these rocks contain small quantities of phosphorus and manganese also (Walker 1914, cited by Krishnan, 1964). The assemblage of minerals in these ironstones suggest their precipitation in environments which were slightly alkaline as well as reducing. This and other lines of evidence have led to the suggestion of a marine invasion into some of the Gondwana basins of India during the Permian (Bose and Sengupta 1993).

VOLCANICLASTIC ROCKS

Sedimentary rocks constituted of fragments of volcanic origin are called volcaniclastic rocks. These rocks contain the products of weathering and erosion of pre-existing volcanic rocks or pyroclastic debris. Pyroclastic is the name given to rocks composed of the ejecta from volcanic vents.

Volcaniclastic sediments are broadly divided into two groups according to grain-size - the coarser volcanic *breccia* (angular fragments) and *agglomerates* (rounded fragments), and the finer *tuffs*. Breccia and agglomerates consist of volcanic bombs and rock fragments. Tuffs are composed of fine crystals, rock fragments and volcanic glass. These are named according to their principal constituent - crystal tuffs, lithic tuffs and vitric tuffs (Fig. 3.20).

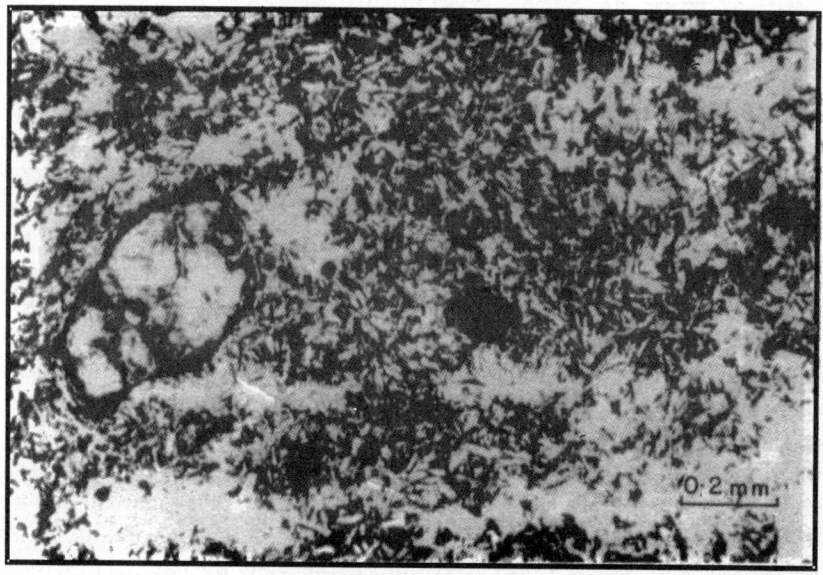

0.2 mm

Fig. 3.20: Photomicrograph of a tuff (plane polarized light). Locality near Padberg, Germany.

The origin of most volcaniclastic deposits is ascribed to eruption of rhyolitic lava which, being highly viscous, with a high gas content, is particularly explosive. The finer ash is carried to the upper atmosphere and transported for long distances. Deposition of ash may lead to extensive pyroclastic beds ('*ash fall*'). Clouds of hot gas ('glowing cloud' or *nuée ardente*), produced out of some acidic volcanic eruptions, may also be responsible for sedimentation of an unusual nature. The basal flows resulting from these eruptions, contain ash, lapilli and blocks which are kept in suspension by escaping gases. The mass may flow downstream as a turbulent mixture. The deposits formed from this mixture are called ash-flow tuffs, welded tuffs, or *ignimbrites*. These poorly sorted, poorly bedded units have compositions comparable to rhyolitic lavas. A single deposit of ignimbrite may be as much as 100 m thick. The ash-fall deposits, on the other hand, are well sorted according to fall velocity of the grains, with mean thickness and grain-size decreasing away from the source. Hence they can be used as palaeocurrent indicators (Scheidegger and Potter 1968).

Silicified tuffs, called *porcellanites*, reported from the Proterozoic Vindhyan sediments by early workers of the Geological Survey of India, contain fragments of quartz and feldspar of vitroclastic shapes. Wedge-shaped fragments of feldspar indicate their pyroclastic origin (Auden 1933, p. 154). These volcanigenic deposits contain bombs, lapillis and huge ash falls (Ghosh 1972). Lava flows, tuffs and pyroclastic breccia around 'crater-like structures' have been reported from the Proterozoic sediments of the Godavari valley, India (Saha and Ghosh 1987). These tuffs show distinct colour lamination, parallely arranged lapilli, flat-ended and stretched pumice fragments (Fig. 3.21). These have been assigned to syntectonic volcanism in Proterozoic island arcs. Ash-beds, interbedded with late Quaternary sediments, have been reported from the Son and Narmada valleys of India. This ash contains angular glass shards and some accessory minerals. In the absence of any evidence of Quaternary igneous activity in the areas, this ash-bed has been tentatively assigned to neotectonic activity along the Son-Narmada lineament, known to be an area of high heat flow (Basu, Biswas and Acharyya 1987).

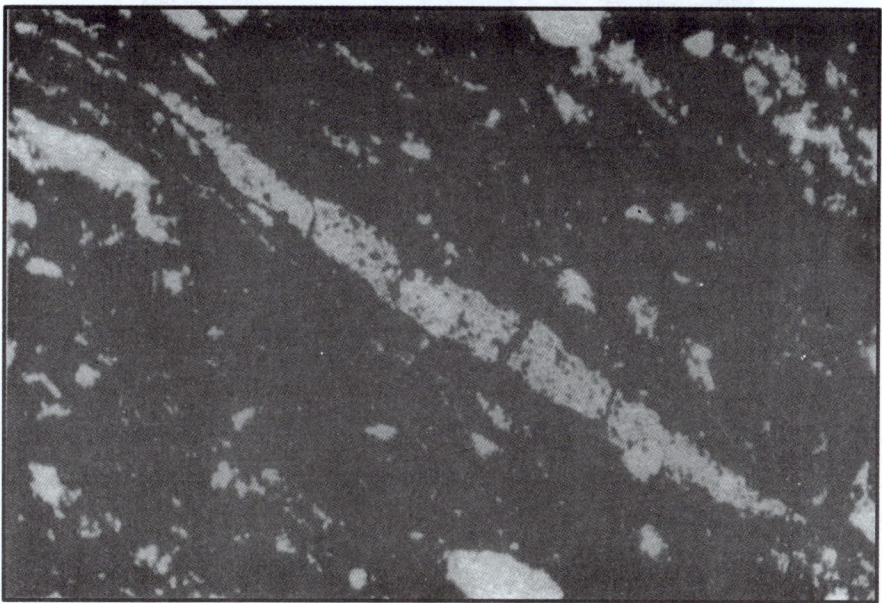

Fig. 3.21: Photomicrograph of a deformed tuff-lava, Sommanpalli, Godavari valley, India. Length of the stretched pumice fragment is about 0.76 mm (courtesy of D.K. Saha).

DIAGENESIS

The changes which sediments undergo between deposition and lithification under normal pressure-temperature conditions are called diagenesis. More recently, the parameters of diagenesis have been extended to depth (4-7 km) and temperature (120°-220° C) conditions where incipient metamorphism sets in. Near-surface (*telegenetic*), shallow depth (*eogenetic*) and deeper level (*mesogenetic*) changes are included in this definition (Burley, Kantorowicz and Waugh 1985).

Diagenesis of Siliceous Sediments

Near-surface diagenetic changes are caused by the interaction of minerals with interstitial pore fluids. This may lead to dissolution of the existing minerals and production of new (*authigenic*) minerals. Authigenic growths ('*overgrowths*') of quartz and feldspar are common features at this stage. The newly formed quartz is added to the detrital quartz grains maintaining the same crystallographic orientation (Fig. 3.22). Authigenic precipitation of feldspar takes place at a higher order of pressure and temperature than that provided by near-surface conditions. Excess potassium in the pore fluid is an essential condition for development of authigenic feldspar. Dissolution features in feldspars and production of kaolinite as pore-filling material are common at this stage.

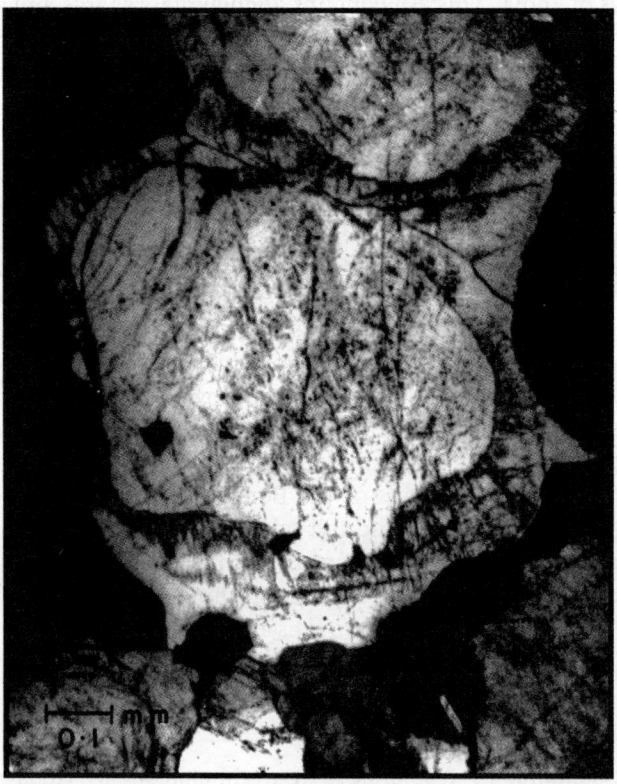

Fig. 3.22: Photomicrogrph showing authigenic overgrowth on a quartz grain in quartz arenite from the Upper Proterozoic Chandarpur Formation, Raipur, India (thin section courtesy of S.S. Das and C.N. Rao).

At a shallow depth, in a warm, humid, non-marine environment, bacterial activity may impart a distinctly black colour to fine-grained sediments. Lateritisation takes place in warm humid climate whereas calcrete forms in warm arid to semi-arid conditions. In a warm humid climate, when the pore fluids are alkaline and rich in calcium, sodium and silica, but deficient in carbonates, a variety of *zeolites* may be produced authigenically. When the climate is hot an dry ferric oxides are produced by oxidation of ferruginous sediments. These oxides impart a red colouration to sediments at shallow depths. In marine conditions a variety of clay minerals, illite, smectite, interbedded illite-smectite, may be authigenically produced. Quartz, feldspar and clay authigenesis is often followed by carbonate precipitation under marine conditions.

At greater depth mechanical processes play important roles in diagenesis. Fracturing, crushing and squashing of detrital grains and wrapping of mica flakes around quartz grains are common mesogenetic changes. Reduction of porosity, partly due to compaction with increasing depth of burial but largely due to chemical processes, takes place. With increasing depth the porosity of sand changes only marginally while that of shale decreases logarithmically. Reduction of intergranular pore space is caused by increase of solubility at the contact of solid clastic grains under pressure. This phenomenon is called '*pressure solution*' (stylolite—see Chapter 6 for details). Pressure solution may change the nature of grain contact from tangential to sutured through concavo-convex (see Fig. 4.13) with increasing depth of burial.

A minimum depth of 1 to 1.5 km is necessary for pressure solution to be effective. Limestones and calcareous sandstones need lesser temperature and pressure for welding of grains than arenites. Finer grained rocks are affected more because they provide larger surface areas for grain contact. Thin films of clay around quartz grains promote pressure solution, but larger quantities of clay deter the process by 'cushioning' (Tada and Siever 1989). Optical microscopes are of little use in distinguishing between overgrowths and pressure solution effects such as concavo-convex contacts. The cathodo-luminescence technique often helps in such cases because the nature of the luminescence differs for detrital quartz and overgrowths (Sippel 1968).

With increasing (mesogenetic) depth of burial, clay minerals such as smectite and kaolinite are replaced by more stable products, illite and chlorite. With increasing depth of burial in the Neogene sediments of the Bengal Basin, Bangladesh, a progressive decrease in the proportion of kaolinite and calcite and an increase in the proportion of illite by *illitization* of illite-smectite were noted. Within the same sediments the following order of diagenesis was reported: chlorite authigenesis, siderite crystallization, development of illite rims, quartz overgrowths and production of late Fe-calcite cement (Imam and Shaw 1984).

Source of Silica Cement

On land silica may be taken into solution during weathering of rock-forming silicates (see Chapter 2). *Kaolinitization* of feldspar and diagenetic alteration of clay minerals may also release some silica. In the early stage of diagenesis the pores of quartz sand may be filled up by silica produced from weathering. Opal and chalcedony cements in a desert environment represent a transition between weathering and diagenesis.

Solubility of both amorphous silica and quartz increases with an increase in temperature and alkalinity of the medium. A sharp increase in silica solubility is noted when pH exceeds 8.5. Hot springs and hydrothermal waters, which satisfy these conditions, might serve as sources of silica in sediments. Precipitation of secondary silica by hydrothermal solution is a common feature. Solutions saturated with silica at high pressure-temperature conditions at depth become further concentrated when they are pushed up to cooler zones by compaction.

In deeply buried sandstones some silica may be produced by pressure solution. The silica thus produced migrates into clay-free, fine-grained layers where authigenic overgrowth can take place readily (von Engelhardt 1977). Silica released by clay minerals during diagenesis of shales might migrate into adjacent sandstones to be deposited in voids.

In a marine environment the silica seems to be of primary origin, generated at the sea-bottom due to dissolution of diatom tests. This leads to concentration of silica in the pores of deep-sea clay and silt. Escape of silica saturated water from the pores of clay or silt to the intergranular space of sand might account for much of the silica saturation in sandstones of marine origin. In some cases, as in the *Buntsandstein sandstones* of Germany, silica cementation has actually been found to increase towards shale (Füchtbauer 1974).

Diagenesis of Carbonates

Replacement of aragonite and high-magnesian calcite by calcite and consequent reduction of pore space are the two most important effects of diagenesis on carbonate sediments of marine origin. These processes are to be studied with reference to the physicochemical conditions prevailing at sea. In a coralline island for example, four different environments prevail. These are : (*i*) a marine *phreatic zone*, where the pore spaces are filled with saline water (*ii*) a freshwater phreatic zone immediately below the exposed part of the island ; (*iii*) a mixed zone with brackish water in the pore space and (*iv*) a *vadose zone* occupying the space between the surface and the water table. The pore space here contains water as well as air (Fig. 3.23).

In the marine phreatic zone, where salinity and Mg/Ca content of the sea-water are fairly high, aragonite is secreted by corals (Fig. 3.15). The porosity of freshly deposited aragonite may be as high as 55%, but micritisation and submarine cementation inside the grains may reduce this porosity to about 40%. Intergranular cementation may further reduce the porosity to 5-10% only. The binding material here is provided by coralline algae which secretes magnesian calcite.

A lowering of sea-level following this primary stage of cementation translocates the carbonate to the freshwater phreatic zone. Decrease of salinity and of Mg/Ca ratio in this zone causes sparry calcite to precipitate within the shells and tests, thereby reducing the porosity to about 10%. Interestingly, the original shapes of the shells are not altered during this conversion. This is due to the presence of a micritic envelope which allows the calcitic pore fluid to pass through it, without destroying the shape of the outer shell. The envelope itself is not involved in this reaction (Friedman and Sanders, 1978). In the case of cavity filling larger crystals are formed at the centre and the smaller ones remain confined to the inner rim of the cavity (see Fig. 3.17). In some cases cementation takes place round the outer boundary of the pre-existing grains in the form of syntaxial rims. The process has been termed *aggrading neomorphism* (Folk 1974).

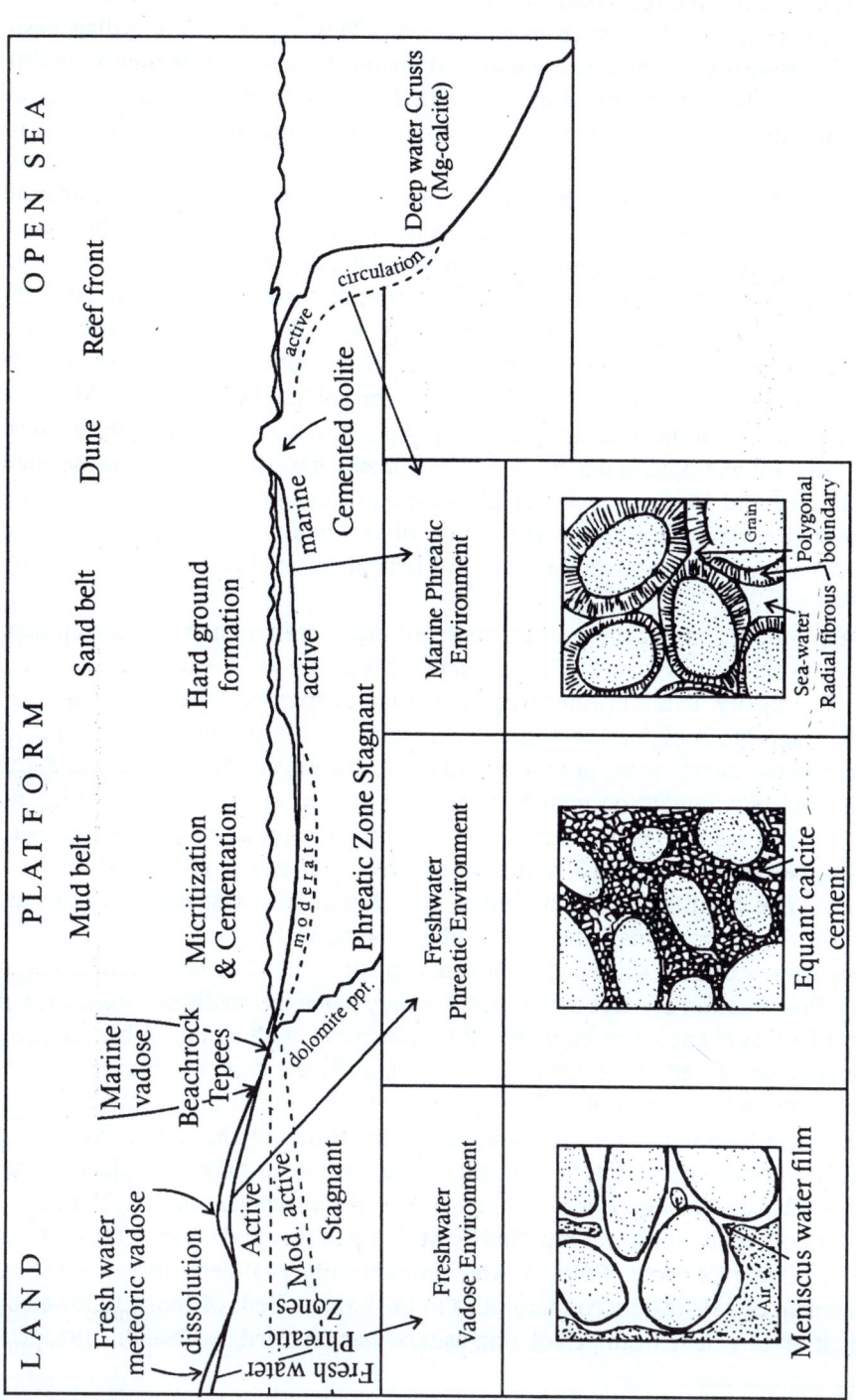

Fig. 3.23: Diagenesis of carbonates in freshwater and marine environments (modified and redrawn after a compilation by Sellwood 1986, with permission).

Neomorphism involves recrystallization of the fine carbonate mud into coarser aggregates ('*microspars*'), 5-10 microns in diameter. This happens during diagenesis because fine crystals of carbonate are thermodynamically unstable. Magnesium ions released to the surface of micrites during diagenesis tend to inhibit crystal growth, but they are often flushed out by meteoric or deep connate water which are deficient in magnesium content.

When the carbonate is left in the freshwater zone for a long time, leaching of aragonite and Mg-calcite takes place by meteoric water since these minerals are unstable under freshwater conditions. Although porosity may increase temporarily at this stage, neomorphism of calcite which follows, fills up the moulds, thereby reducing the porosity to about 5-10%. A vuggy porosity may develop, however, if the limestone is brought up to the vadose zone and retained there for a sufficient length of time. The secondary porosity developed in this way may provide space for holding hydrocarbons. Much of the reservoir porosity of the oil-bearing 'Bombay High' structures located off the western coast of India, for example, is due to the vuggy porosity developed in the vadose zone (Roychoudhury and Deshpande 1982). Dissolution of carbonate and feldspar in the post-cementation phase, leading to generation of secondary porosity, has also been reported from the Neogene sediments of the Bengal Basin, Bangladesh (Iman and Shaw 1984).

Although vadose zone leaching makes an important contribution to the development of porosity, prolonged exposure of a carbonate rock in the vadose zone may cause a reduction in porosity. This is effected by precipitation of sparry calcite within the vugs. When the aragonite is of marine origin, bicarbonate with a positive value of δC^* is produced because the mean δC of marine carbonate is positive. On the other hand, the humus-derived bicarbonate occurring in the near-surface vadose zone has a highly negative δC. The reaction between these two bicarbonates produces a pore fluid with δC ranging between -6 and -10, values not as highly negative as the vadose zone bicarbonate. The condition is ideally suited for precipitation of sparry calcite in the vadose zone. The process may drastically reduce the porosity.

The occurrence of both silica and carbonate cement in the same sandstone has been a mystery. Precipitation of carbonate is possible only when the medium is alkaline, but solubility of silica is known to increase with an increase in alkalinity. When an algal mat growing over a carbonate sediment is periodically exposed to atmosphere, the interstitial water attains exceptionally high alkalinity, leading to supersaturation, and eventually precipitation of silica on carbonate. This is one of the special conditions wherein simultaneous precipitation of silica and carbonate may take place (Baas-Becking, Kaplan and Moore 1960). In India, a case of replacement of early diagenetic rhombic carbonates by silica was reported by Mukhopadhyay and Chanda (1972) from the Iron Ore Group of rocks of Orissa. Another interesting post-depositional sequence of replacement of calcite by secondary silica in the Lameta Beds of India, followed by reprecipitation of calcite from percolating waters, was reported by Chanda (1963).

---

*δC represents deviation, measured in parts per thousand (per mil or ‰), from the standard. The standard is taken to be a belemnite from the Cretaceous, Pedee Formation in North Carolina, USA.

APPENDIX

Techniques of Sedimentary Petrography

Scheme for Description of Sandstones under the Microscope

A. Framework Grains
1. *Mineralogy*

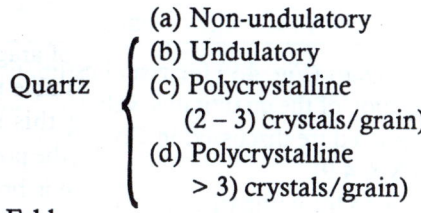

Quartz
- (a) Non-undulatory
- (b) Undulatory
- (c) Polycrystalline
  (2 – 3) crystals/grain)
- (d) Polycrystalline
  > 3) crystals/grain)

Feldspars
Rock Fragments – specify
Micas – muscovite/biotite
Heavy Minerals – specify
Other Minerals – specify

The proportions of quartz, feldspars and rock fragments should add up to 100. Use Fig. 3.24 for visual estimation of proportions. Also determine proportions of quartz grains of different varieties separately, so that they add up to 100.

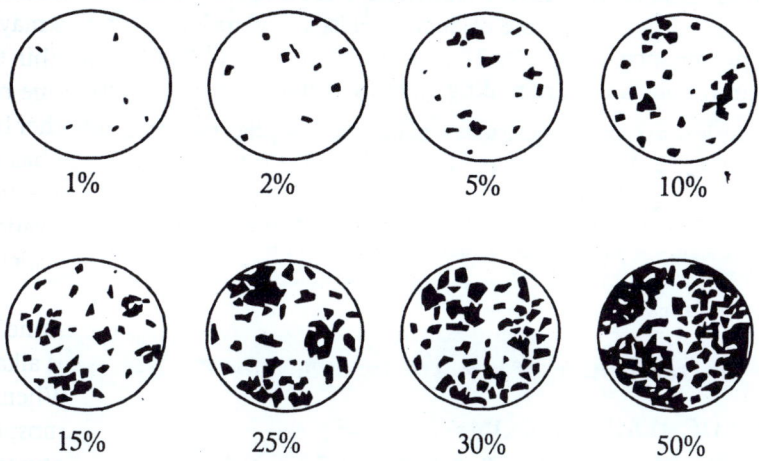

| 1% | 2% | 5% | 10% |

| 15% | 25% | 30% | 50% |

Fig. 3.24: Charts for estimating percentage composition of minerals under the microscope. Prepared by M.S. Shvetsov (based on Terry and Chilingar 1955). With permission of SEPM (Society for Sedimentary Geology).

2. *Texture*
   (i) Grain-size   –   use pre-calibrated eyepiece micrometer for size determination. See Table 4.1 for Wentworth nomenclature of size classes. Designate proportions also in qualitative terms: abundant, frequent, few, rare, e.g., medium sand-size particles abundant.

(*ii*) Sorting   –   use the following scheme of classification:

Sorting	Matrix content	Grain-size variation
Well sorted	Low	Most grains lie within one or two Wentworth size classes
Moderately sorted	Moderate (~ 10%)	Moderate
Poorly sorted	High	Large

(*iii*) Roundness   –   Use Powers' chart (Fig. 4.9) for visual estimation.
      &   For computation of the quantitative value of roundness
  Sphericity   follow the procedure discussed in Chapter 4
                       (see Figs 4.8 & 4.9).

(*iv*) Nature of grain contact—see Fig. 4.13 for comparison.

B. CEMENT- identify type—siliceous/calcareous/ ferruginous/any other.

C. MATRIX - determine proportion (greater or less than 12.5%). Name the rock (wacke or arenite) accordingly.

D. CLASSIFICATION AND NOMENCLATURE - use revised Dott's (1964) scheme (see Fig. 3.3), e.g., Arkosic Wacke.

E. COMMENTS ON THE POSSIBLE NATURE OF THE SOURCE ROCK
Use proportions and types of quartz, feldspar and rock fragments (see discussions on provenance in Chapter 2). Plot 'a', 'b', 'c' and 'd' types of quartz on Fig. 2.4 to draw conclusions regarding igneous or metamorphic nature of source rock.

Note: An easy way to classify a sandstone is to determine whether the total matrix content is greater or less than 12.5%, i.e. respectively. whether the matrix fills one-half of one quadrant of the field of the microscope. The rock may be named arenite or wacke depending on whether the matrix content is smaller or greater than 12.5% respectively. Proportions are to be determined in three stages : (*i*) proportion of matrix (*ii*) proportion of the three important constituents— quartz, feldspars and rock fragments (*iii*) proportions of four varieties of quartz : undulatory, non-undulatory etc.

Scheme for Description of Limestones under the Microscope

A. ALLOCHEMICAL COMPONENTS
    1. Intraclasts   Determine proportion and size
    2. Fossils   (sand size or larger) of each of
    3. Oolites   these and tabulate in Folk's
    4. Pellets   scheme (Table 3.9).

B. ORTHOCHEMICAL COMPONENTS
    Micrite (microcrystalline ooze, 0.001-0.004mm)
    Sparite (sparry calcitic or dolomitic mosaic, 0.02 - 0.1 mm)

Decide name of the limestone by using appropriate combination of A and B. Use Folk's scheme of classification (see Table 3.9). Add 'rudite' as suffix if the majority of the allochemical grains are larger than sand size.

4

SEDIMENTARY TEXTURE

The term texture refers to the pattern formed within a sedimentary rock by the constituting grains. This can best be studied under a microscope. In sedimentary rocks of mechanical origin the framework grains are joined together, either by finer fragmental particles or by chemically precipitated cement. The former is called *matrix* and the latter, *cement*. The pattern formed by framework grains, matrix and cement is called *clastic texture*. Sandstones show this kind of texture.

The texture of a rock of chemical origin is produced wholly by chemically precipitated materials. No intergranular space is to be found in such a rock. Limestones of orthochemical origin show this kind of texture. Allochemical limestones, produced by fragmental pieces of calcareous material joined together either by fine carbonate mud or by calcite spars, show a clastic texture (see Chapter 3 for definition of terms).

Clastic textures are defined mainly by size, shape and roundness of the framework grains. Other features which play important roles in clastic texture are the nature of grain contact (grain-to-grain relationship) and grain packing. Porosity and permeability of a clastic rock depend on these parameters.

GRAIN-SIZE

Concept of Particle Size

The definition of particle size is related, at least in part, to the technique used for measuring it. When a particle is large enough to allow direct measurement of its diameters, say by calipers or by a micrometer scale fitted to the microscope oculars, size can be expressed in terms of the diameters (long, short and intermediate). A problem arises when a grain is too irregular to allow direct measurement of its diameters. In such cases the size is measured indirectly, firstly by measuring its volume, by the process of displacement of water, and then by working out the diameter of the sphere of equivalent volume by the following formula:

$$V = 4/3\pi r^3,$$

where V and r are the volume and radius of the sphere respectively. The diameter thus obtained was named '*true nominal diameter*' by Wadell in 1932.

Udden-Wentworth Size Scale

Fragmental particles constituting sedimentary rocks cover a wide range of grain-sizes—

from boulder, which by definition is greater than 25cm, to clay which is less than 0.004 mm in diameter (less than 0.002mm in Friedman and Sanders 1978, classification). Since handling of such a wide range of grain-size is a difficult problem, a geometric scale is used instead of an arithmetic one. In the so-called *grade scale* proposed by Udden in 1914, the consecutive figures in the scale increase by multiples of 2 on the higher side and decrease by multiples of ½ on the lower side. The various size ranges within the grade scale were named by Wentworth (1922) as follows :

Boulder	Cobble	Pebble	Granule	Sand	Silt	Clay
256	64	4	2	1/16	1/256 (mm)	

The Udden-Wentworth system not only provided an appropariate scale for representation of grain-size but also standardized the sedimentological nomenclature. Adoption of this system of nomenclature by a committee of geologists and hydrologists dispelled the confusion created by arbitrary assignment of names to different parts of the grade scale by various authors. The Udden-Wentworth grade scale and nomenclature system are universally used today.

Phi Scale

In the early days of sedimentology, when mechanical or electronic computing facilities were not readily available, handling of fractional values of the Udden grade scale involved tedious computations. As a solution to this problem, Krumbein (1934) introduced a method of logarithmic transformation as follows:

$$\phi = -\log_2 d,$$

where d = diameter of the grain in millimetres. Conversion of phi (ϕ) to grain diameter (d) is attained as follows:

$$d = 2^{-\phi}$$

The use of 2 as the base of the logarithm allows direct conversion of Udden's grade scale to Krumbein's phi scale. Moreover, inversion of sign converts the fractions to positive, round integers, thereby making easier computations of the finer grains (sand, silt and clay), which constitute the bulk of sediments. Log scale condenses the large range of grain-sizes of sedimentary particles thereby helping graphical representation. These advantages popularized the phi scale throughout the world for nearly half a century. However, with the advent of modern electronic computing devices, handling of fractional values being no longer a problem, the phi scale is losing its popularity. Sedimentologists are returning back to the millimetre scale.

The combined Udden-Wentworth-Krumbein grain-size scale is shown in Table 4.1.

Grain-Size Masurement

The techniques commonly used for grain-size analysis are: (*i*) direct measurement (in macro- or microscale), (*ii*) sieving and (*iii*) settling. Grains larger than sand can be measured directly by means of calipers, while an optical device, varying from a hand lens to a microscope, is needed for measuring finer particles. Sieving is generally used for particles ranging from granule to fine sand. Settling techniques are used for sizes finer than sand (silt and clay). More sophisticated methods such as image analysis of

SEM pictures (without intermediary photographs), electrical resistive pulse and light-blockage counters, disc centrifuge and laser scattering devices are available now (see Tables 4.2 and 4.3). The principles behind the techniques traditionally used for grain-size measurement are briefly discussed below.

Table 4.1: Grain-size classification and nomenclature

Wentworth nomenclature	Boulder	Cobble	Pebble	Granule	Sand	Silt	Clay
		256	64	4	2	1/16	1/256mm
Udden's grade scale		(2^8)	(2^6)	(2^2)	(2^1)	(2^{-4})	(2^{-8}) mm
Krumbein's phi-scale		−8	−6	−2	−1	+4	+8 (ϕ)

	Sand				
	V. Coarse	Coarse	Medium	Fine	V. fine
−1	0.0	+1.0	+2.0	+3.0	+4 (ϕ)

Grain-size measurements conducted under a microscope are influenced by the shape as well as the orientation of the grains. When the grains are oriented at random, the 'maximum horizontal intercepts' through the grains are measured. The sizes thus obtained are obviously smaller than the actual sizes because seldom does the line of traverse pass through the grain centre. Theoretical studies as well as experimental observations have shown that the data obtained by this method are some 24% smaller than the actual in most cases. Thus the average radius of a sphere, computed on a random section by the techniques of maximum intercept measurement, is 0.736 of the actual radius of the spheres (Krumbein and Pettijohn 1938, p. 130).

To take care of the shape variation of irregular grains, Wadell (1935) recommended computation of the area of the grain by a planimeter on an enlarged picture. The nominal sectional diameter, which is the diameter of a circle of the same area, can be obtained from this measurement. For elongate grains the three diameters – a, b, and c, must be measured separately and the mean diameter then computed by one of the following formulae:

arithmetic mean, $d_a = (a + b + c)/3$

harmonic mean, $d_h = (3\ abc)/(ab + bc + ca)$

geometric mean, $d_g = \sqrt[3]{abc}$

Computation of the arithmetic mean suffices in most cases. On average, the harmonic mean lies within about 6% of the arithmetic mean.

Sieving Technique

Sieves are screens of standard size. In the commonly used sieves specified by the American Society for Testing Materials (A.S.T.M.), the openings increase uniformly in multiplies of $\sqrt[4]{2}$, starting from an opening of 0.0029 inch (200 mesh). The millimetre equivalents of these openings lie quite close to the Wentworth grade limits and can be used for sedimentological purposes. A list of the standard ASTM sieves together with their mm and phi equivalent is given in Table 4.6 in Appendix.

Table 4.2: Techniques commonly used for grain-size analysis (modified after McCave and Syvitski 1991)

Altered Atterberg	Modified Udden-Wentworth	Classical method of analysis	Modern method of analysis
Gravel	Gravel	Count	Field seieve
— 2 mm —			
Sand (3 grades)	Sand (5 grades)	Sieve	Settle Count Scatter light
— 63 µm —			
Silt (3 grades)	Silt (5 grades)	Settle	Count Settle Scatter light
— 2 µm* —			
Clay	Clay	Settle	Settle Centrifuge
— molecular —			

* Modified silt-clay boundary after Friedman and Sanders 1978.

Table 4.3: Techniques of grain-size analysis applicable to different size ranges (modified after McCave and Syvitski 1991)

Range			Principle			
			Settle from		Laser scatter	Photon correlation spectroscopy
	Count	Sieve	Top	Suspension		
Gravel	/a	/	X	X	X	X
Sand	/b	/	/f	X	/i	X
Silt	/c	/e	/g	/h	/i	X
Clay	X_d	X	/g	/h	X^i	/j

Symbols: / = technique applicable;

X = technique of limited applicability or impossible;

a = Measurement on outcrops and photographs;

b = Measurement on thin section or grain mount either directly or by image analyzer;

c = Image analysis of SEM photographs or by pulse and light blockage counters;

d = Electrical resistive pulse method measures down to ~0.5 µm;

e = Commonly measures down to 44 µm, in special cases to 20-25 µm or even down to 5 µm by special Fritsch sieve shaker;

f = Settling tubes measure weight, differential pressure on pan, visual accumulation or light attenuation method of sensing;

g = Disc centrifuge method. Sediment is introduced via a density gradient and sensed by light attenuation. Discrimination down to 0.01 µm is claimed;

h = Homogeneous settling suspension. Measurement by simple pipette method, sedimentation balance, withdrawal from bottom, light or X-ray attenuation techniques.

i = Laser forward scattering devices;

j = Coulter and many other similar counters.

In the sieving technique a known weight of sediment sample consisting of loose grains is placed on top of a stack of screens arranged from bottom to top in order of increasing sieve diameter. When the stack is shaken, the particles divide into several fractions, depending on the number of sieves in the stack. The fraction lying on each sieve is collected separately and weighed.

The theory of sieving assumes that the weight of material lying on a particular screen is proportional to the weight of the original material of that particular size in the sample. Moreover, the sieve diameter of a grain is assumed to be the same as its nominal diameter. The theory of sieving, based on these assumptions, has many drawbacks. An elongate grain can pass through a square or round hole screen vertically. The smallest diameter of the grain involved in this case, may not even be remotely related to the nominal diameter. Moreover, a grain having a long diameter 1.4 times the sieve opening can pass through a square mesh sieve diagonally, the diagonal of a square of unit length being $\sqrt{2} = 1.41$. Common manufacturing defects of sieves also add to the problem. Square mesh sieves often stretch into rhombic shapes after prolonged usage, thereby increasing the effective diameter. Moreover, in this method, the grains are sorted according to their shapes as well as sizes. Flakes of mica, for example, lie flat on screens far bigger than their nominal diameter, while elongate grains, passing vertically, accumulate on the finer sieves.

Interpretation of the results of sieving is also not straight forward. The conventional techniques of statistical analysis demand measurement of frequency (number counts), whereas in sieving the total weights of grains lying on each sieve are measured instead of numbers. This weight is represented by a complex parameter, frequency (n) multiplied by the mass of each grain on the sieve (M), or

$$W = n.M,$$

Where $M = (4/3 \, \pi r^3) \, \rho$, $4/3\pi r^3$ being the volume of a spherical grain of radius r and density ρ. This leads to additional complications because density of the grains influences size measurements. Moreover, r, the sieve opening, is *assumed* to be the same as the nominal grain diameter.

In view of these complications, parameter W, the total weight of grains on each sieve, is designated as '*weight frequency*' to distinguish it from true frequency (n). Appropriate techniques for analysis of weight frequency data not being available, the conventional techniques for analysis of frequency data are used for computing statistical parameters of grain-sizes from sieve data. Some attempts have been made to formulate new techniques of computation of mean and variance of the weight frequency data (Adhikari, Roy and Sengupta 1981), but the methods developed are too sophisticated to be discussed here.

Settling Techniques

In this method the particle concerned is allowed to fall through a column of water and the size of the grain is worked out from its fall velocity (v), using *Stokes' equation* :

$$v = \frac{2(\rho_1 - \rho_2)\, g\, r^2}{9\eta}$$

where ρ_1 and ρ_2 are the densities of the grain and the fluid respectively, η = coefficient of viscosity, g = acceleration due to gravity and r = radius of the grain concerned. The Stokes' equation can therefore be simplified as follows:

$$v = C \cdot r^2,$$

all parameters, except the grain diameter (r) and fall velocity (v), being constant for a particular experiment. Hence the radius of the particle can be worked out from its settling velocity.

The type of equipment commonly used for measuring grain settling velocity is illustrated in Fig. 4.1. The particles concerned are allowed to fall through a tall (about 2m) column of water and the weight of the grains accumulating on the balance pan immersed in water is recorded automatically against time. The advantage of this system is that each grain of the sample is allowed to fall through the whole column of water and hence the curve obtained is a direct record of the cumulative weight percentage of each size fraction. Various other models of *settling tubes*, measuring fall velocities of fine particles from suspension densities, are available today. Modern versions of settling tubes use pressure-sensitive devices or light attenuation techniques for measurement. Computerized systems in modern settling tubes take care of such variables as water temperature, grain shape, grain density, providing grain-size distribution patterns of the samples directly.

In spite of these developments the settling method cannot be used for measuring grains of all sizes because of the preconditions necessary for Stokes' law to be valid. The more important of these are: (1) the particle concerned is spherical, (2) the fluid medium is homogeneous, (3) the medium is of unlimited extent, (4) a constant settling velocity is attained during fall of the particle through the fluid column and (5) the settling velocity of the grain is not too high.

The lower size limit for which the above-mentioned conditions are valid lies between 0.15 and 0.5 µm. On the coarser side, 60 µm is believed to be the limit (some would extend it to 80 µm). Free fall of grains larger than this size is hindered by generation of wakes (low pressure zones), thereby invalidating Stokes' law. These limitations confine the applicability of Stokes' law to the silt and clay size grains only. Equations were developed by Oseen in 1910 and Rubey in 1933; to extend the limit of settling analysis to coarse silt. The rates of settling of grains coarser than this size vary as the square root of grain diameter (see Krumbein and Pettijohn 1938, p. 101).

In view of the difficulties involved in converting settling time to grain diameter, theoretical or empirical curves relating settling velocity to grain diameter are often used for conversion (see, for example, Fig. 4.2).

A comparative study of the different techinques of grain-size measuement is presented in Table 4.4. In spite of its drawback, the sieving method is still very popular for grain-size measurement because the equipment is relatively inexpensive and the results are replicable, that is, overall reproducibility is good.

Some techniques for conversion of sieve ('weight frequency') data to thin section (number frequency) data are available (Friedman 1958, 1962; Sahu 1964). Friedman (1962) empirically estblished a linear correlation between plots of thin-section quartile parmeters and sieving quartile parameter (Fig. 4.3). This provides an easy conversion of one to the other. Mathematical models for comparison of sieve and thin-section data are also available (Burger and Skala 1976).

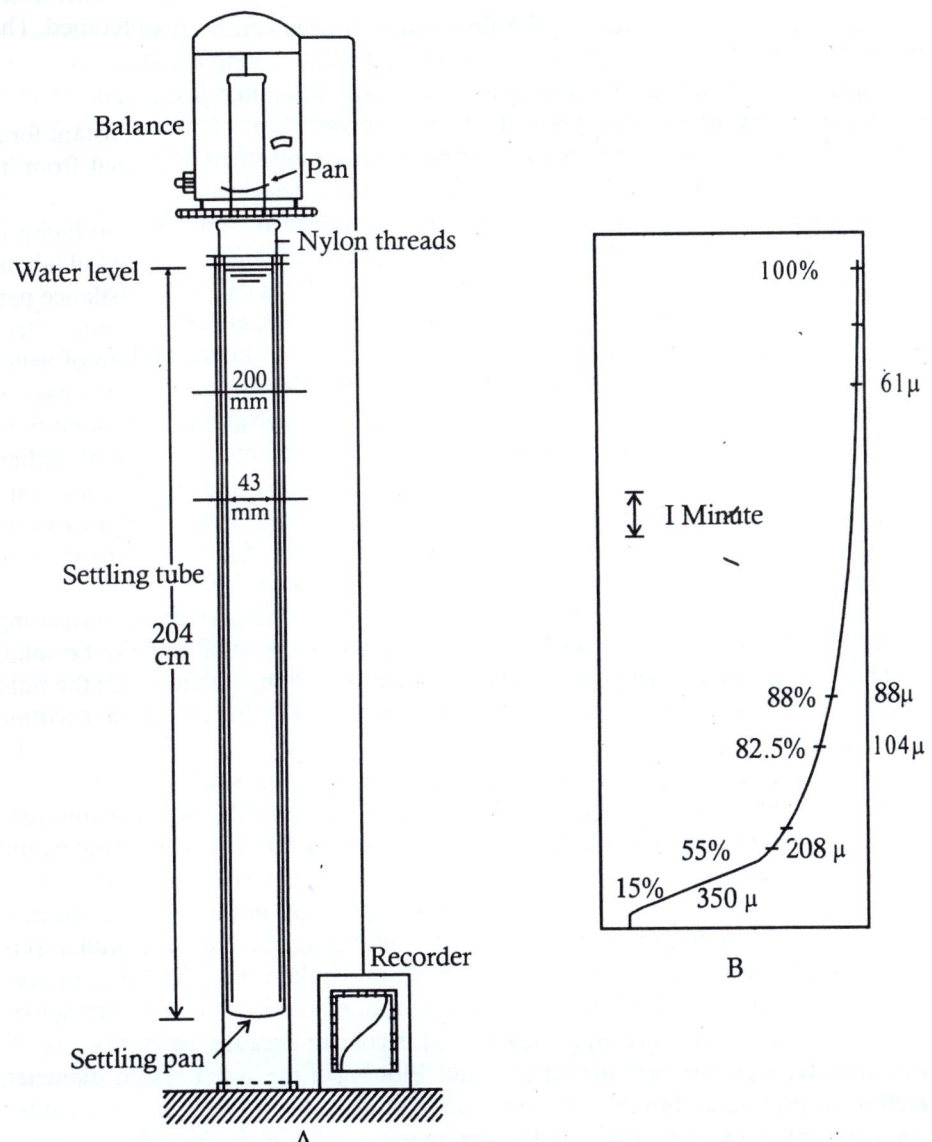

Fig. 4.1: (A) Schematic representation of the Groningen sedimentation balance for grain-size measurement by settling technique. (B) Sketch of a grain-size distribution chart obtained from the Groningen balance (modified after Sengupta and Veenstra 1968 with permission of Elsevier Science Publishers B.V.).

Analysis and Interpretation of Grain-Size Data

Grouped data of grain-sizes, obtained by any of the methods described above, are represented in tabular form. The attributes of grain-size frequency distribution (central tendency, dispersion, symmetry and peakedness) are then computed, either graphically or by using the technique of moment measures. Moments are preferred to graphic plots because they take care of every part of the frequency distribution. Following the procedure given in the Appendix, the first four moments can be readily calculated using a desk calculator (see Tables 4.7-4.11 in Appendix). Using a suitable software, the results can be obtained in seconds on an electronic computer. It is a good practice to plot the results also as histogram, frequency polygon and cumulative frequency curve (Fig. 4.4). The visual impression created by these plots often helps in arriving at geologically meaningful conclusions. Before the modern electronic computers came into use it was a common practice to compute the statistical parameters in terms of phi-units with the help of the percentile values read off the cumulative frequency curve. The 'inclusive graphic measures' proposed by Folk and Ward (1957) were very popular for computational purposes for many years (Table 4.8. in Appendix).

The particles produced out of the natural processes of weathering and erosion of a source rock are further sorted during transportation and deposition. The mechanism of grain sorting, which involves selective elimination of certain grain-sizes, varies from environment to environment. Grain-size parameters of sediment populations from various modern environments have been studied intensively for more than half a century with the hope of establishing a direct correlation between the two. The techniques used in most cases involved plotting of one of the parameters of grain-size distribution against another with the intention of discriminating between different environments (Friedman 1961, 1967; Passega 1957, 1964; Sahu 1964a; Moiola and Weiser 1968). Friedman tried combinations of various statistical parameters and found plots of third moment ('mean cubed deviation') against standard deviation (square root of second moment) to be one of the most efficient methods for environmental discrimination (Friedman 1979).

The principle involved in this discrimination can be demonstrated by referring to Friedman's earlier publications (Friedman 1961; also see Figs. 4.5 and 4.6 of this text). River sand shows poorer sorting than beach sand because of the lower energy conditions prevailing in rivers. River sands are also distinguished by positive skewness because of the presence of fines. The finer particles are washed away in the beach, leading to negative skewness. A perfect discrimination between beach and river sands can thus be obtained by plotting standard deviation (sorting) against skewness (Fig 4.5). In figure. 4.6 the (eolian) dune sand is finer than the beach sand, and is also positively skewed because the fines are retained in it. Beach sand is coarser because the competence of the ocean wave is greater than that of the wind. Plotting of mean size against skewness therefore, discriminates beach from eolian dune sand.

In another line of approach popular among sedimentologists, the different modes of sediment transportation are identified by plotting the cumulative frequency distributions of grain-sizes on *probability paper* (Fig. 4.7). A perfectly symmetrical ogive, which is the cumulative plot for a normal distribution, plots as a straight line on log-probability paper. When several such straight lines show up on the log-probability plot, as in figure. 4.7, it is assumed that the sediment population has been transported by different mechanisms, each of which has led to a normal distribution. The net distribution is then divided into traction load (coarsest fraction), suspension load (finest fraction) and saltation load (medium fraction). Using this technique, Visher (1969) identified a variety of transportational-depositional conditions on the assumption that specific combinations of traction, suspension and saltation load characterize particular environments.

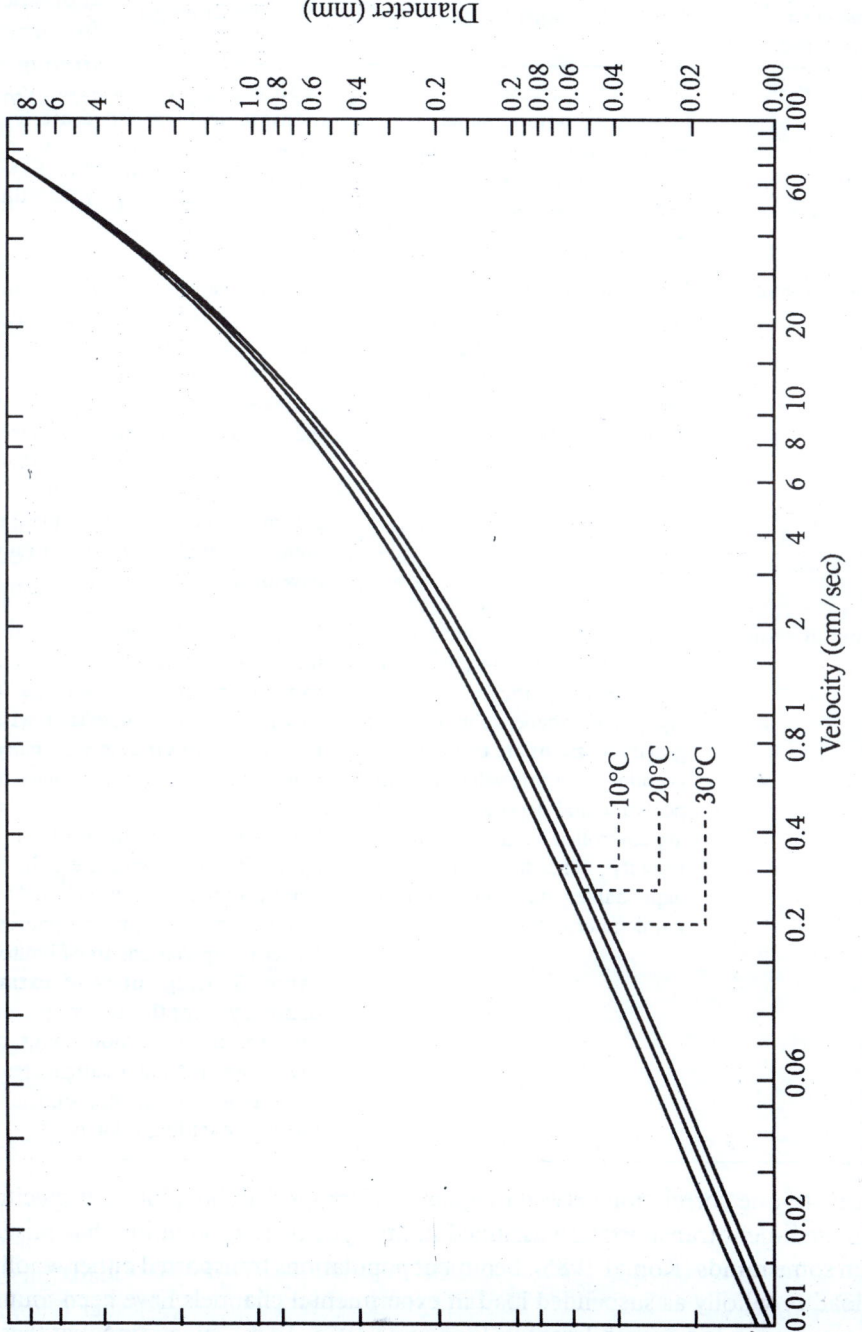

Fig. 4.2: Terminal fall velocities of quartz spheres of different diameters in water at different temperatures, as determined experimentally by the U.S. Department of Agriculture, Soil Conservation Service (modified and redrawn after Vanoni, Brooks and Kennedy 1961).

Table 4.4: Advantages and disadvantages of various methods used for grain-size measurement

Technique of measurement	Advantages	Disadvantages
Direct method calipers or microscope (ocular micrometer)	Observations recorded as frequency (number counts). Hence statistical computations are easy. Shape and roundness can also be measured together with grain–size.	Tedious and time-consuming. Measurements obtained are less (by about 24%) than actual value. Conversion to 'weight frequency' is required for comparison with sieve data.
Sieving method	Easy to use. Equipment inexpensive. Overall reproducibility of results is good. This technique is widely used. Hence comparison with results obtained elsewhere is possible.	Particle shape and density influence measurements. Square hole sieves create special problems. Manufacturing defects are common in sieves. Procedure, though not as tedious as microscope measurements, is time–consuming. Measures 'weight frequency', a parameter for which the statistical mehod of analysis is not yet fully developed.
Settling method	Rapid. Easy to use. Continuous record of settling time (velocity) available automatically by modern equipment. Shape, volume, specific gravity of grains are taken into account in settling velocity. Transportation and deposition in nature are controlled by grain settling velocity ; hence this is the most dependable parameter for hydrodynamic interpretation.	Shape and density of grains influence settling velocity but not exactly in the same way as in sieving. Hence comparison with results of sieving is not easy. Conversion of settling velocity to grain-size has to be done by formulae which have limited applicability, or with the help of empirically derived curves based on observations on quartz spheres. Hence interpretations are of limited value. Settling tubes of extraordinary length are required for handling the whole range of sediment size. Natural samples tend to fall as one unit, thereby artificially increasing settling velocity.

The one-to-one correlation between segments of log-probability plots and specific modes of sediment transportation assumed in this type of interpretation, has raised doubts in some minds (Komar 1986). Sediment populations transported either wholly as bed load or wholly as suspended load in experimental channels have been found to be segmented in log-probability plots (Sengupta 1975, 1979), suggesting that these are not necessarily the result of transportation by more than one mechanism. The degree of log-normality attained by a sediment population is perhaps a measure of the flow parameters rather than depositional environment. In another study segmented log-probability plots, commonly interpreted as mixtures of several log-normal polulations, have been found to fit log-hyperbolic better than log-normal distributions (Christiansen, Blæsild and Dalsgaard 1984). The recent finding that log-normality may develop out of a log-hyperbolic distribution when the sorting process is nearly completed by a high velocity current, has provided new clues to the interpretation of grain-size data (Sengupta, Ghosh and Mazumder 1991; Sengupta, Das, Maji, 1999).

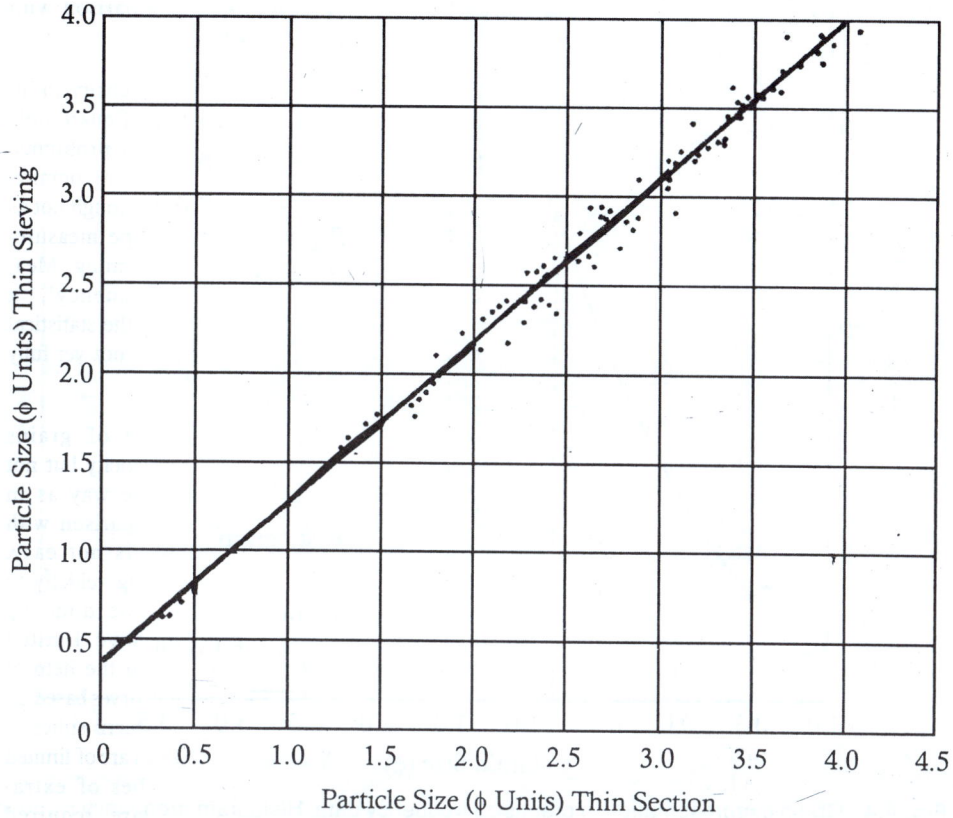

Fig. 4.3: Single overall regression line for determining sieve-size distribution from thin section analysis (after Friedman 1958, with permission of the author and The Unviersity of Chicago Press).

ROUNDNESS AND SHAPE

Roundness and shape of clastic particles are two separate parameters, independent of each other. Roundness implies rounding of the edges and corners of a clastic grian. Shape is expressed as the ratio between the three measurable diameters of an irregular grain : length, breadth, and width. Other measures of shape, including sphericity and flatness, are also used.

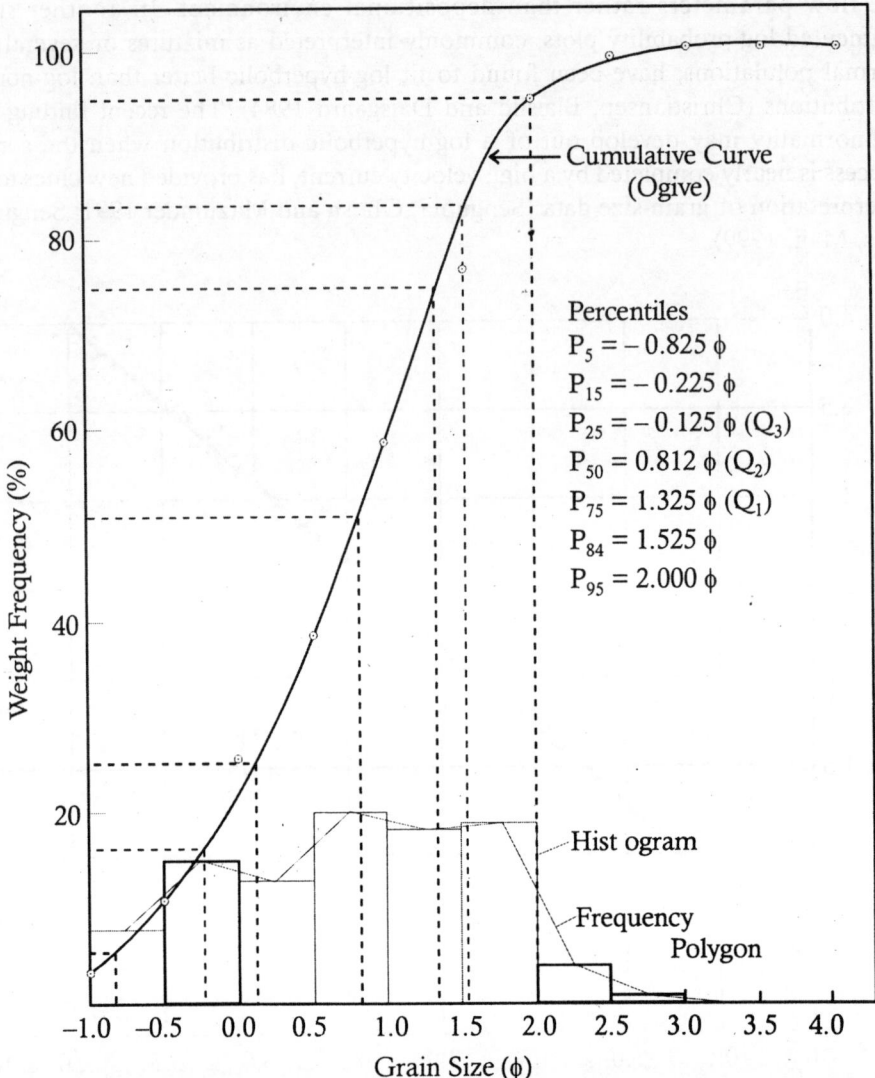

Percentiles
$P_5 = -0.825\ \phi$
$P_{15} = -0.225\ \phi$
$P_{25} = -0.125\ \phi\ (Q_3)$
$P_{50} = 0.812\ \phi\ (Q_2)$
$P_{75} = 1.325\ \phi\ (Q_1)$
$P_{84} = 1.525\ \phi$
$P_{95} = 2.000\ \phi$

Fig. 4.4: Graphic representation of grain-size frequency data. Histogram, frequency polygon and cumulative curve for the data shown in Table 4.7 in Appendix.

Roundness

Following Wadell, roundness is defined as the ratio of the average radius of curvature of the edges and corners of a grain to the curvature of the maximum inscribed sphere. Measurements on spheres being difficult, two-dimensional projections of grains are generally used. Roundness in such a case is expressed as:

$$\text{Roundess} = \left(\sum_{1}^{N} r_i\ /\ N \right) /\ R$$

where $r_i\ s$ are the radii of the corners of the particles, N = the total number of corners measured, and R is the radius of the largest possible inscribed circle (Fig. 4.8).

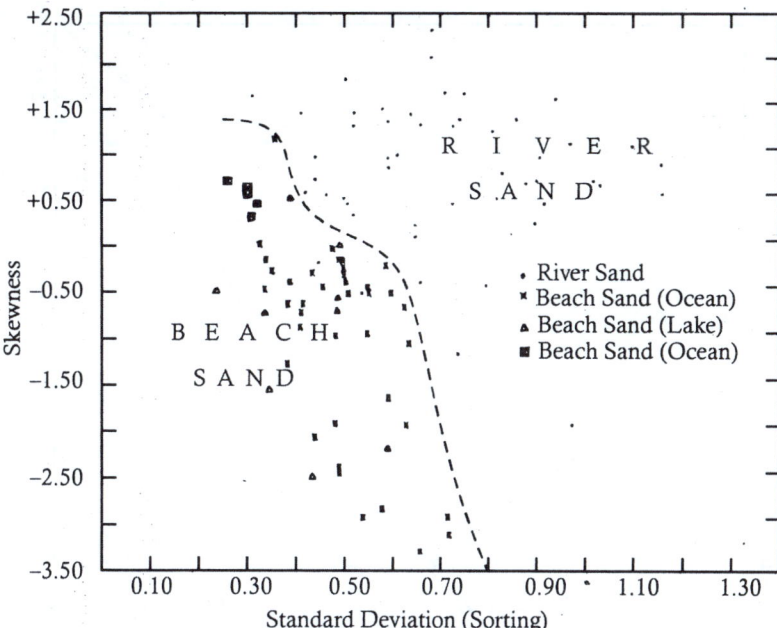

Fig. 4.5: Plots of sorting vs. skewness of grain-size distribution for discrimination of river and beach sands (from Friedman 1961). With permission of the author and SEPM (Society for Sedimentary Geology).

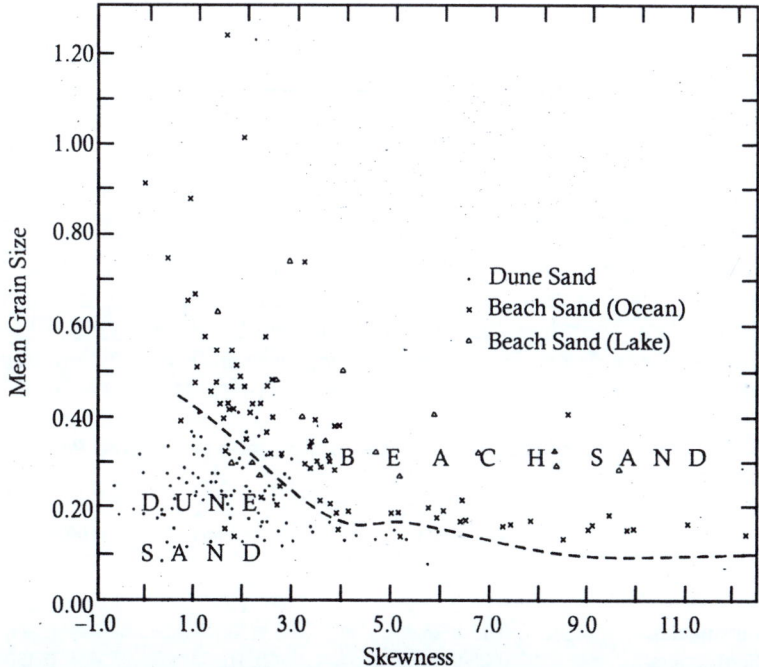

Fig. 4.6: Plots of skewness vs. mean grain-size for discrimination of beach and (eolian) dune sand (from Friedman 1961). With permission of the author and SEPM (Society for Sedimentary Geology).

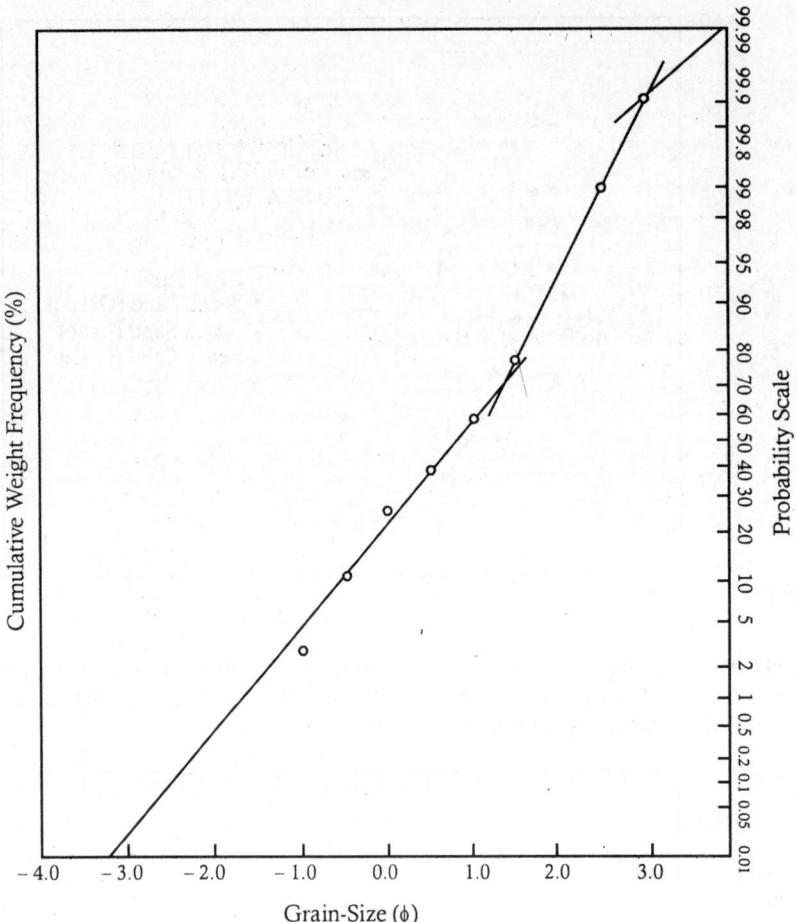

Fig. 4.7: Probability plot of grain-size distribution of sand Sample No. USR 23-A from the Usri River, India (data shown in Table 4.7 in Appendix).

According to Wadell's definition, the roundness of a sphere is 1.0. The nearer a natural particle is to this value, the more rounded it is. A chart giving photographs of pebbles of different roundness values was published by Powers (1953) to provide a quick aid to roundness determination by visual comparison (Fig. 4.9). Verbal expressions of roundness as given by various authors are shown in Table 4.5.

Table 4.5: Comparison of roundness values of Pettijohn (1975), and Powers (1953).

| Roundness terms | Pettijohn (1975) | | Powers (1953) |
	class limits	midpoints	midpoints
Very angular	–	–	0.145
Angular	0.00-0.15	0.125	0.210
Subangular	0.15-0.25	0.200	0.300
Subrounded	0.25-0.40	0.315	0.420
Rounded	0.40-0.60	0.500	0.595
Well rounded	0.60-1.00	0.800	0.850

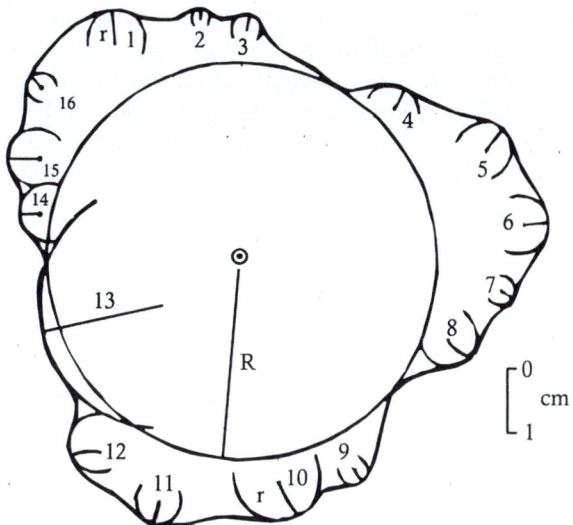

Fig. 4.8: Diagram illustrating technique of measurement of grain roundness on an enlarged projection. Roundnss = $\left(\sum_{1}^{N} r_i / N \right) / R$ = (81.1/16) mm/34.5 mm = 0.15.

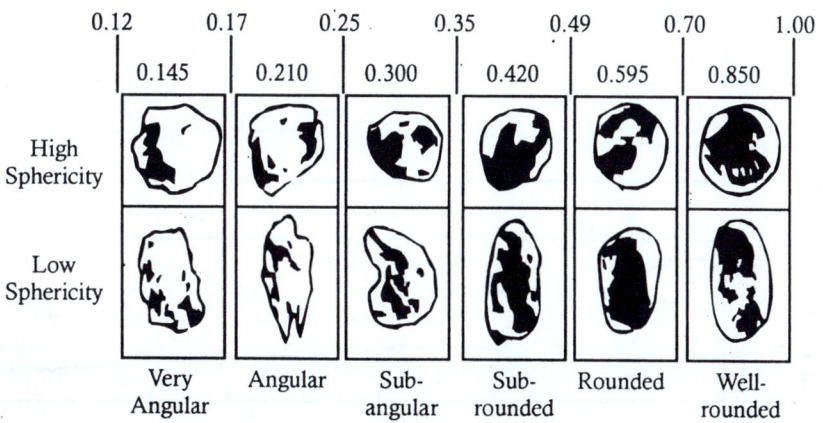

Fig. 4.9 : Chart for estimation of roundness of sand grains of high and low sphericity by visual comparison (redrawn from Powers 1953). With permission of SEPM (Society for Sedimentary Geology).

Rounding of sand-size particles of quartz and feldspar is negligible during fluvial transport. Whatever rounding is attained during transportation is confined only to the first few kilometres and that, too, observable only in the pebble-size grains (Fig. 4.10). Beach action is an effective agent for rounding of pebbles but not for sand. Rounding of a sand grain is believed, on the basis of experimental studies, to be essentially due to eolian action. The effect of solution in the rounding of quartz grains has been shown to be negligible (Kuenen 1960 a, b and 1964).

For grains larger than sand, size and roundness generally show a positive correlation. This is so because the smaller grains, which are produced by chipping of the corners of a large pebble, are more angular than the remaining central part. The fact that pebble roundness is controlled by size of the pebble rather than distance of transportation puts a serious limitation to the use of pebble roundness as a palaeocurrent indicator. Size and roundness relation in sand sizes is more complicated. For example, Banerjee (1964) showed a negative correlation between size and roundness of sand and explained this as being the result of mixing of first-cycle and second-cycle sand grains.

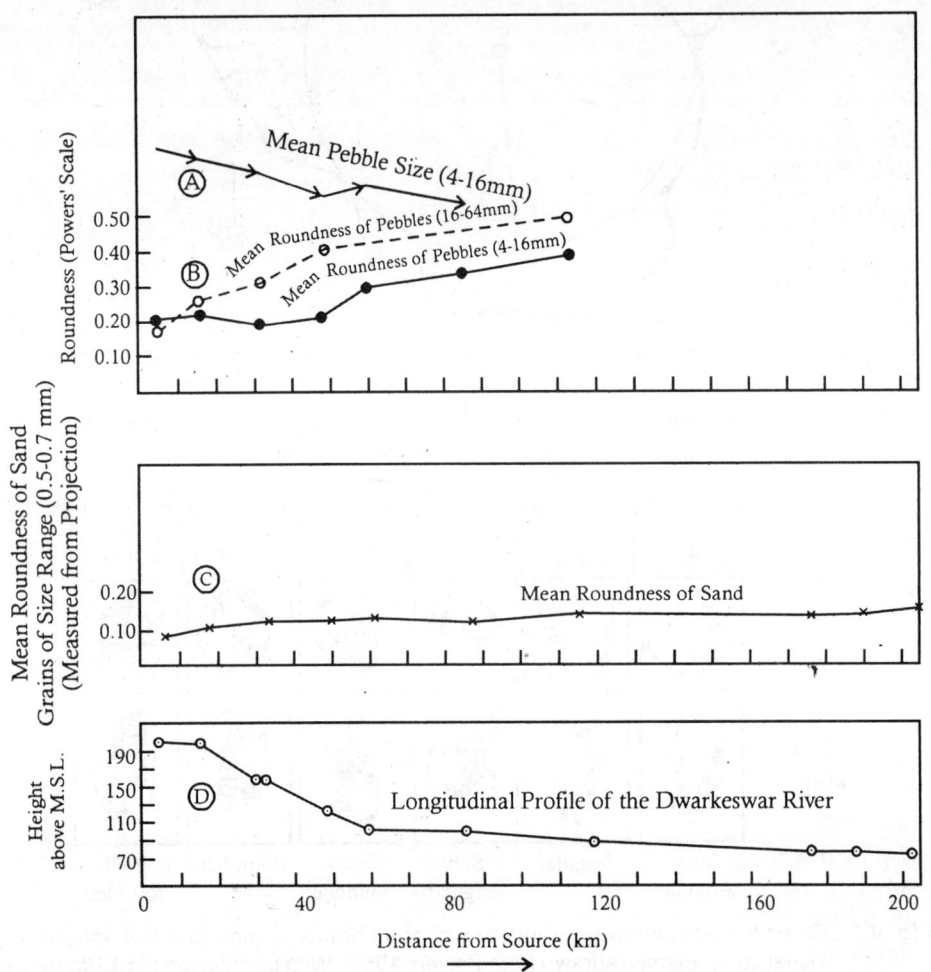

Fig. 4.10: Downstream change in roundness of pebbles and sand in the Dwarkeswar River, India (after A.K. Mukhopadhyay 1988, unpublished M.Sc. thesis, IIT, Kharagpur).

Shape

Shapes of sedimentary particles can be defined in many ways. A simple shape classification, introduced more than half a century ago (Zingg 1935), used the raito of breadth to length (b/a) and thickness to breadth (c/b) of the particle concerned. Four different shape nomenclatures (tabular, equant, bladed and prolate) were proposed on the basis of these ratios (Fig. 4.11).

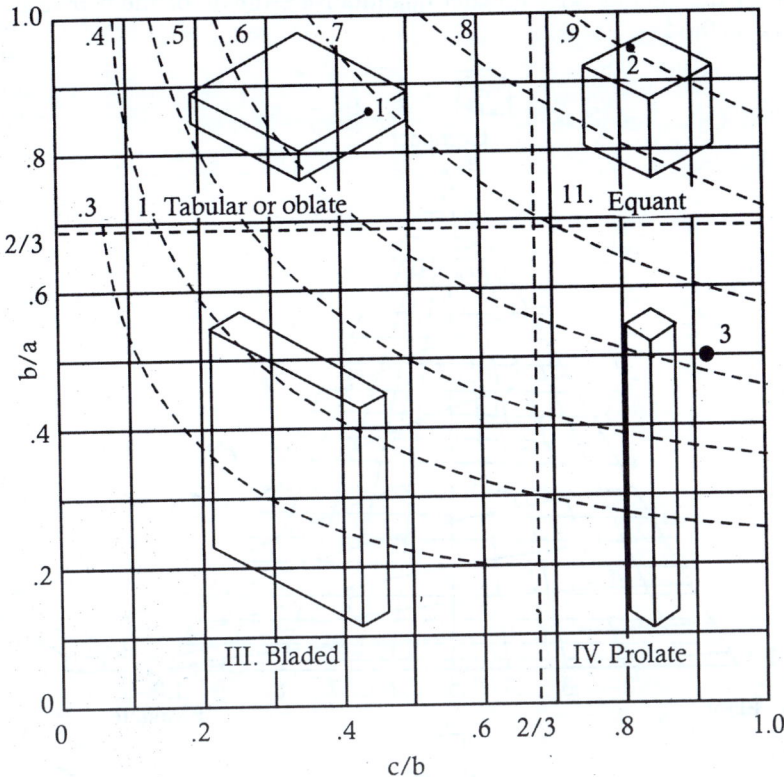

Fig. 4.11: Curves of equal sphericity superimposed on Zingg's (1935) scheme of classification of pebble shapes. Note that pebble 1 (oblate) and pebble 3 (prolate) fall in the same sphericity class (complied from Pettijohn 1975, Figs. 3.18 and 3.19).

In another approach to shape measurement the closeness of a grain to the shape of a sphere (sphericity) is determined by the ratio s/S, where 's' is the surface area of a sphere of the same volume as the particle having a surface area of 'S'. For an ideal sphere this ratio is 1. For practical purposes, following Wadell, the diameter of a sphere (d_n) having the same volume as the particle in question is compared to the diameter (D_s) of the smallest sphere circumscribing the particle of irregular shape. The Wadell sphericity index in that case is represented as d_n/D_s. Since measurements of spheres are difficult, the diameter of a particle of irregular shape and that of the circumscribing sphere are compared with the help of two-dimensional projections. It is difficult to establish a direct relationship between Wadell's sphericity index and Zingg's shape for a sedimentary particle. Pebbles of entirely different Zingg shapes (oblate and prolate) may have the same or similar sphericity.

A more modern measure of equidimensionality of grains was introduced by Sneed and Folk (1958) for analysis of forms of particles. The triangular diagram (Fig. 4.12) used for this purpose has three end members: compact or spherical (all axes equal), platy or oblate spheroid (two long and one short axes) and elongate or prolate spheroid (one long and two short axes). A given particle, with all three axes unequal, can occupy any field within this triangle, depending on (L – I)/(L – S) ratio, where L, I and S represent the long, intermediate and short diameters respectively. The form classes obtained by this method give a better quantitative estimate of the pebble shape than the Zingg method.

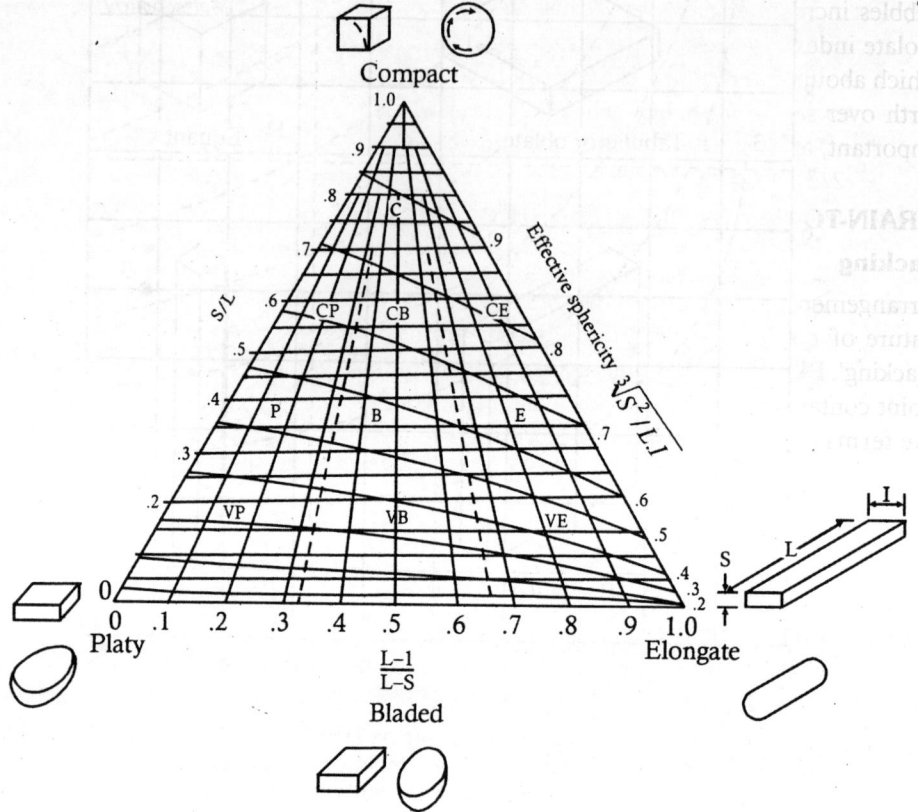

Fig. 4.12: Graphic method of determination of sphericity after Sneed and Folk (1958). Procedure : compute S/L and (L-I)/ (L-S). The sphericity of the point at the intersection of these two parameters is obtained by interpolating between the curved lines. L, I, and S represent the long, intermediate and short diameters of the grain respectively. (With permission of The University of Chicago Press). Shapes of the grains of the end members are after Fritz and Moore 1988.

A new shape property, *'rollability'*, was proposed by Winkelmolen (1971). Rollability is measured by the time taken by a grain, put into a slightly inclined, revolving drum, to roll down to the bottom of the drum. The average rollability value of a given sample is calculated per sieve fraction. The shape distribution character (SDC) obtained by plotting the relative rollability values against the size fractions, is claimed by Winkelmolen (1972) to have genetic significance. Shelf, shore and alluvial channel deposits of the Lower Oligocene of Belgium were shown to have distinctly different SDC characteristics.

The shape of a mineral or rock fragment, often controlled by its cleavage or fracture, is rarely modified by prolonged transportation. Analysis of particle form led Sneed and Folk (1958) to conclude that sphericity depends mainly on the inherent abrasional properties of different rock types and is a function of size as well as distance of transportation. It is little affected by selective sorting. In the course of another study,

Dobkins and Folk (1970) concluded that roundness of isotropically wearing basalt pebbles increases from rivers to beaches, while sphericity decreases and the oblate-prolate index becomes progressively negative, i.e., disc-like. The production of discs, which abound on beaches, is the result of abrasion caused by pebbles sliding back and forth over sand or smaller pebbles in the surf zone. Selective shape sorting is also important, at least in some cases, as sandy beaches are capable of trapping the discs.

GRAIN-TO-GRAIN RELATIONSHIP

Packing

Arrangement of the framework grains within a clastic sedimentary rock and also the nature of contact between the adjacent grains, control what is commonly termed 'packing'. Packing of the grains in turn controls porosity and also, indirectly, *permeability*. Point contact, long contact, concavo-convex contact and sutured contact are some of the terms commonly used in describing the nature of grain contacts (Fig. 4.13).

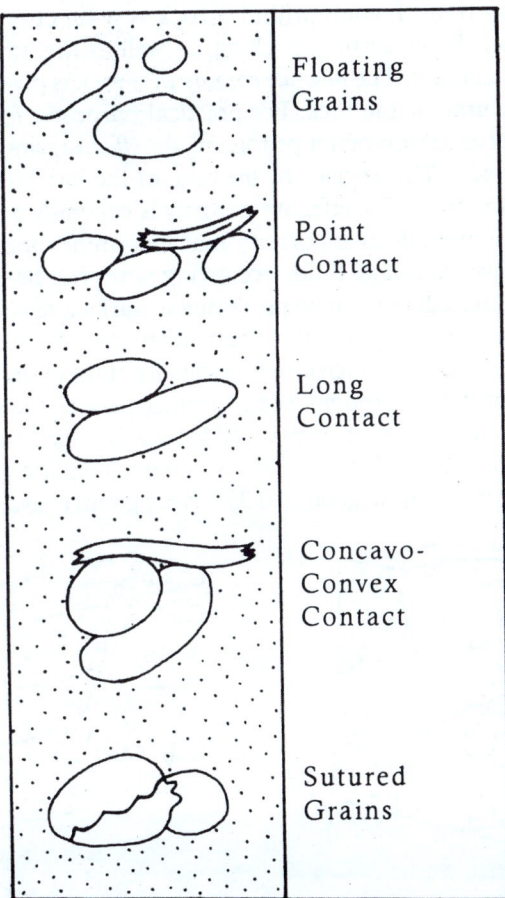

Fig. 4.13: Common types of grain contacts found in clastic sedimentary rocks (based on Pettijohn, Potter, and Siever 1987).

When the matrix proportion is very high, grains simply 'float' in the matrix. Concavo-convex contact generally implies solution effect, whereas solution, together with pressure, produces sutured grain contacts. In most other cases, when the rock is unaffected by solution or pressure, point, tangential or long contacts are found.

The index of packing of the framework grains may be determined by counting the number of grain-to-grain contacts in a randomly cut section of the rock. The result may be expressed either as '*packing proximity*' or as '*packing density*'. The former is defined as the ratio of number of grain-to-grain contacts to the total number of grains counted in the traverse. The latter is the ratio of the sum of all grain intercepts to the total length of the traverse. This, as Pettijohn (1975, p. 75) has pointed out, will have an inverse relationship to porosity. In this method of computation it is assumed that there is no matrix or cement present in the rock. A comparison by Pettijohn shows that packing (number of grain contacts) is higher in orthoquartzites than subgreywackes. Loose sand and calcareous oolites have a much smaller packing index than most sandstones.

Porosity

Porosity or the proportion of voids within a rock is a function of all the textural parameters, grain-size, shape, sorting, packing, as well as the amount of void-filling materials, matrix and cement. The *absolute porosity* of a rock is given by the ratio - total pore volume/bulk volume of the rock. For practical purposes, for example to assess the usefulness of a rock as a reservoir for petroleum, the *effective porosity* is more important than the absolute porosity. This is given by the ratio of the interconnected pores to the total bulk volume of the rock. The effective porosity is generally 5 to 10% less than the absolute porosity. It is the effective porosity that determines the space occupied by extractable fluids. However, not all fluids occupying pores can be extracted because in very fine pores the liquid adheres to the pores due to surface tension.

Porosity is determined to some extent by the manner of *packing* of the grains within a rock. An arrangement of packing which increases stability of the grains invariably decreases porosity. For this reason a stable rhombohedral arrangement of the grains within a rock produces a porosity of only 26%, whereas an unstable cubic arrangement produces a porosity of 47.6% (Fig. 4.14). Porosities of reservoir rocks for petroleum are graded as follows : 5-10% (poor), 10-15% (fair), 15-20% (good), 20-25% (very good) (Levorsen 1958, p. 98).

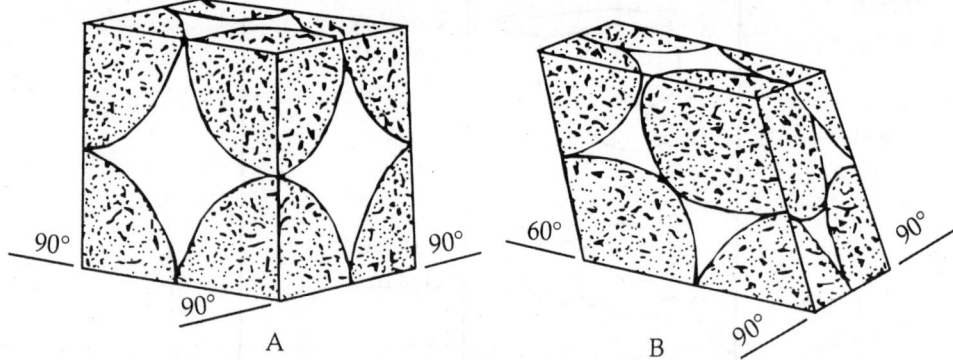

Fig. 4.14: Influence of grain packing on porosity: (A) cubic packing (porosity = 48%), (B) rhombohedral packing (porosity = 26%) (after Graton and Fraser 1935). With permission of The University of Chicago Press.

Porosity can be determined under the microscope in thin sections of rock specimens. Cores or well cuttings collected from boreholes are invariably studied under a binocular microscope for estimation of porosity. The concerned rock is then qualitatively classified as tight, dense, porous or cavernous. Porosity may also be estimated from such parameters as speed of drilling. A sudden decrease in drilling speed may indicate an increase in porosity of the rock formation because loosely packed, porous materials can only be drilled slowly. *In situ* determination of porosity can be done in the subsurface using various geophysical logs. These methods have been discussed in the section on wireline logging in Chapters 9 and 10.

Permeability

Permeability is the property allowing passage of fluid through a rock. The unit of permeability is a *darcy*. According to the American Petroleum Institute, a porous medium is said to have a permeability of one darcy 'when a single phase fluid of one centipoise viscosity that completely fills the voids of the medium will flow through it under conditions of viscous flow at a rate of one centimeter per second per square centimeter of cross-sectional area under a pressure of equivalent hydraulic gradient of one atmosphere (76.0 cm of Hg) per centimeter'.

The rate of flow of a fluid through a porous medium is dependent on hydraulic gradient. All rocks become permeable to some extent when the pressure difference is sufficiently high or the fluid viscosity is sufficiently low. Temperature therefore exerts an influence on permeability. Higher temperature lowers fluid viscosity thereby making an otherwise non-permeable rock permeable. Grain shape and packing are the two other important factors controlling permeability of a formation. A decrease in grain-size lowers permeability by increasing capillary pressure. Moreover, permeability in many cases is an anisotropic property controlled by features such as bedding planes, grain orientation and fractures.

In the laboratory permeability is determined by forcing a liquid through a core sample of the rock concerned by means of a pump and measuring the pressure drop across the sample on a manometer. The equipment used is called a *permeameter*. *In situ* measures of permeability can be obtained in boreholes by means of logging techniques.

Uncemented gravel and sand with a permeability of $10\text{-}10^5$ darcy are good aquifers. Those with lower darcy values are poor aquifers. Clays with darcy values of less than 10^{-5} are impervious. In the petroleum industry the unit of permeability used is a millidarcy (md = 0.001 darcy). Rocks are classified as follows according to their permeability : fair (1.0-10 md), good (10-100 md), very good (100-1000 md) (Levorsen 1958).

APPENDIX

Techniques of Granulometric Analysis

Determination of Grain-Size Frequency ('Weight Frequency') Distribution by Sieving

1. Grain-size frequency ('weight frequency') distributions of loose sand-size grains can be determined by the following method. Well-cemented sandstones must be disaggregated (without breaking the framework grains) before sieving. Sandstones with carbonate cement can be easily disaggregated by boiling chips of the rock in dilute HCl. Ferruginous cement in sandstones can be dissolved by boiling chips of the rock in dilute HCl together with granules of tin (Sn). Nascent Cl⁻ liberated by the process converts oxides of iron into soluble ferric chloride. Sandstones with siliceous cement cannot be disaggregated without damaging the framework grains. Grain-size distributions of such rocks must be determined under the microscope using a calibrated eyepiece micrometer.

2. Sand samples must be thoroughly dried in hot air oven before sieving. To eliminate organic materials treat the sample with H_2O_2.

3. When the original sample size is too large for sieving a subsample of desired weight must be obtained by splitting. The cone-and-quartering method of sample splitting usually works well but a mechanical sample splitter should preferably be used. Too large a sample clogs the sieves; too small a sample causes delays because of the extreme care needed for collection and weighing.

   Recommended initial weights of samples to be placed on the topmost sieve (after Vanoni *et al.*, 1961):

Sand size	Frame diameter of sieves	
	8-inch (20.3 cm)	3-inch (7.6 cm)
Medium sand size	25-50 gm	3-7 gm
Fine sand size	10-20 gm	1.5-3 gm

4. A list of the standard A.S.T.M. sieves together with their mm and phi equivalents is given in Table 4.6.

5. The sieves should be thoroughly cleaned and arranged in a stack from bottom to top in order of increasing sieve diameter in a mechanical ('Ro-Tap') sieve shaker. Ten minutes of shaking is recommended normally but the time may be increased when the sample size is large.

6. All the materials accumulated on each sieve are to be collected and weighed. Gentle tapping and sweeping by a camel hair brush may be necessary for collection of the grains.

7. The results of sieving analysis are to be tabulated in the manner shown in Table 4.7. Net weight of material corresponding to each size class, weight % and cumulative weight % are to be calculated.

8. The results of sieving analysis are to be presented graphically and statistical parameters of grain-size distributions are to be worked out as follows.

Graphic Presentation

This involves presentation of the results of sieving in the form of a histogram, frequency polygon and cumulative curve (see Fig. 4.4). The cumulative curve is the best-fitting smooth curve through the cumulative weight frequency plots. Successive points on the graph must not be joined by straight lines.

Computation of Statistical Parameters

This is done either (*i*) graphically or (*ii*) analytically.

(*i*) In *the graphic method* percentiles are computed from the cumulative curve of weight frequency and the values obtained are substituted in appropriate formulae, e.g., the Folk and Ward formulae (Table 4.8) to obtain the graphic measures for mean (M_Z), standard deviation (σ_1), skewness (S_{K_1}) and kurtosis (K_G). The results corresponding to Folk and Ward measures for the sample No. USR 23-A are shown in table 4.9.

(*ii*) The *analytical method*: Computation of moment measures

The procedures for computation of moments are given in Table 4.10 and 4.11. The statistical parameters of grain-size distribution are obtained as follows:

The first moment (arithmetic mean), $\overline{X} = \mu_1 \Sigma(fm)/\Sigma f$, where m and f denote midpoint of the class and weight frequency of the class respectively.

The second moment, $\mu_2 = \sigma^2 = \{\Sigma(fm^2)-(\Sigma fm)\mu_1\}/\Sigma f$ measures variance (see. Table 4.10). Standard deviation (σ) is the square root of variance. The third (μ_3) and fourth (μ_4) central moments are functions of skewness and kurtosis of the distribution respectively. Simpler approximations of skewness and kurtosis are given by α_3 and α_4 (see Table 4.11)

$$\alpha_3 = \{\Sigma(m_i - \mu_1)^3 \cdot f_i\}/N\sigma^3$$

$$\alpha_4 = \{\Sigma(m_i - \mu_1)^4 \cdot f_i\}/N\sigma^4$$

Values of μ_1, σ, α_3 and α_4 for the sample No. USR 23-A are shown in Tables 4.10 – 4.11.

Tests for Normality of Distribution

(*i*) *Analytical Procedure*: whether or not the grain-size frequency is 'normally' distributed can be tested either analytically or graphically. In normal distribution β_2 attains the value of 3 and γ_2 is zero, where β_2 and γ_2 are as follows :

$$\beta_2 = \mu_4 / \mu_2^2,$$

$$\gamma_2 = \beta_2 - 3,$$

where μ_4 = 4th central moment; μ_2 = 2nd moment = variance. Elaborate computations are necessary for obtaining β_2 and γ_2. A simpler method is to ascertain whether α_3 and α_4 correspond to zero and three respectively.

(*ii*) *Graphic Procedure*: sedimentologists are accustomed to testing the normally of distribution by plotting the cumulative grain-size frequency on a 'probability paper' (see Fig. 4.7). A symmetrical ogive, which is the cumulative curve corresponding to a normal frequency distribution, has a perfectly linear appearance on a probability plot. The better the fit of a straight line through the plots, the closer the match to a perfectly normal distribution. When the grain-sizes are plotted on a logarithmic (phi) scale, a straight line fit indicates that the distribution is log-normal.

Table 4.6: Specifications of standard A.S.T.M sieves

A.S.T.M Mesh No.	Sieve opening (mm) $\sqrt[4]{2}$ scale	Phi size (ϕ)	Nomenclature
			Pebble (-2 to -6 phi)
5	4.00	−2.00	
6	3.36	−1.75	
7	2.83	−1.50	Granule
8	2.38	−1.25	
10	2.00	−1.00	
12	1.68	−0.75	
14	1.41	−0.50	V. Coarse Sand
16	1.19	−0.25	
18	1.00	0.00	
20	0.841	0.25	
25	0.707	0.50	Coarse Sand
30	0.595	0.75	
35	0.500	1.00	
40	0.420	1.25	
45	0.354	1.50	Medium Sand
50	0.297	1.75	
60	0.250	2.00	
70	0.210	2.25	
80	0.177	2.50	Fine Sand
100	0.149	2.75	
120	0.125	3.00	
140	0.105	3.25	
170	0.088	3.50	V. Fine Sand
200	0.074	3.75	
230	0.063	4.00	
270	0.053	4.25	Silt
325	0.044	4.50	

Table 4.7: Results of sieving analysis. Sample No. USR 23-A. Initial weight of sample = 50.292 gm

ASTM No.	Sieve opening (mm)			(ϕ)	Mid point (ϕ)	Gross (wt) (gm)	Empty watch glass wt. (gm)	Weight f (gm)	Weight (%)	Cumulative weight frequency
10	>2.00		<−1.00		−	30.950	29.526	1.424	2.823	2.823
14	2.00	1.41	−1.00	−0.50	−0.75	18.721	14.763	3.958	7.847	10.670
18	1.41	1.00	−0.50	0.00	−0.25	37.015	29.526	7.489	14.847	25.517
25	1.00	0.71	0.00	0.50	0.25	36.013	29.526	6.487	12.861	38.378
35	0.71	0.50	0.50	1.00	0.75	39.562	29.526	10.036	19.897	58.275
45	0.50	0.35	1.00	1.50	1.25	38.622	29.526	9.096	18.033	76.308
60	0.35	0.25	1.50	2.00	1.75	38.959	29.526	9.433	18.701	95.009
80	0.25	0.177	2.00	2.50	2.25	31.563	29.526	2.037	4.039	99.048
120	0.177	0.125	2.50	3.00	2.75	15.201	14.763	0.438	0.868	99.916
170	0.125	0.088	3.00	3.50	3.25	14.794	14.763	0.031	0.061	99.977
230	0.088	0.062	3.50	4.00	3.75	14.770	14.763	0.007	0.014	99.991
PAN	<0.062		>4.00		−	14.767	14.763	0.004	0.008	99.999

Total weight   50.440 gm
Initial weight of sample   50.292 gm
Gain in weight   0.148 gm
         Error = + 0.294%

Table 4.8: Folk and Ward (1957) grain-size measures

Parameter	Formula
Graphic mean	$(M_z) = \dfrac{\phi_{16} + \phi_{50} + \phi_{84}}{3}$
Inclusive graphic standard deviation	$(\sigma_1) = \dfrac{\phi_{84} - \phi_{16}}{4} + \dfrac{\phi_{95} - \phi_5}{6.6}$
Inclusive graphic skewness	$(SK_1) = \dfrac{(\phi_{84} + \phi_{16} - 2\phi_{50})}{2(\phi_{84} - \phi_{16})} + \dfrac{(\phi_{95} + \phi_5 - 2\phi_{50})}{2(\phi_{95} - \phi_5)}$
Graphic kurtosis	$(K_G) = \dfrac{(\phi_{95} - \phi_5)}{2.44(\phi_{75} - \phi_{25})}$

Table 4.9: Statistical parameters of grain-size distribution. Sample No. USR 23-A

	Mean (ϕ)	Standard deviation (ϕ)	Skewness	Kurtosis
Graphic measures (Folk and Ward 1957):				
M_Z	0.704			
σ_1		0.865		
SK_1			– 0.172	
K_G				0.965
Moment measures :				
μ_1	0.777			
σ		0.860		
α_3			– 0.095	
α_4				2.133

Table 4.10: Computation of first and second moments. Sample No. USR 23-A

Size class (ϕ)		Midpoint m_i (ϕ)	Weight frequency f_i	$f_i m_i$	$f_i m_i^2$
< –1.00		–	(1.424)	–	–
– 1.00	– 0.50	– 0.75	3.958	– 2.968	2.226
– 0.50	0.00	– 0.25	7.489	– 1.872	0.468
0.00	0.50	0.25	6.487	1.622	0.406
0.50	1.00	0.75	10.036	7.527	5.645
1.00	1.50	1.25	9.096	11.370	14.212
1.50	2.00	1.75	9.433	16.508	28.889
2.00	2.50	2.25	2.037	4.583	10.312
2.50	3.00	2.75	0.438	1.204	3.311
3.00	3.50	3.25	0.031	0.101	0.328
3.50	4.00	3.75	0.007	0.026	0.098
>4.00		–	(0.004)	–	–
			$N = \Sigma f$ = 49.012	$\Sigma f\, m$ = 38.101	$\Sigma f\, m^2$ = 65.895

Note: Figures in parentheses are omitted in computation

First moment (arithmetic mean): $\mu_1 = \Sigma\,(fm)/\,\Sigma f = 38.101/49.012 = 0.777\ \phi$

Second moment (variance): $\mu_2 = \sigma^2 = \{\,\Sigma\,(fm^2) - (\Sigma fm)\,\mu_1\}\,/\Sigma f = 0.740\ \phi$

Standard deviation : $\sigma = 0.860\ \phi$

Table 4.11: Computation of α_3 and α_4. Sample No. USR 23-A

Size class (ϕ)	Midpoint m_i (ϕ)	'Weight Frequency' f_i	$(m_i - \mu_1)$	$(m_i - \mu_1)^3$	$(m_i - \mu_1)^3 \cdot f_i$	$(m_i - \mu_1)^4$	$(m_i - \mu_1)^4 \cdot f_i$
< -1.00		(1.424)	–	–	–	–	–
-1.00 -0.50	-0.75	3.958	-1.527	-3.560	-14.093	5.437	21.519
-0.50 0.00	-0.25	7.489	-1.027	-1.083	-8.112	1.112	8.331
0.00 0.50	0.25	6.487	-0.527	-0.146	-0.949	0.077	0.500
0.50 1.00	0.75	10.036	-0.027	0.000	0.000	0.000	0.000
1.00 1.50	1.25	9.096	0.437	0.106	0.963	0.050	0.455
1.50 2.00	1.75	9.433	0.973	0.921	8.689	0.896	8.455
2.00 2.50	2.25	2.037	1.473	3.196	6.510	4.708	9.589
2.50 3.00	2.75	0.438	1.973	7.680	3.364	15.153	6.637
3.00 3.50	3.25	0.031	2.473	15.124	0.469	37.402	1.159
3.50 4.00	3.75	0.007	2.973	26.277	0.184	78.123	0.547
> 4.00		(0.004)	–	–	–	–	–
		$N = \Sigma f_i$ $= 49.012$			$\Sigma (m_i - \mu_1)^3 \cdot f_i$ $= -2.970$		$\Sigma (m_i - \mu_1)^4 \cdot f_i$ $= 57.192$

Note : Figures in parentheses are omitted in computation.
Computation of α_3 and α_4 :

$$\alpha_3 = \{\Sigma (m_i - \mu_1)^3 \cdot f_i\}/(N \cdot \sigma^3) = -0.095$$

$$\alpha_4 = \{\Sigma (m_i - \mu_1)^4 \cdot f_i\}/(N \cdot \sigma^4) = 2.133$$

5

HYDRAULICS, SEDIMENT TRANSPORTATION AND STRUCTURES OF MECHANICAL ORIGIN

HYDRAULICS OF OPEN CHANNEL FLOWS

Running water transports much of the sediments in nature. To follow the process of sediment transportation therefore, one must have a clear idea of the dynamics of flowing water. The mechanism of sediment transportation by wind is also comparable in many respects to that of running water. The basic principles of hydraulics are discussed in this section to provide an insight into the mechanism of sediment transportation under aqueous and eolian conditions.

The most fundamental relationship between water discharge and flow velocity in an open channel is given by the *continuity equation*:

$$Q = w.d.v. = Av.$$

where Q = total flow discharge, w = channel width, d = water depth, v = flow velocity, A = cross-sectional area of the channel. It follows from this equation that discharge and water level remaining constant, the velocity of flow through an open channel varies inversely as the channel width. All open channel flows, including natural streams, are guided by this fundamental relationship. An important concept in fluvial hydrology is *hydraulic radius,* which is obtained by dividing the cross-sectional area of the channel by the wetted perimeter.

The physical state of flow through an open channel is controlled by several factors, of which flow velocity and fluid viscosity are more important. The term *viscosity* (also called molecular or dynamic viscosity) indicates the rate of deformation (du/dy) a fluid will undergo under a given stress (τ). In the equation: $\tau = \mu \, (du/dy)$, μ is the coefficient of molecular viscosity. The dimensions of molecular viscosity are $ML^{-1}T^{-1}$. In SI units it is kg/m.s or (N.s/m^2). It's unit is *poise*. μ is inversely proportional to temperature. The reciprocal of viscosity is *fluidity* that is, viscosity decreases (and fluidity increases) with increase of temperature. At 20°C the viscosity of water is 0.01 poise. Of the common fluids, glycerine is most viscous (nearly a thousand times as viscous as water).

Physical States of Flow

Within a body of fluid, a velocity gradient may generate due to gravity. It may also be produced by an externally applied pressure. Such flows are called *gravity waves*. Density differences within a body of fluid caused by changes in sediment concentration or temperature variation may also produce a flow. Such currents are called *density currents*.

When the resistance offered by viscosity of the medium is high compared to the flow velocity, the fluid particles move in straight, parallel paths in the form of laminae. Such a flow is called *laminar flow*. The resistance offered by fluid viscosity is overcome when the flow velocity increases. The fluid particles under such a condition move in a haphazard fashion. Multidimensional secondary motions are superimposed on the primary flows and eddies generate (Fig. 5.1). The flow at this stage is called a *turbulent flow*. Osborne Reynolds expressed the relative effects of flow velocity and viscosity on the physical state of flow in 1883 by the following equation:

$$R = v.l / v$$

where v = mean flow velocity, l = a linear measure like depth of water, grain-size, hydraulic radius, v = kinematic viscosity of the fluid. *Kinematic viscosity* is obtained by dividing the coefficient of viscosity (μ) by density (ρ).

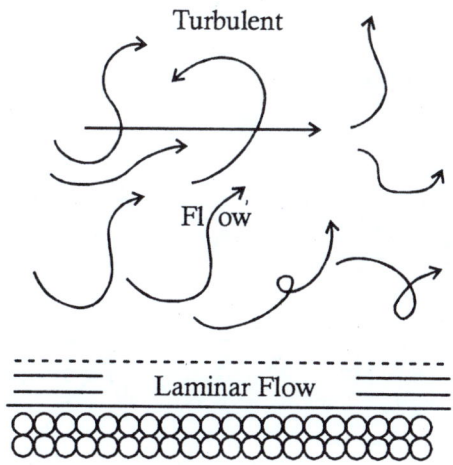

Fig. 5.1: Laminar and turbulent flows over a fine-grained bed (based on Fritz and Moore (1988). With Permission of W. J. Fritz.

Reynolds Number, a dimensionless parameter, is an expression for the balance between two contrasting forces acting within the fluid. The inertial forces shown in the numerator allow the fluid mass to accelerate, while the viscous force in the denominator resists deformation. For water at 20°C the changeover from laminar to turbulent flow takes place at the critical Reynolds number of 500. This changeover, however, is not sharp. Even when the temperature is constant, a zone of transition marks the conversion from laminar to turbulent flow (Fig. 5.2, also see Chow 1959).

While the relationship between inertial and viscous forces determines whether a flow will be laminar or turbulent, the ratio between the inertial and gravity forces determines whether the flow will be tranquil (*subcritical*) or rapid (*supercritical*). The following equation named after Froude (pronounced frood) shows the relative effects of velocity and gravity waves in determining flow condition:

$$F = v / \sqrt{g.d}$$

where v = flow velocity, d = water depth, g = acceleration due to gravity. F, a dimensionless parameter, is called the *Froude Number*. When flow velocity equals the velocity of propagation of the gravity wave, that is, $F = 1$, the flow is said to be critical.

At $F < 1$, the flow is subcritical (also called *tranquil*). At $F > 1$ the flow is rapid or supercritical. In the subcritical stage, the water level is high but the flow velocity is low. This is called the *Lower Flow Regime* of the stream. In the supercritical stage, the water level is low but the flow velocity is high. This is the *Upper Flow Regime* of the stream. The changeover from higher to lower regime is often abrupt. When it takes place, some of the kinetic energy of the supercritical flow is suddenly transformed into turbulent energy and rollers appear within the flow. The phenomenon is known as *hydraulic jump*. The fields of critical Reynolds Number and critical Froude Number might overlap in nature in spite of the fact that Reynolds and Froude Numbers are fundamentally different (see Fig. 5.2 stippled field).

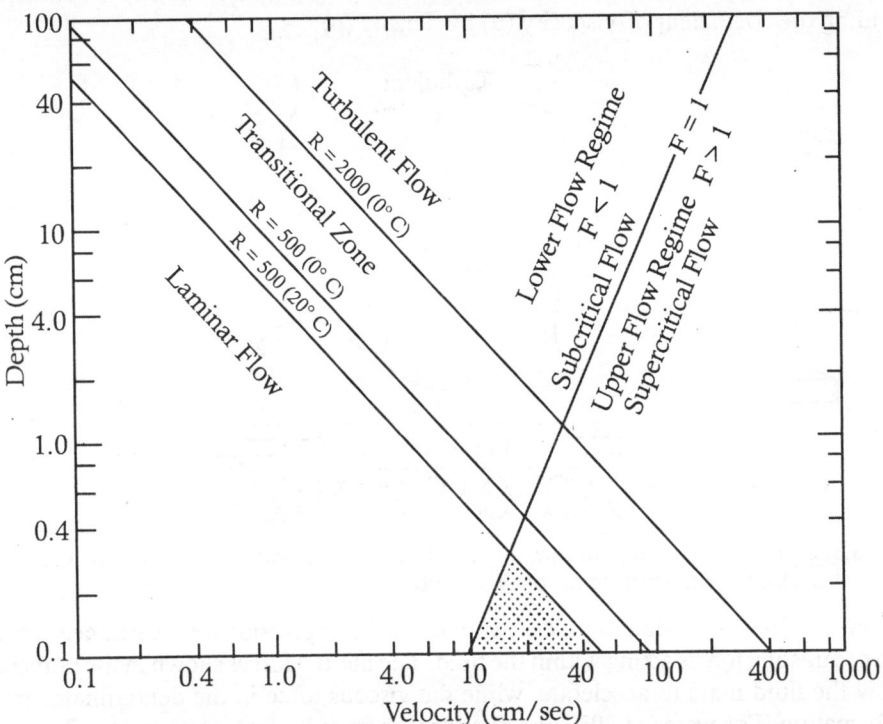

Fig. 5.2: Hydraulic conditions at various combinations of water depth, flow velocity and temperature (after Sundborg 1956). With permission of the author.

Structures of Turbulent Flow

Turbulent flows are characterized by eddies having vortices. Fluid eddies produced by the vortices are responsible for redistributing mass as well as momentum within the flow. Turbulent eddies are three-dimensional. Hence, the flow, while moving essentially in the x-direction, has y- and z-components also. Researches conducted over the last 30 years or so have provided insights into the physical nature of turbulent motion. It is now known that a turbulent flow is characterized by 'lumps' of fluids, which move a) from outer to inner flow region ("sweeps"), and b) from inner to outer flow region ("bursts"). These movements take place in the horizontal (x, y) plane, at the bottom of the flow. Near the base of the vertical (x, z) plane slow- and fast-moving streaks ("lanes") of fluid alternate. Higher up they lose their identity. In three-dimensions, the vortex structures have the shapes of hairpins whose legs are constituted of low-speed streaks. The hairpin vortex eventually give rise to longitudinal ridges and generate bed defects. Sweeps amplify bed defect heights. This is followed by flow separation, which generates downstream erosion and initiates ripple propagation on the sediment at the bed (Best 1992, see Fig. 5.3). Spacing of these streaks, and also, the rate of bursting of the low-speed streak decrease with increase of shear velocity. The picture is ever changing because new sreaks are generated and regenerated across the flow. Cogent accounts of these recent findings can be found in Leeder 1999, Mazumder, 2000.

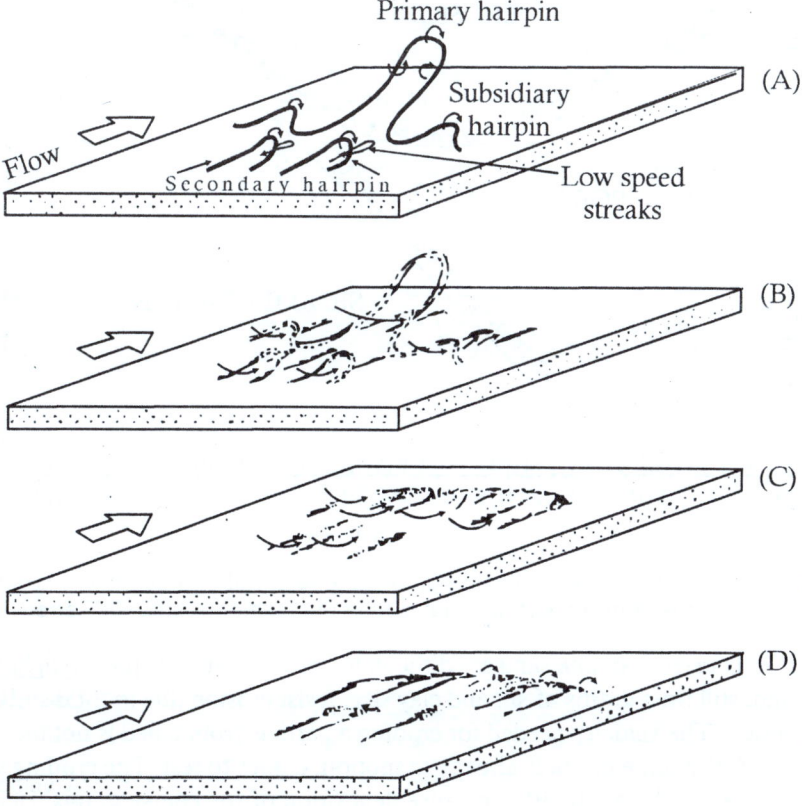

Fig. 5.3: Turbulent boundary layer structures leading to sediment entrainment by flow. (A) Hairpin structures and low speed streaks generate at the boundary layer. (B) Multiple streaks and sweeps generate ridges parallel to flow. (C) Bed defects are produced. (D) Flow separation causes downstream erosion and ripple propagation. Modified and redrawn from Best, J. on the entrainment of sediment and initiation of bed defects: insights from recent developments within turbulent boundary layer research. *Sedimentology* (1992) **39**, 797-811. Publisher, International Association of Sedimentologists (With permission of Blackwell Publishing).

SEDIMENT TRANSPORTATION

When the velocity of water flowing over a bed of non-cohesive sediment exceeds a certain threshold value, the grains are put into motion (entrained). At this stage, the flow overcomes the resistance offered by the grain due to its mass, cohesion and other forces. Properties like shape and position of centre of gravity of the grain play important roles in determining the degree of resistance offered. A diagram complied by Hjulström (1935) revealed the relationship between *critical erosion velocity* and grain-size (Fig. 5.4). The field above the curve indicates the condition suitable for entrainment of grains. It is clear from these diagrams that the particles most easily eroded from a bed are of fine-sand size (0.3 to 0.6 mm).

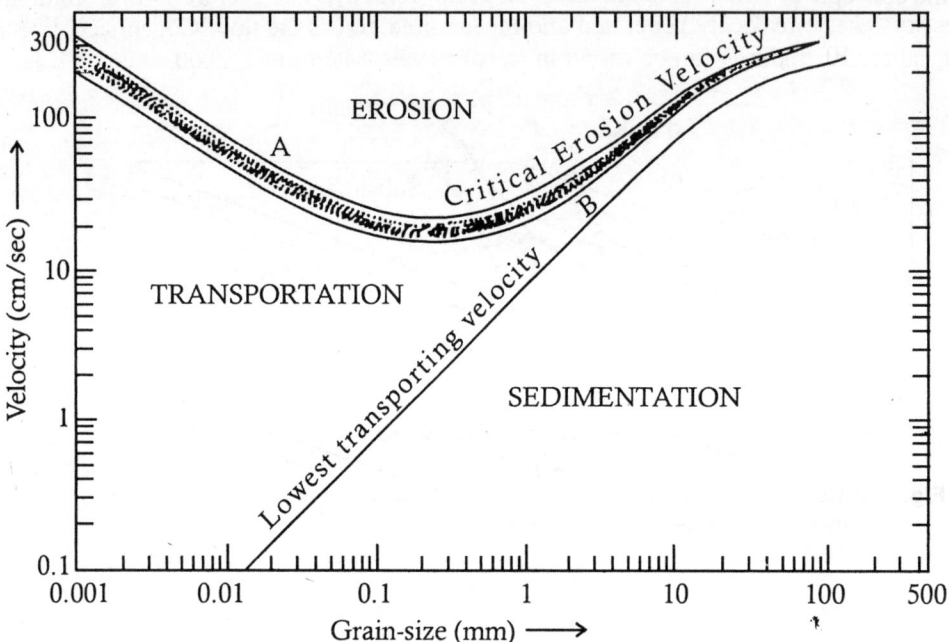

Fig. 5.4: Relationship between flow velocity (at a height of 1 m above bed), grain-size and erosion-transportation-deposition of sediments (after Hjulström 1935).

Particles coarser than fine sand are difficult to erode because of their weight. On the other hand, still finer grains of silt and clay size, resist erosion due to their smoothness and cohesion. The velocity needed for eroding a particle from a bed is not the same as that at which the same grain, if already in motion, comes to rest. The eroding velocity must be decreased by about 30% to cause deposition of the coarse grains. Hence, the critical erosion velocity (CEV) in the Hjulström diagram is a zone, not a line. Clay and fine silt, once entrained, may remain in suspension even at a low velocity. Curve B in this diagram corresponds to the lowest transporting velocity. Hjulström's studies, although of fundamental importance, have their limitations. For example, later workers have demonstrated that for the very coarse sediments the Hjulström curve underpredicts the entrainment velocity (Novak 1973).

The influence of the drag force (τ_*) of the fluid on the sedimentary particle, and the resistance offered by the grain in terms of turbulence (R_*) at the boundary on the critical erosion velocity was experimentally determined by Shields in 1936 (Fig. 5.5). Shields' curve is valid for grains transported by water as well as wind. To move from water to air one only needs to change the parameters of the equations. Experiments conducted by Bagnold (1954) showed that in the case of wind also fine sand is most easily eroded. The surface provided by silt and clay is aerodynamically smooth and hence difficult to erode. Experiments have shown that for water as well as wind a certain amount of form resistance is essential for initiation of grain movement. This resistance is provided either by some extraneous bodies lying on the bed surface, or by some irregularities projecting out of the bed into the laminar sublayer. These findings have given rise to the concepts of smooth and rough boundaries.

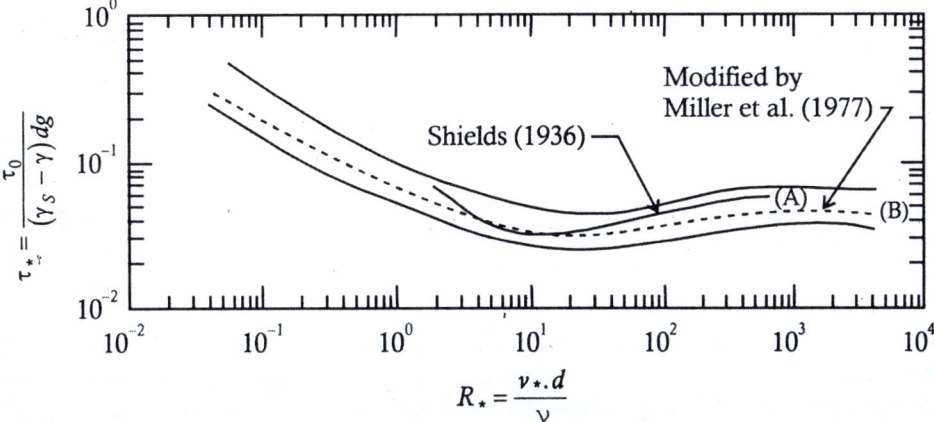

Fig. 5.5: Curves for sediment transportation as determined (A) by Shields 1936, and (B) modified and extended by Miller *et al.* 1977. $v*$ = shear velocity, d = particle diameter, v = kinematic viscosity, γ = specific gravity of the fluid, γ_s = specific gravity of the particle, τ_0 = average boundary shear. With permission of International Association of Sedimentologists.

Smooth *boundaries* develop when the sedimentary particles of a bed lie wholly concealed within the viscous sublayer. On the other hand, *rough boundaries* may be produced when the particles are not fully concealed. In the latter case the particles projecting out into the laminar sublayer may shed off eddies whose sizes are directly proportional to those of the particles. In the intermediate stage, when the grains are only partly concealed, the boundary is called *transitional*. The flow velocity immediately above a sediment bed is retarded due to friction at the bed surface. The flow loses a major part of its kinetic energy here. The thin layer of fluid adjacent to the bed where this retardation takes place is called the *boundary layer*. Within this layer the flow can be either laminar or turbulent. Even when the flow is turbulent, a very thin film, immediately in contact with the bed, shows laminar motion. This *laminar sublayer* generates over extremely smooth beds offering no resistance to flow. Close to the wall of the channel also there is a layer of liquid, which retards movement due its surface attraction. This is called the *adsorbed layer.*

The turbulent boundary layer can be divided into two distinct zones: an 'inner' (lower) zone, close to the bed and an 'outer' (upper) zone extending up the free flow surface. The former, which is the zone of vigorous turbulent activity, contains the low- and high-speed streaks of fluid. When the shear velocity is high, the low-speed streaks are lifted up to the base of the upper boundary. The '*burst cycles*', initiated at the lower boundary of the overlying layer, dissipate, while being transmitted through the upper layer, and may show up at the surface as boils.

The nature of form resistance provided by a rough bed is given by a number of well-established equations given below. (Rouse 1938, p. 279; Chow 1959, p.99)

The Chézy equation, $\quad V = C\sqrt{R_h . S.}$

The Manning equation, $\quad V = \left[1 \cdot 49\left(R_h\right)^{2/3} . S^{1/2}\right] / n.$

The Darcy equation, $\quad \sqrt{\dfrac{8}{f}} = R_h^{1/6} / 3.8\,n,$

where V = uniform flow velocity; R_h = hydraulic radius, S = channel gradient, C = Chézy friction coefficient, n = Manning roughness coefficient; f = Darcy's friction factor. The expression for Manning equation given above is in f.p.s units. In metric units the constant 1.49 is replaced by unity.

The velocity profile of the body of water flowing through an open channel is logarithmic with respect to height except at the very bottom where it is linear (Fig. 5.6). This velocity profile, deduced from the *similarity theorem* of von Kármán varies, depending on whether the bed concerned is smooth or rough. The constants in the Kármán equations were experimentally determined by later workers. After inserting these constants the well-known Prandtl-Kármán velocity distribution equations read as follows (Sundborg 1956):

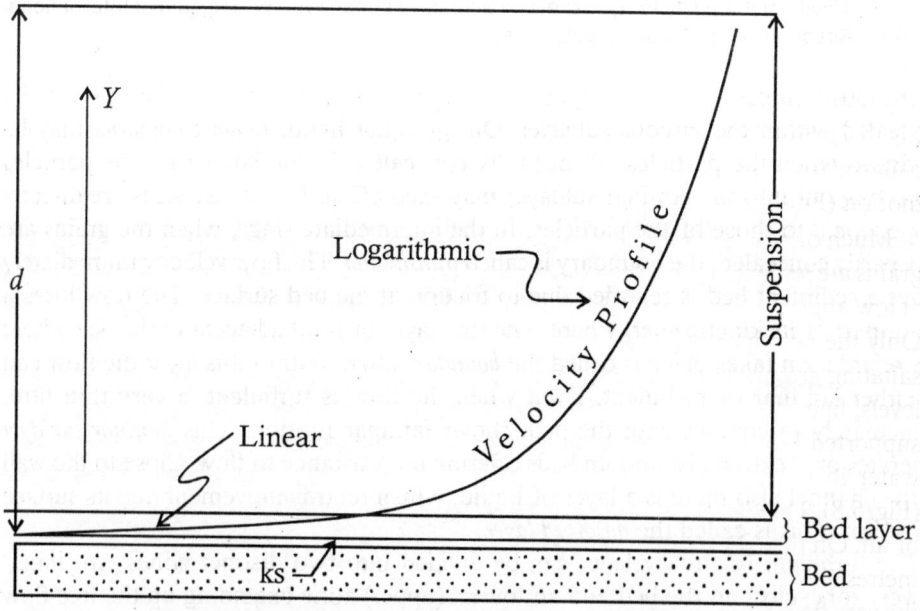

Fig. 5.6: Velocity distribution over a non-cohesive sand bed.

$$u/u_* = 5.75 \log_{10} (zu_*/\nu) + 5.5 \quad \text{for smooth boundary}$$

$$u/u_* = 5.75 \log_{10} (z/\kappa) + 8.5 \quad \text{for rough boundary}$$

where u is the average velocity at a distance z from the boundary surface, u_* is the shear-stress velocity, ν is the kinematic viscosity, and κ is the diameter of sand grain on a sandy surface.

Sediment Transportation by Water and Wind

Once entrained, running water in four different ways transports sediments: sliding, rolling, saltation and suspension. The first two, confined to the stream bed, contribute to what is commonly called the *bed load or traction load*. In natural streams rolling is more important than sliding. A grain rolls forward in response to the impact of fluid ('velocity pressure') against it. Larger particles offering greater surface areas move more easily than smaller grains. This movement is opposed by the viscosity pressure whose magnitude depends on the fluid's coefficient of viscosity, density and velocity. The total drag (Fc) is due to the pressure difference between the two forces.

In an accelerated flow, the Bernoulli law demands a reduction of pressure transverse to the direction of fluid motion. This causes a reduction of pressure on the upper surfaces of the grains moving in the flow. A grain is lifted up into the flow whenever the force created by the pressure difference between its upper and the lower surfaces exceeds its submerged weight. The upward component of the turbulent flow is also an additional aid to the process of lifting. In fact, for grains of sand and silt size, turbulence might be the controlling factor, because the effect of Bernoulli lift on such grains is negligible.

Tangential forces acting on a particle in motion may occasionally force it to leap forward into a state of temporary suspension until the force of gravity brings it down to the bed. The whole process is then repeated and the particle moves forward in a series of jumps, constituting what is called the *saltation load* (Latin: *saltare*-to dance, see Fig. 5.7). Saltating grains follow low, smooth trajectories and are governed largely by ballastic forces. The velocity of a particle during saltation is less than that of water in motion (Francis 1973).

Much of the sand size particles is transported by wind in saltation while the coarser grains are left behind as lag deposits. The saltating grains impinging on the sand surface at low angles put into motion the grains, which are normally, too large to be entrained. Only the upper few millimetres of a sand bed are affected by this *surface creep* but the saltating grains may themselves be raised up to a height of one metre or so. At higher levels, finer dust constituted of silt and clay size particles, moves in suspension, being supported by the upward component of turbulence. This is true for wind as well as water in motion. Windblown dust extends to many kilometres above desert sand (Fig. 5.8). The turbulence in case of wind is particularly high because of the low viscosity of air. On the contrary, the suspension load in water rarely extends to more than a few metres above the bed. Washload ('dustload' in case air flows) is the name assigned to the very fine (clay size) particles which are transported in permanent suspension (Leeder 1999, p. 134).

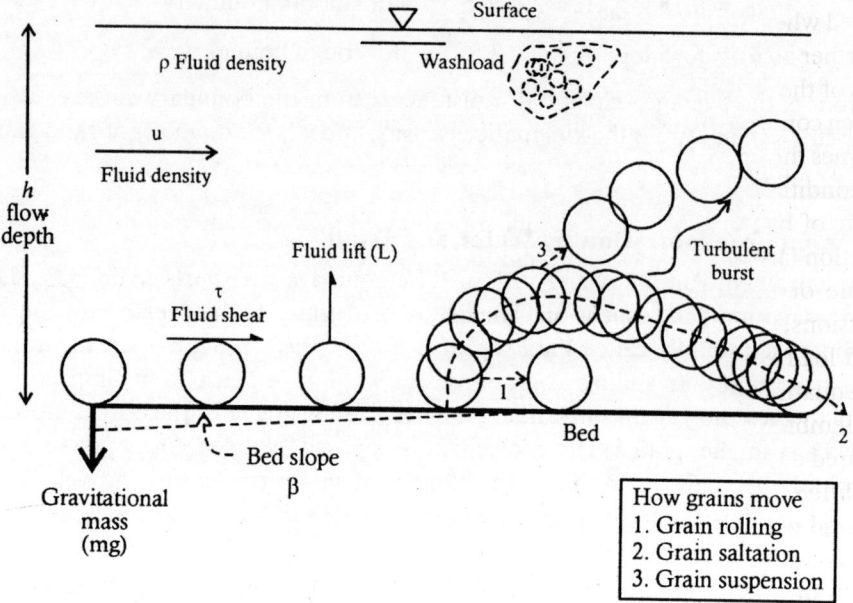

Fig. 5.7: The physical conditions for grain entrainment, saltation and suspension (from Leeder 1999). With permission of Blackwell Science Ltd.

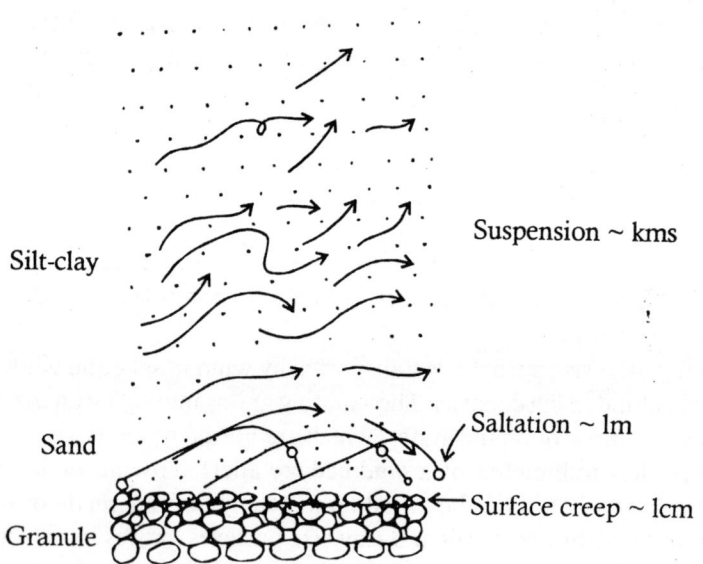

Fig. 5.8: Eolian sediment transportation (based on Fritz and Moore 1988). With permission of W.J. Fritz.

Estimation of Bed Load

An early attempt at estimation of bed load was made by H.A Einstein in 1942. In the modified version of his equation, Einstein (1950) assumed that particle movement is initiated when the hydrodynamic lift force exceeds the submerged weight of the particle. He further assumed that the probability of this particle being redeposited is equal at all points of the bed and the average distance travelled by the particle as bed load is constant between consecutive points of deposition. Einstein's assumption that this distance (about 100 times the diameter of the particle) is independent of the rate of transport, bed- and flow conditions, was questioned by later workers. Einstein's bed-load equation relates the rate of bed-load transport (ϕ^*) to the properties of the grain and the flow (ψ^*). A correction factor (θ) was introduced in 1953 by Einstein and Chien for materials which are fine or non-uniform (see Grade and Ranga Raju 1977, p. 145 for detailed discussions).

Bed load equations were later developed by many authors. Shen (1978) reviewed these equations and recommended the use of Einstein's methods as modified by Colby and Hembree (1955) for estimation of bed and suspended loads relative to some measured data when these are available. In India, Goswami (1988) applied as many as eight different formulae to estimate the bed-load transportation rate in the Brahmaputra River and obtained widely varying results, suggesting that no one of these equations can fully accommodate all the natural variables. Laursen's (1958) equation, according to Goswami, provides the most acceptable result in the case of the Brahmaputra.

Estimation of Suspension Load

A suspension (suspended) load is defined as one in which the weight of the suspended particles is continuously supported by the fluid. The force necessäry to keep the particles heavier than water or wind in suspension is provided by the vertical component of turbulence. As this is a fluctuating phenomenon, attempts to compute in absolute terms the amount of suspension load in a turbulent flow have led to extraordinarily complex exercises. On the other hand, equations have been developed for estimation of relative concentration of suspended load in open channel flows (Rouse 1938, p. 341):

$$C_y / C_a = [\{(d-y)/y\} \cdot \{a/(d-a)\}]^Z.$$

where $z = v_s/\chi u_*$ C_a and C_y are sediment concentrations at distances a and y above the bed respectively, d is total depth of water, v_S is the settling velocity of the grain, u_* is the shear velocity. χ, the von Kármán constant, is 0.40. A simpler expression of the exponent z has been given by Sundborg (1956, p. 214):

$$Z = 6.25c \ln (d/z_0)/U_{max}.$$

where U_{max} is maximum velocity in profile; c is settling velocity of the grain; z_0 is a measure of the roughness of the bed given by $k_S/30$, where k_S, the representative grain diameter of the bed material, is given by that sieve size of which 65% of the mixture (by weight) is finer. The equation provides a simple method for estimation of suspension concentration at any level (C_y), provided the concentration at the reference level (C_a) is known.

Sundborg (1956) constructed several modified curves for critical erosion velocity, cessation of grain movement and relative concentration in suspension. Ljunggren and Sundborg (1968) also combined the curves for transportation of minerals of different densities in a composite diagram (Fig. 5.9). It must be noted that all these diagrams are based on experiments with (grain-size wise) homogeneous materials. In natural

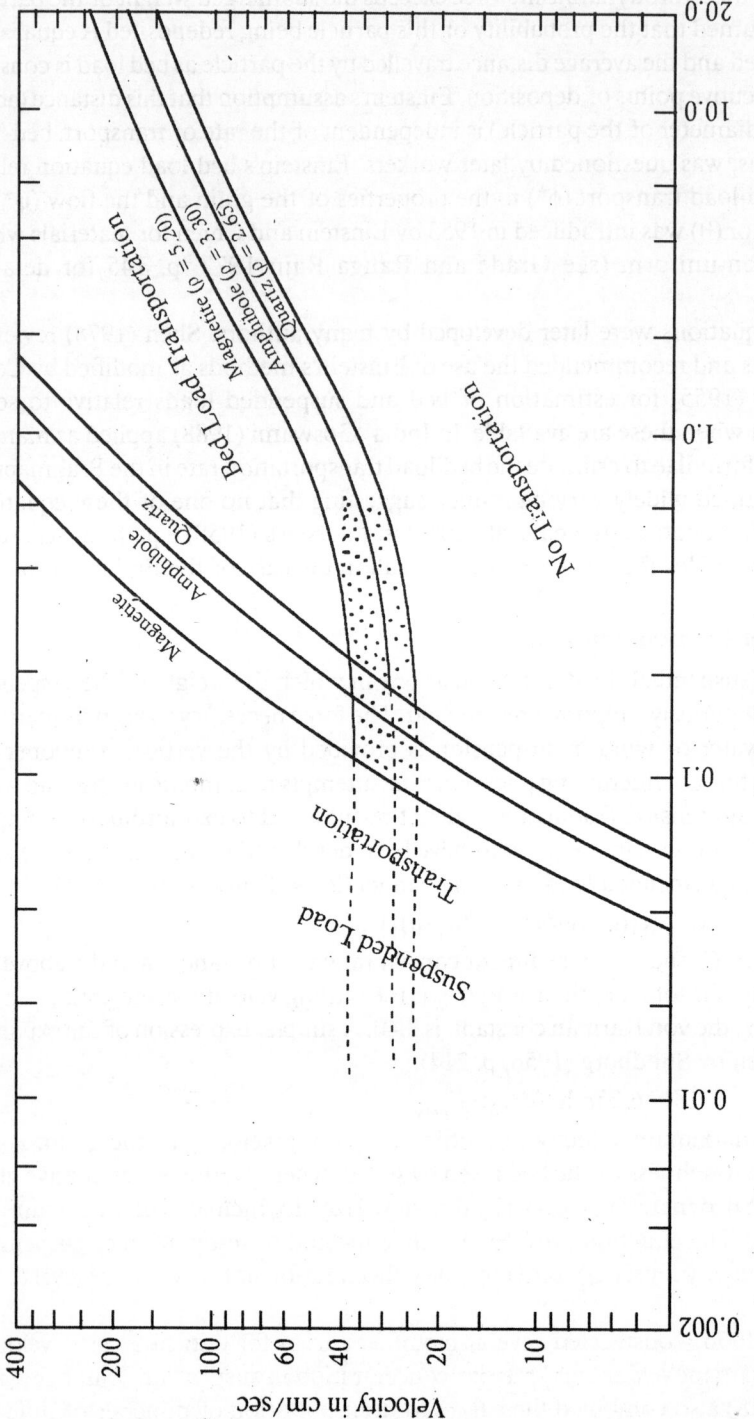

Fig.5.9: Relationship between grain-size, flow velocity and transportion of minerals of different specific gravities. Stippled area indicates environment suitable for heavy mineral enrichment (after Ljunggren and Sundborg 1968). With permission of Å Sundborg.

conditions, the process of sediment transportation is much more complex. This is so because a large range of grain-size is involved in nature. Only a few attempts have been made so far to study the mechanism of grain sorting in heterogeneous sediment loads (Sengupta, 1975, 1979, Sengupta *et al.* 1991, Ghosh *et al.* 1981, and 1986, Sengupta *et al.* 1999).

Capacity and Competency

Gilbert (1914) defined the total load a stream can transport as capacity and the size of particles it can carry as competency. In case of transportation by suspension current, capacity is expressed in terms of grams per litre of suspension. *Competency* is represented by the coarsest grain (or by the size of the coarsest 1% of the grain-size distribution) transported. Experiments have shown that the ratios of grams per litre in suspension at two different flow velocities vary with the size class involved. Hence a current's capacity to carry a load in suspension depends on the coarsest size it can carry, that is, on its competency. Grains belonging to a particular size class in suspension are deposited when the capacity of the current to hold materials of that size class in suspension is exceeded (Kuenen and Sengupta, 1970).

BED CONFIGURATION DURING SEDIMENT TRANSPORTATION

Water flowing over a plane, non-cohesive sediment surface throws it into a series of rhythmic patterns called bedforms. These are initiated at points of bed irregularities by processes shown in Fig. 5.3 (Best, 1992). The process of evolution of a variety of bedforms, starting from a simple, plane bed was experimentally investigated by a number of researchers. The results of one such pioneering study by Simons, Richardson and Nordin (1965, see Fig. 5.10) demonstrate the influence of grain-size and flow-intensity on the morphology and magnitude of bedforms. Stream power ($\tau_0 . \bar{v}$) was used in this diagram instead of velocity because the latter varies as logarithm of height. $\tau_0 . \bar{v}$, which has the dimension of power (work done over unit area in unit time) represents the erosive power of flow on the bed. Fall diameter, which indicates the hydraulic behaviour of a grain, is defined as the diameter of a sphere having a specific gravity of 2.65 and the same terminal uniform settling velocity as the particle concerned. The flows responsible for generation of bedforms of different types were divided into three categories according to Froude number: lower flow regime ($F < 1$), transitional flow regime ($F \sim 1$), and upper flow regime ($F > 1$).

The first to develop in the lower flow regime are the small sand ripples. With increasing stream power and fall diameter of the bed material, larger bedforms (*dunes*) appear. Long, flat sand waves mark the transitional zone ($F \sim 1$). With increasing steam power, these are replaced by plane beds of the upper flow regime and eventually by antidunes. At this stage, the water waves break in a direction opposite to the flow and the sediment transport direction is reversed. The antidunes are in phase with the waves generated at the water surface, whereas the ripples and dunes in the lower flow regime are not. More recently low-amplitude bedforms have been discovered on the upper-stage plane beds. These are out of phase with the overlying flows (Leeder 1999, p. 156).

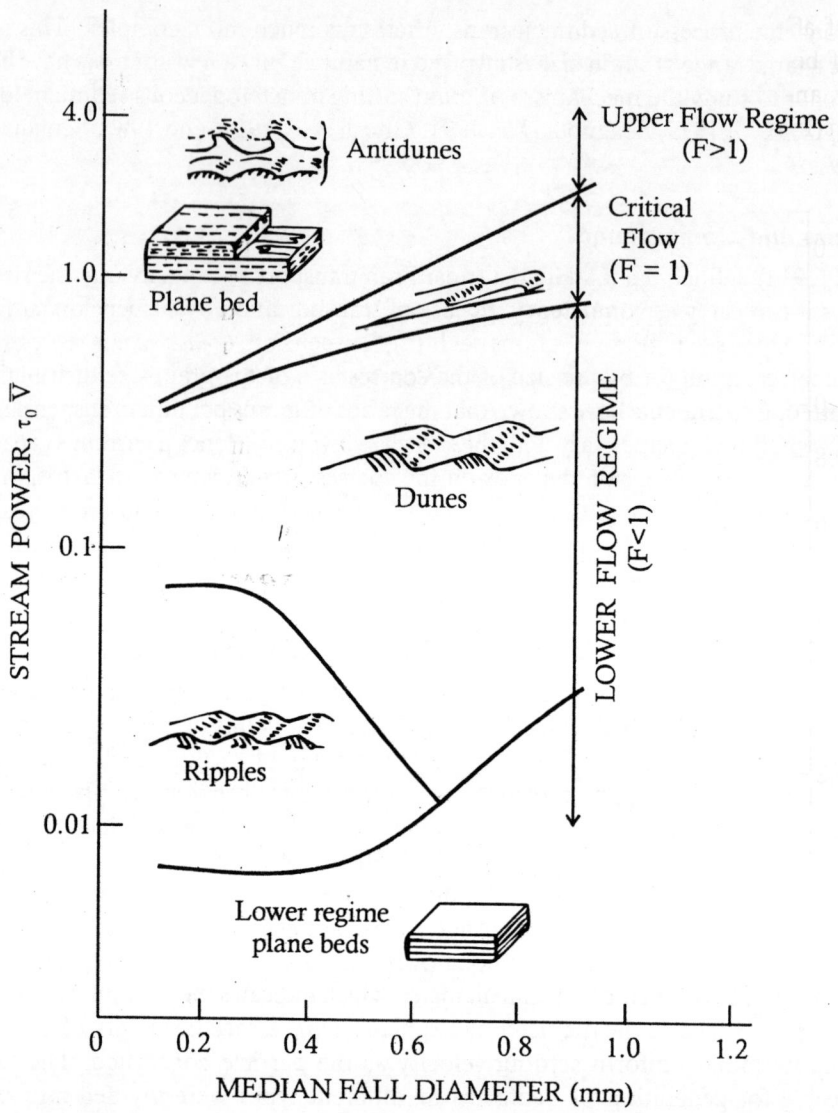

Fig. 5.10: Bedforms generated in experimental channels at various combinations of fall diameter and stream power (after Simons, Richardson and Nordin 1965).

Repetition of these studies in different laboratories during the last three decades has led to modification and refinement of the original Simon curve. One of the later results, based on the experiments conducted by Southard and Boguchwal (1990) is shown in Fig. 5.11. Grain-size plays an important role in the passage of one bedform to the other. Only when the grains are of medium sand size do all the four different types of bedforms appear at increasing velocities. When the grain-size is very small (< 0.15 mm), ripples grade directly into upper plane beds without passing through an intervening dune stage. Similarly, small ripples do not generate at all in coarse sand. Another important factor in the development of bedform is depth of water. A minimum depth of about 4 cm of water is essential for the appearance of all the four different types of bedforms. The influence of the three different parameters, grain-size, flow velocity, and water depth on the nature of bedforms is shown in a three-dimensional diagram (Fig. 5.12). A similar diagram was also constructed by Rubin and McCulloch 1980.

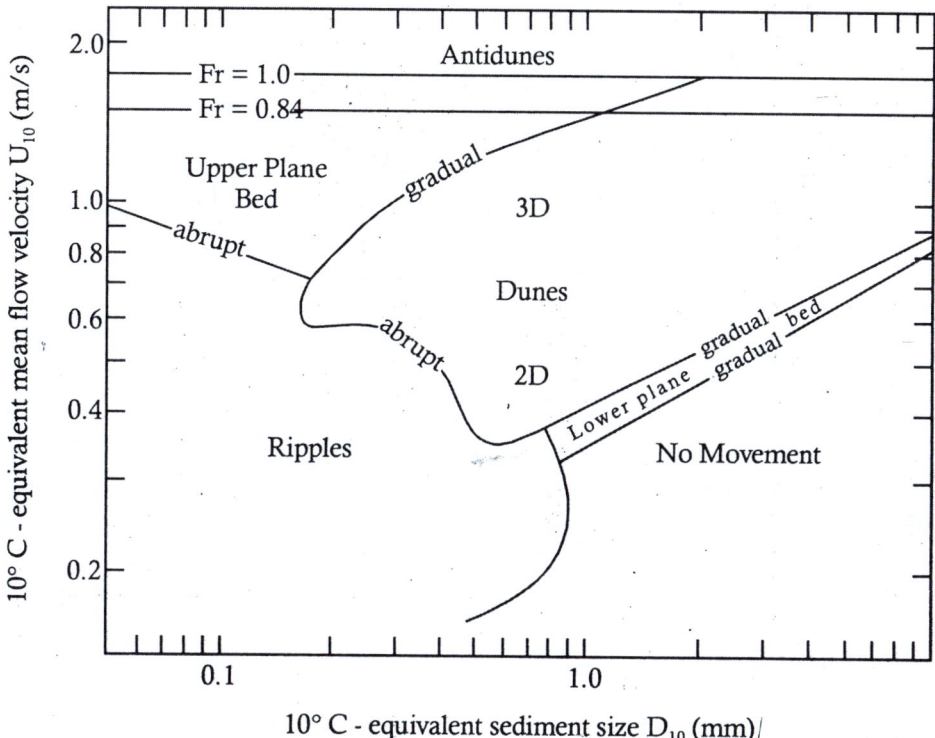

Fig. 5.11: Schematic representation of stability fields for bedforms of different types at various combinations of flow velocity and grain-size for flow depths = 0.25-0.40 m. Froude Number is indicated by Fr. (From Southard and Boguchwal 1990). With permission of SEPM (Society for Sedimentary Geology).

Introduction of the concept of flow regime in the study of bedforms was prompted by the idea that an assemblage of bedforms implies a state of hydraulic equilibrium that is hard to define. A number of variables, not only fall diameter and stream power, but even parameters like temperature and the amount of clay suspension, which affect viscosity, play important roles in defining this equilibrium. The concept of flow regime in a closed system like a laboratory channel again, is not the same as that applicable to nature. In the latter case, variables like rates of sediment supply and deposition, and also non-availability of the time needed for attaining hydraulic equilibrium, may cause complications which are not faced in laboratory experiments (Blatt, Middleton and Murray 1972, p. 123-124).

The simplest type of repetitive bedform, generated transverse to unidirectional flows, is a sand ripple (Fig. 5.13). The steeper face of a ripple dipping in the downcurrent direction is called the *lee face*. The gently sloping face dipping in the opposite direction is the stoss. The crest of the ripple is named the *summit point*. The vertical height of the crest above the base is the *height* (H) of the ripple. The horizontal distance from one crest to the next is the length or wavelength of the ripple (L or λ). The *ripple index*, also called the *vertical index*, is given by the ratio of length to height (L/H).

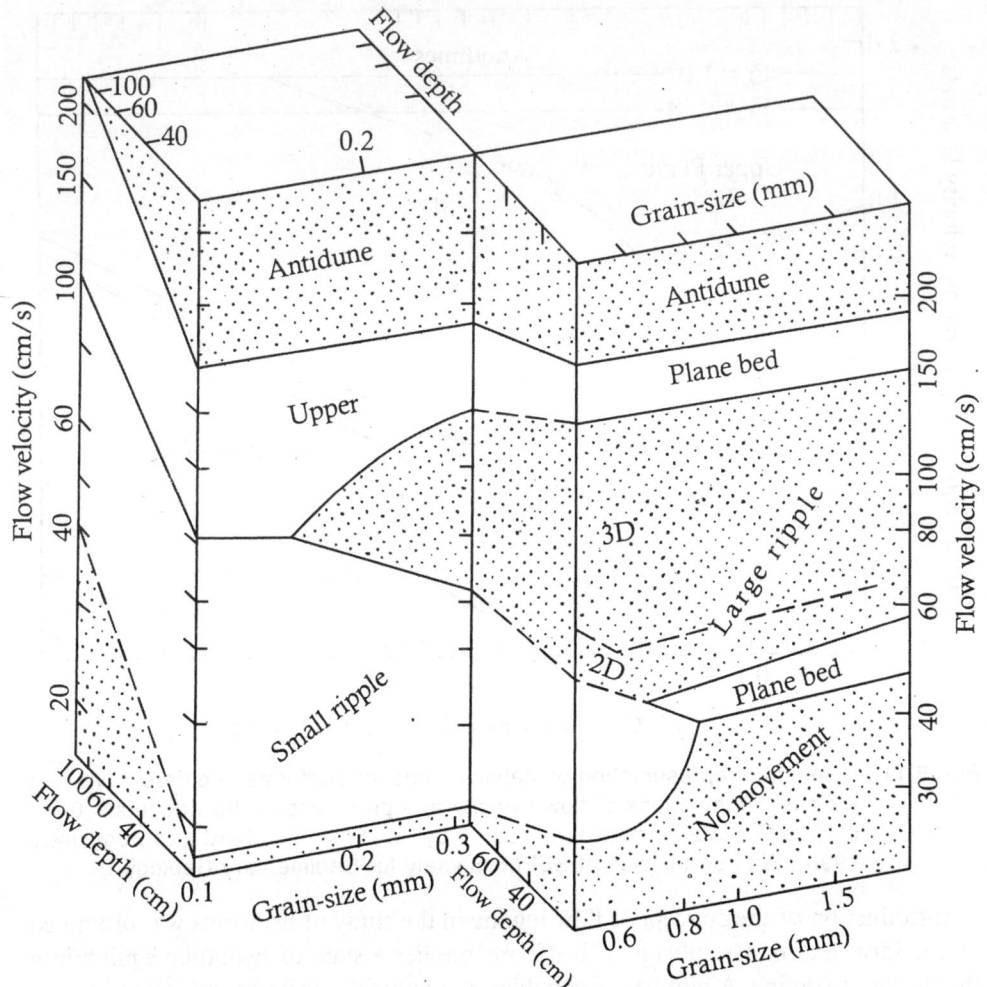

Fig. 5.12: A three-dimensional representation of stability fields of bedforms at various combinations of water depth, flow velocity and grain-size.

The stream flow lines associated with an asymmetric sand ripple are shown in Fig. 5.13. Erosion of sediments from the stoss face of the ripple and deposition of these sediments on the lee side causes the ripple to 'migrate' downstream. The internal stratification developed out of this process produces *cross-stratification* on the lee face.

On the lee side of the ripple, immediately below the ripple crest, a change in the direction of flow causes instability (termed *Kelvin-Helmholtz instability*) and develops a system of vortices rotating on horizontal axes. This is an important mixing mechanism.

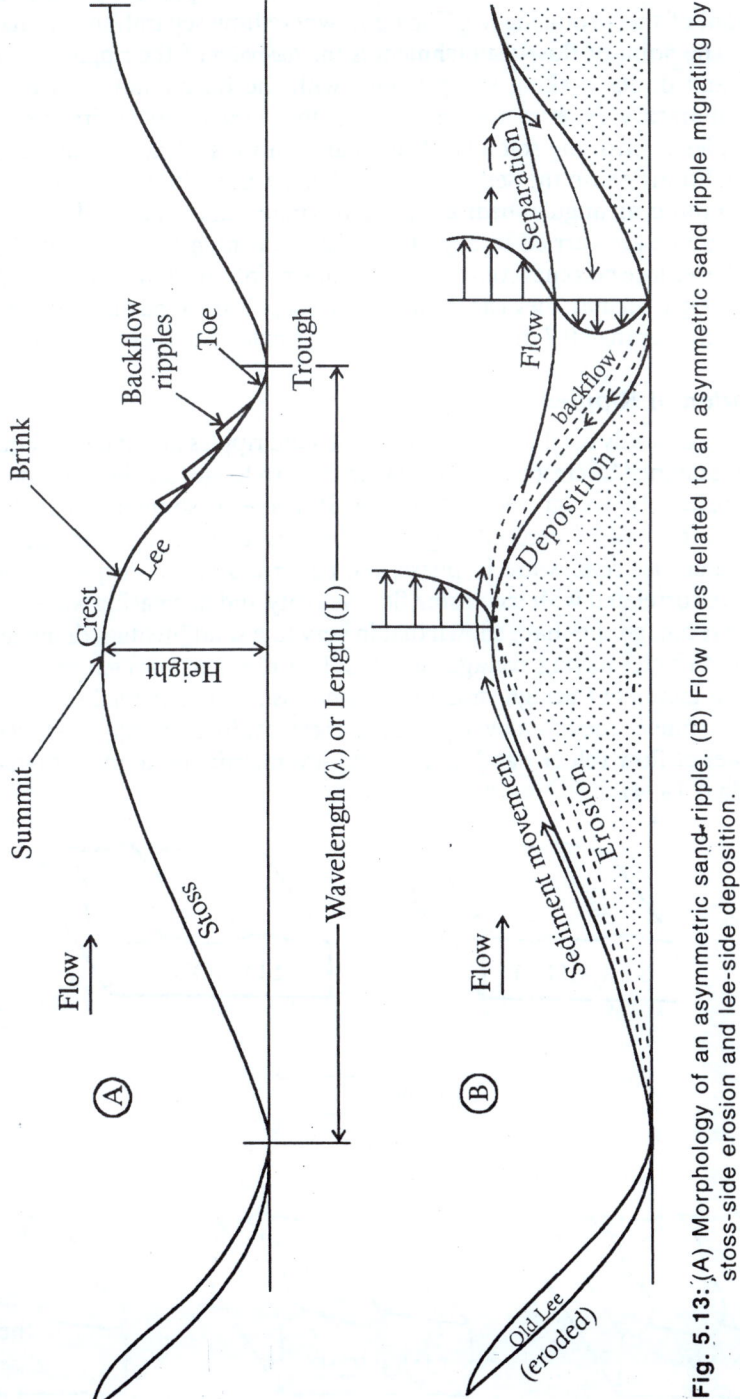

Fig. 5.13: (A) Morphology of an asymmetric sand-ripple. (B) Flow lines related to an asymmetric sand ripple migrating by stoss-side erosion and lee-side deposition.

As the pressure increases in the downflow direction the layer of fluid at the bottom is forced upstream by a process called *flow separation*. The process generates *backflow ripples* in front of the advancing lee. The point where flow separation is initiated is the *brink point*. The point of flow-reattachment is the *toe-point* of the ripple.

Ripples and dunes making sharp angles with the horizontal are called *angular bedforms*. Laboratory experiments have shown that largely the sediments on the lee sides of ripples control the stability of angular bedforms. When grain-size and flow velocity conditions cause the sediment to avalanche near the ripple crest, to make a concave-up profile, an angular bedform pattern is maintained. Flat beds generate when the coarser grains are carried to a greater distance at high flow velocity. Under this condition, the lee face becomes convex and the brink point vanishes. Coarse sands (d > 0.7 mm) do not ripple because they cannot maintain a slip face (Hand and Bartberger 1988). In fact, sand coarser than 0.7 mm is believed not to ripple at all (Leeder 1999, p. 150).

Classification of Ripples

The criteria commonly used for classification of sand ripples are shape and magnitude. The former is defined by crestline and the latter, by wavelength and height. Four different shapes of ripples were recognized by Allen (1968). These are straight, linguoid, catenary, and lunate (Fig. 5.14). Linguoid ripples are arranged in en echelon (out-of-phase) patterns. When the ripples are in phase, the pattern is called cuspate. These ripple patterns were correlated with increasing flow velocity and decreasing water depth (Fig. 5.15). More recent studies have shown that in very fine sand (median diameter = 0.095 mm) linguoid ripples having average height of 13.1 mm and wavelength of 115.7 mm develop from incipient through straight and sinuous forms provided sufficient time is allowed. The time needed for development of these equilibrium ripples is related to an inverse power of flow velocity and ranges from several minutes to more than hundreds of hours (Baas 1994).

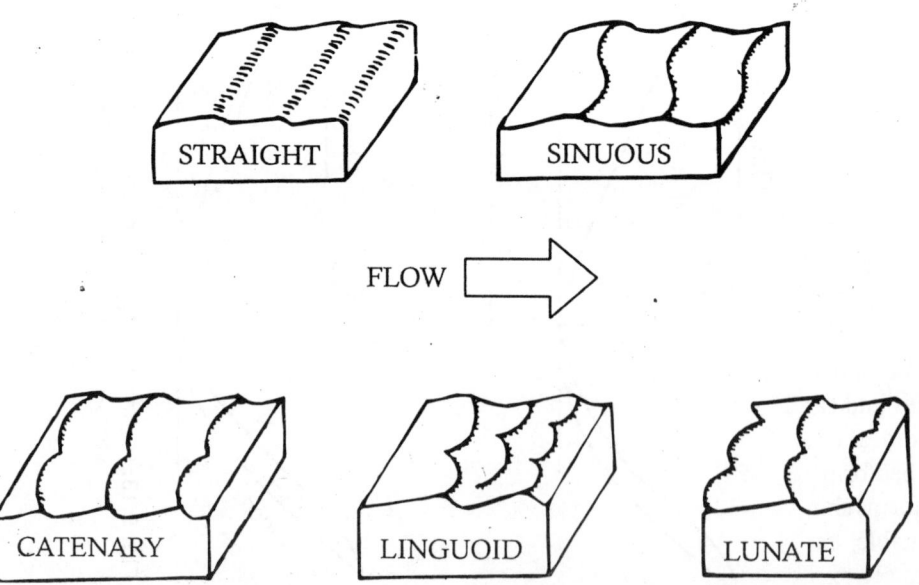

Fig. 5.14 : Classification of ripple forms according to crestline (after Allen 1968).

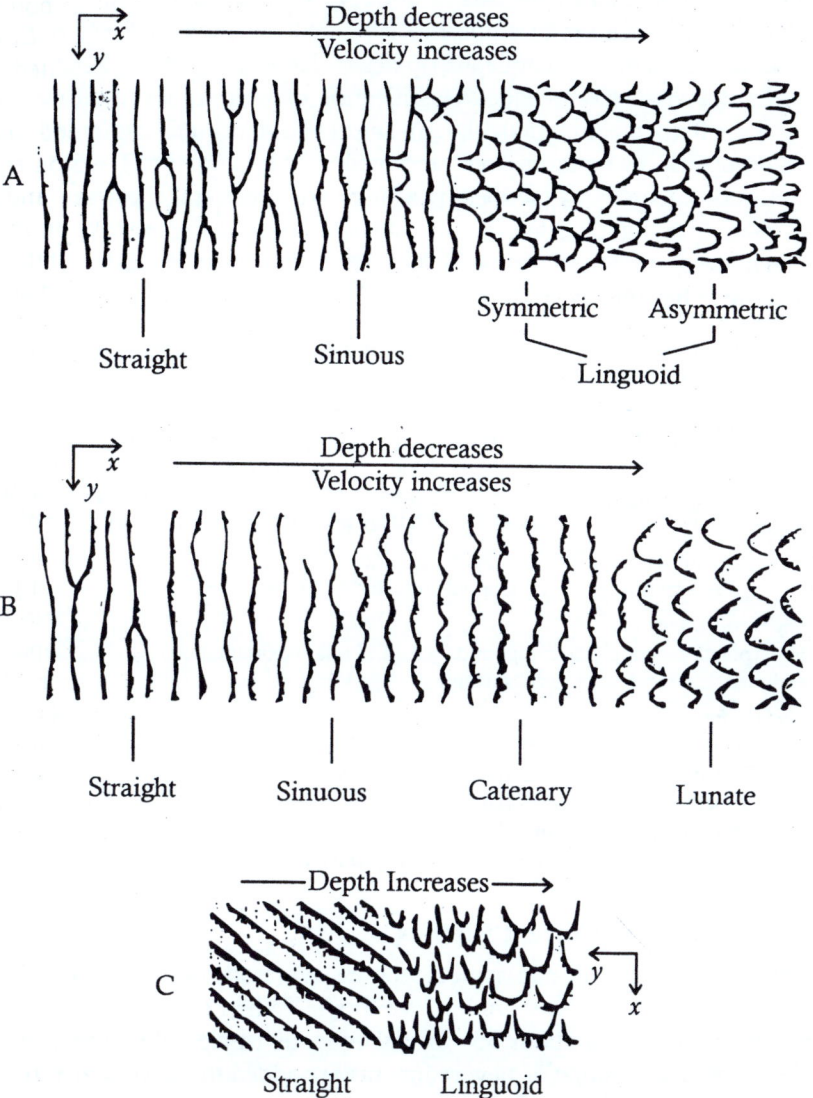

Fig. 5.15: Influence of flow velocity and/or water depth on ripple form. A = small scale ripples, B = large scale ripples, C = small scale ripples transverse to flow. Lee sides are stippled (from Allen 1968a with permission).

Magnitude-wise ripples have been traditionally classified into ripples, megaripples, dunes and sandwaves, but lack of uniformity in the definition of these terms led to a good deal of confusion in the past. Attempts were made from time to time to review the bedform nomenclature to resolve the confusion. According to current thinking, bedforms should be classified according to descriptive parameters, which are objectively definable, easily measurable and have genetic implications. The classes should be mutually exclusive and their names should unambiguously define the nature of the bedform (Ashley 1990).

A committee appointed by the SEPM Society for Sedimentary Geology reviewed the problem of classification of ripples in 1987 and recommended use of the term 'dune' for eolian as well as subaqueous ripples of large (L > 0.6 m) dimension. The same Committee also recommended classification of sand dunes into four groups : small (L = 0.6-5.0 m), medium (5.0-10.0 m), large (10-100 m) and very large (> 100 m). Only two groups of shape terms were accepted by the Committee. When the dune geometry can be adequately described by a single transect parallel to the flow, that is, when the dune is straight crested, it is designated as 'two-dimensional'. Dune forms characterized by scour pits and curved lee faces are called 'three-dimensional'. The other secondary criteria recommended by the SEPM Committee are listed in Table 5.1.

Table 5.1: Classification scheme for subaqueous ripples and dunes (based on SEPM recommendations in Ashley 1990)

I. Essential Characters	Ripples	Dunes			
		Small	Medium	Large	Very large
Size : Spacing (L or λ)	=0.1- 0.6m	0.6-5 m	5–10 m	10–100 m	> 100 m
Height (H)	~ 0.04 m	0.075–0.4m	0.4–0.75 m	0.75–5 m	> 5 m

Shape : 2– dimensionl (2–D) Straight crest line
3–dimensional (3–D) Curved crest line

II   *Important Characters*
Superposition : simple or compound (sizes and relative orientation)
Sediment characteristics (size, sorting)

III   *Useful Characters*
Bedform profile (stoss and lee slope lengths and angles)
Full beddedness (fraction of bed covered by bedforms)
Flow structure (time-velocity characteristics)
Relative strengths of opposing flows
Dune behaviour-migration history (vertical and horizontal accretion)

Structures Related to Current Ripples

As current flows the stoss side of a ripple is progressively eroded and the eroded materials continuously accumulate at the brink of the ripple. When the lee slope exceeds the angle of repose, which is about 35° for subaqueous ripples, the eroded materials avalanche down the lee slope. If new sediments are added to the system at this stage, the ripple not only migrates forward but also builds upward. The net direction of ripple movement now is described by an angle of climb, θ (Fig. 5.16A). The lamination produced by this process of accumulation is called *ripple-drift cross-lamination*. Jopling and Walker (1968) recognized three different types of ripple-drift cross-laminations. When the grain movement is restricted to the bed, the ripple stoss is eroded almost completely and climbing sets develop on the lee side (type-A ripple-drift, Fig. 5.16). When fallouts from suspension are also involved, sinusoidal ripple lamination, consisting of continuous series of sine-wave profiles (type-B) is generated (Fig. 5.16B). Intermediate between these two is the 'type C' of Jopling and Walker (1968). The type-C ripple laminae consist of climbing sets of lee-side faces together with stoss-side laminae. Alignment of ripple crests in type-B climbing ripples may sometimes produce the illusion of cross-stratification ('pseudostratification'; also see Fig. 5.17).

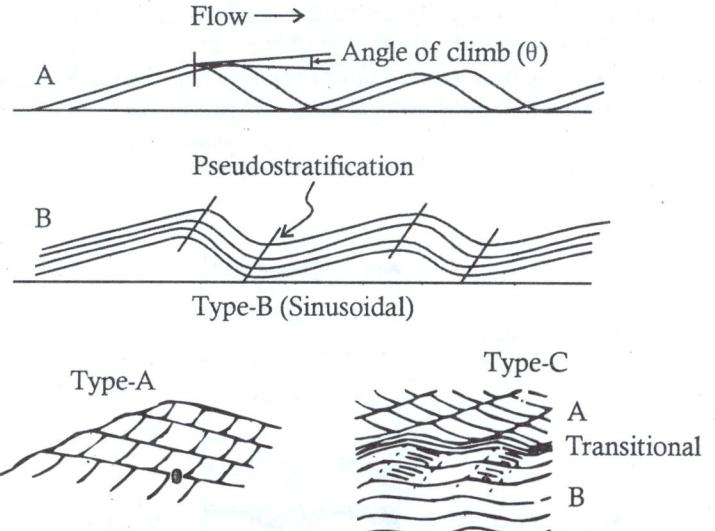

Fig. 5.16: (A) The mechanism of generation of ripple drift lamination (based on Harms *et al.* 1982). B) Different types of ripple drift laminations (based on Jopling and Walker 1968).

Fig. 5.17: Type-A and Type-B ripple drift lamination as seen in a lacker peel of channel island deposit. Ganga River near Bhagalpur, India. Scale in cm (courtesy of A. Chakrabarti).

When the depositional system contains mud as well as sand, intermittent breaks in the current flow may lead to preservation of streaks of mud within the ripple troughs. The resultant structures have been termed *flaser bedding* (Reineck and Wunderlich 1968). With increasing proportions of mud, *wavy beddings* are produced. When the proportion of mud is very high compared to that of sand, only isolated lenticles of sand may be preserved within thick layers of mud, leading to what has been termed *lenticular bedding* (Fig. 5.18).

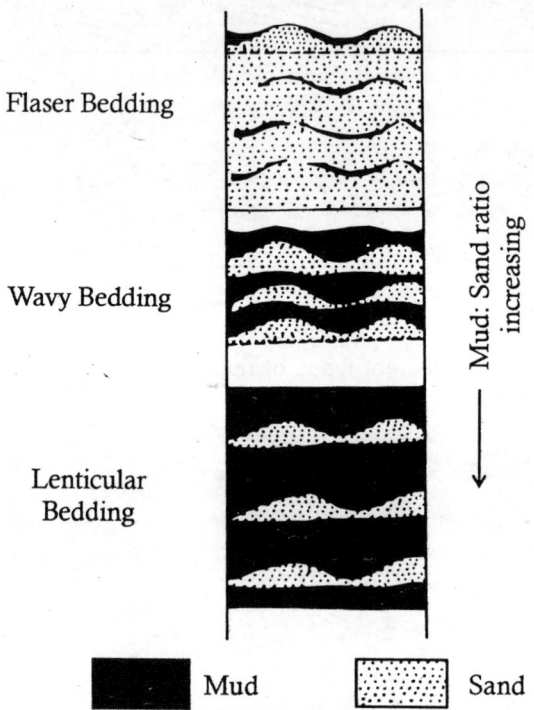

Fig. 5.18: Influence of mud and sand proportions on flaser, wavy and lenticular beddings (after Reineck and Wunderlich 1968).

Contemporaneous cutting and filling by a migrating ripple produces one of the very common sedimentary structures called *cross-stratification*. These structures consist of beds having a higher inclination than the principal surface of accumulation. Cross-stratifications may also generate in two stages—an initial stage of scouring followed by a stage of deposition (scour-and-fill). Emptying of sediments carried by streams into a relatively quite pool of water may also produce cross-bedded sets. These cross-beds have three components: a *topset* lying parallel to the principal surface of accumulation, a middle part (*foreset*) having stratification inclined at a much higher angle than the topset, and a lower bounding surface parallel to the depositional floor (bottomset). An ideal delta depicting all the three cross-bedding components is often called a *Gilbert-type delta*, after G.K. Gilbert who described such features from the delta of Lake Bonneville about a century ago.

Cross-strata may be classified according to their magnitudes and shapes. Those measured in ranges of centimetres are called *cross-beds* (>1 cm), and those measured in millimetres are called *cross-laminations* (< 1 cm). The shape classification uses the character of the lower bounding surface (*simple, planar and trough*) and the shape of sets of cross-strata (*tabular, lenticular and wedge*) as the main criteria (McKee and Weir 1953). Several other less important criteria are also used (see Fig. 5.19 and Table 5.2). Allen (1963) proposed a very comprehensive scheme of classification based on the character of the lower bounding surface and five other morphological features (Fig. 5.20 and Table 5.3).

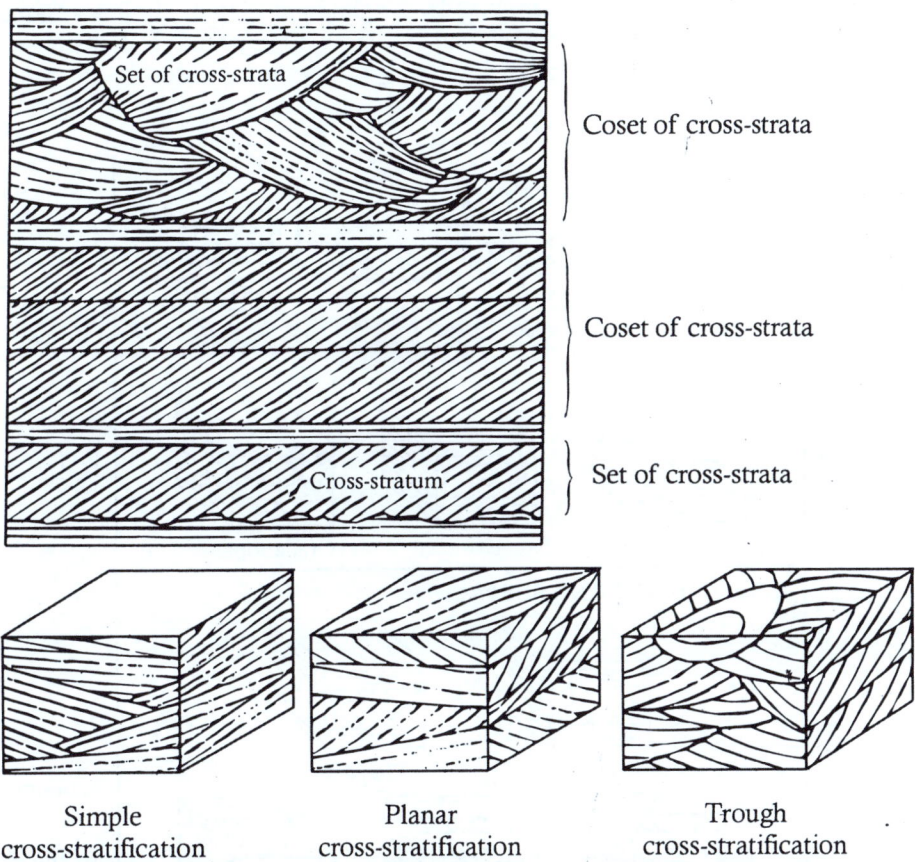

	Simple	Planar	Trough
	cross-stratification	cross-stratification	cross-stratification

Fig. 5.19: Classification of cross-bedding (after McKee and Weir 1953).

Table 5.2: Classification of cross-stratification (abridged from McKee and Weir 1953)

I.	NATURE OF LOWER BOUNDING SURFACE (LBS)	Terminology
	A. LBS - non-erosional	--- Simple
	B. LBS - planar surface of erosion	--- Planar
	C. LBS - curved surface of erosion	--- Trough
II.	SHAPE OF EACH SET OF CROSS-STRATA	
	A. Sets bounded by converging surfaces	--- Lenticular
	B. Sets bounded by planar, parallel surfaces	--- Tabular
	C. Sets bounded by planar, converging surfaces	--- Wedge

Note: Based on criteria I and II cross-bedding sets are termed planar-tabular, planar-wedge or trough-leticular.

Both planar and trough types of cross-stratifications may be produced by dune migration under aqueous and eolian conditions. In *tabular* cross-strata the bedding surfaces are planar or parallel and the foreset laminae meet the lower bounding surface at sharp angles. In plan the traces of the laminae are straight or nearly so. These are produced by migration of two-dimensional, straight-crested dunes (Fig. 5.21 B). A *though type* of cross-bedding is produced by migrating, three-dimensional, sinuous-crested dunes. The set boundaries are curved in longitudinal section. The foreset laminae of trough cross-beds are also curved in plan and have a tangential relationship to the underlying erosional Surface (Fig. 5.21 A).

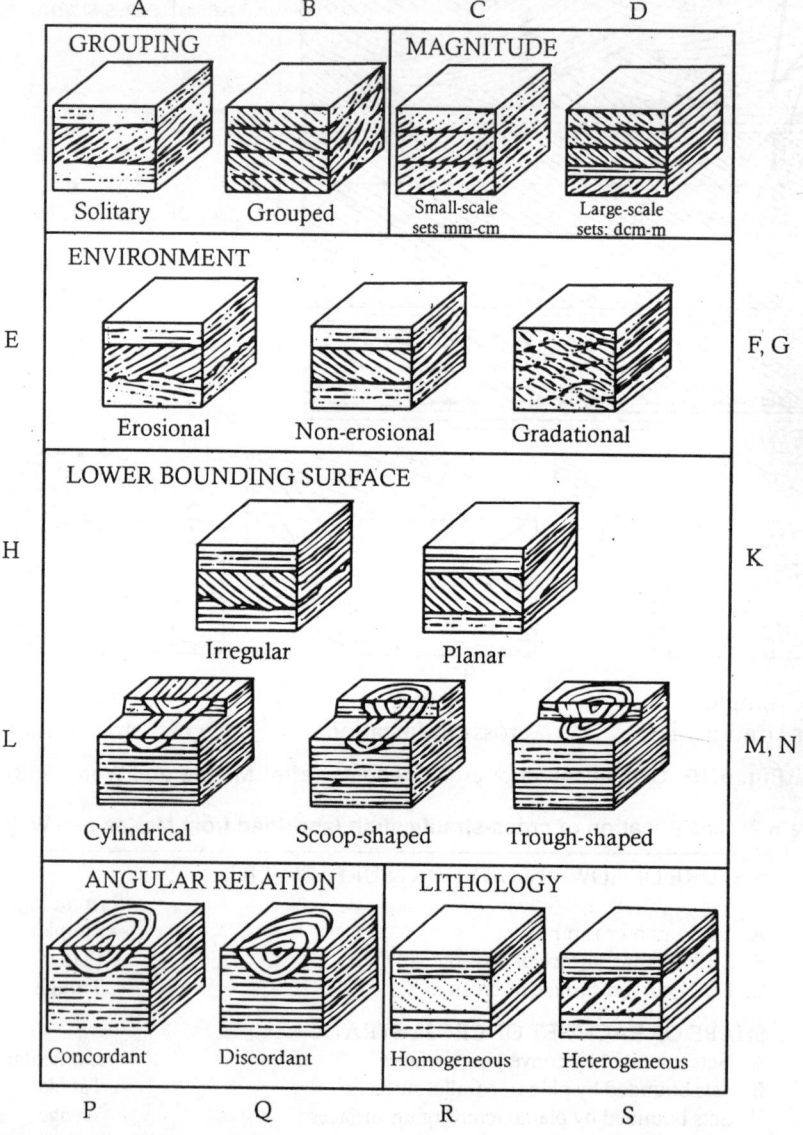

Fig. 5.20: Classification of cross-bedding (after Allen 1963). With permission of the author and Elsevier Science Publishers.

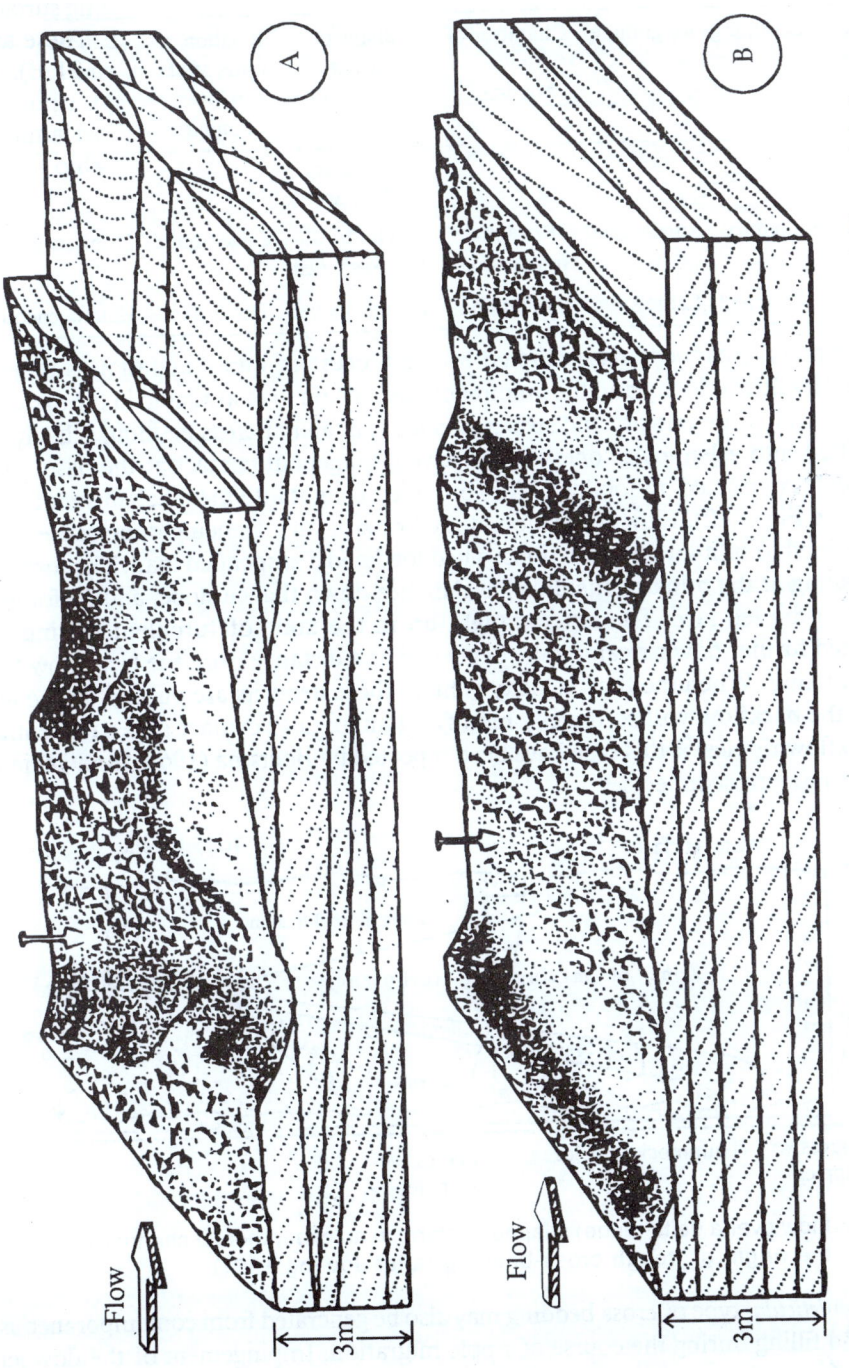

Fig. 5.21: (A) Large-scale trough-lenticular cross-bedding produced by migrating sand dunes, (B) Planar-tabular cross-bedding produced by migrating sand waves (after Harms et al. 1975). With permission of SEPM (Society for Sedimentary Geology).

Table 5.3: Criteria of classification and descriptive terms applicable to cross-stratified units (Allen 1963).

Criteria	Grouping	Magnitude	Character of lower boundary	Shape of lower boundary	Relation of cross-strata to lower bondary	Lithology of cross-strata
				Irregular Planar		
Descriptive terms	Solitary	Small-scale	Erosional Non-erosional	Cylindrical Scoop shaped	Concordant	Homogeneous
	Grouped	Large-scale	Gradational	Trough-shaped	Discordant	Heterogeneous

The type of cross-bedding generated by the process of dune migration is largely controlled by flow velocity and the amount of suspension load available in the system. With increasing flow velocity and suspension load, a tabular cross-bedding is likely to be modified into a lenticular one with a curved, tangential contact at the base. The experiments conducted by Jopling (1965) provided insight into the mechanism of this change (see Fig. 5.22). Flow separation at the brink of an advancing dune causes the sediment load to split into two fractions: a bed load and a suspended load. The former, accumulating at the brink of the dune, avalanches down the lee slope, adding foreset laminae to the lee face. The latter, caught up in the zone of turbulent mixing, is transported further downstream, to be deposited at a distance from the lee, below the zone of mixing. When the flow velocity is high, these deposits are pushed up the lee slope by the backflow vortex, thereby modifying an angular contact into a tangential one. Backflow ripples of this type have been reported by Boersma (1967) from ancient as well as modern sediments.

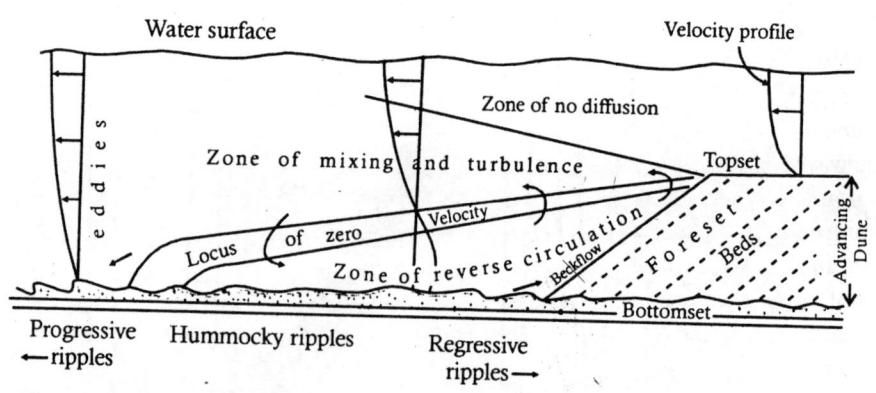

Fig. 5.22: Flow over a delta (dune) shaped bedform in an experimental channel indicating the origin of trough cross-bedding (after Jopling 1961).

A *trough-leticular* type of cross-bedding may also be generated from contemporaneous cutting and filling during the course of ripple migration. Impingement of the flow jet in front of an advancing dune may cause the bed to erode at the point of flow reattachment. When the dune crest is sinuous, the process of erosion may produce a series of spoon-shaped scours at different distances from the advancing dune front. These are soon filled by lenticular, lee-face laminae of the migrating dune, producing a trough-lenticular type of cross-bedding. Large-scale, trough-shaped scours may also be scooped out of stream-beds by eddies (called '*kolks*') swirling with semi-vertical axes. Such eddies are commonly generated at the confluence of stream channels meeting at sharp angels. The filling of these trough-shaped scours by advancing dune laminae may produce trough-lenticular cross-beds of large dimensions. Such features have been described front ancient and modern sediments among others, by Harms and Fahnestock (1965), Coleman (1969) and Sengupta (1974) (see Fig. 5.23B).

The internal stratifications of *eolian* dunes are not exactly the same as those of subaqueous dunes. Four different eolian stratifications have been identified: plane bed lamination, translation ripple lamination, grainfall lamination and sand-flow lamination (Hunter 1977). Thin, parallel laminae (parallel bed lamination) develop on the stoss of eolian dune when the wind velocity is too high to generate ripples. Accretion on the stoss side of dunes produces translation (climbing) ripple lamination. When the angle of climb exceeds the dip of the stoss face, wavy lamination and internal cross-lamination may develop. This is the supercritical stage. When the angle of climb is less than the slope of the stoss side due to a slow rate (subcritical) of accretion, tabular lamination, a few millimetres thick, may develop, and the laminnae may also show inverse grading. Grainfall laminations develop when flow separation by strong winds causes deposition of saltating grains below the dune crest. These may show poorly defined internal structures. When the angle of repose at the dune crest is exceeded, *en masse* avalanching may produce linguoid flow lobes (sandflow lamination) on the lee face of small dunes and cover the pre-existing sedimentary structures (see Chapter 7 and Fig. 7.18).

Structures Related to Wave Ripples

The simplest type of wave motion is caused by movement of wind on the free surface of water. At the water-air interface globules of water move in circles of the same diameter as the wave. At depth the diameters decrease exponentially. Wave ripples are generated on a non-cohesive sediment surface when the oscillatory motion of the water globules is strong enough to move the grains. The wave ripples thus produced are usually straight crested in plan. Only in exceptional cases are they sinuous or three-dimensional. The crests of wave ripples are sharp and the round troughs are often indented. These ripples are symmetrical in longitudinal section except when the forward and backward speeds of water waves are unequal.

It may be difficult to distinguish between wave and current-generated ripples on a purely morphological basis. Current ripples generally have longer lengths than heights. At a *ripple index* (R. I. = ripple length/ripple height) exceeding 15 only current ripples occur (Reineck and Wunderlich 1968). Symmetry in the outward pattern of a current ripple may be produced due to erosional processes also, but the true nature of a ripple in such a case will be indicated by the internal stratifications.

Fig. 5.23: Longitudinal sections of (A) planar-tabular and (B) trough-lenticular cross-stratifications. Trench wall sections in a modern point bar, the Usri River, India. Length of scale is 30 cm (from Sengupta 1974). With permission of Geological, Mining & Metallurgical Society of India.

The type of *oscillation ripple* generated on a bed surface is a function of the orbital speed (U_m) and time period (T) of oscillation. Density, mean grain-size and shape of the sedimentary particles, and density and viscosity of the fluid are some of the other factors controlling the nature of wave ripples. A sequence of oscillation ripples of varying shapes and dimensions may appear as the orbital speed and time period of oscillation wave change (see Fig. 5.24A). The first to generate on a non-cohesive sediment bed at low orbital speed are small, straight-crested ripples with gentle slopes. These have been termed '*rolling-grain ripples*'. With increasing orbital speed these ripples change over to steeper-sided '*vortex ripples*' in which lee-side vortices erode and carry sand each time the flow direction reverses. The vortex ripples are stable over a wide range of orbital speed and time period. At higher orbital velocity and at a small to moderate oscillation period the vortex ripples change over to flatter, post-vortex ripples. During each half of the flow cycle these post-vortex ripples show strongly unidirectional flow. With an increasing time period and at a larger duration of one-way flow, the post-vortex ripples are converted to *reversing-crest ripples*. At this stage the grain transport in each direction levels off the stoss sides of the ripples. When the oscillation periods are very long, both post-vortex and reversing crest ripples change over to flat beds (Harms, Southard and Walker 1982).

The internal stratification of an oscillation ripple is controlled by the rates of migration and aggradation (Fig. 5.24B). When the former is high compared to the latter, ripple troughs erode parallel to the already deposited laminae and the internal strata dip in one direction only. On the other hand, when the aggradation rate is much higher, minor shifts in crest positions cause the laminae to be dovetailed in a chevron-like pattern. At intermediate stages, when some migration takes place in an environment which allows rapid vertical build-up, the climb angle is steep and the entire ripple profile is preserved (Fig. 5.24B). Box cores taken from shallow marine sand (Newton 1968) show complex internal structures within features produced under oscillatory conditions. A large number of laminae dip randomly within such structures.

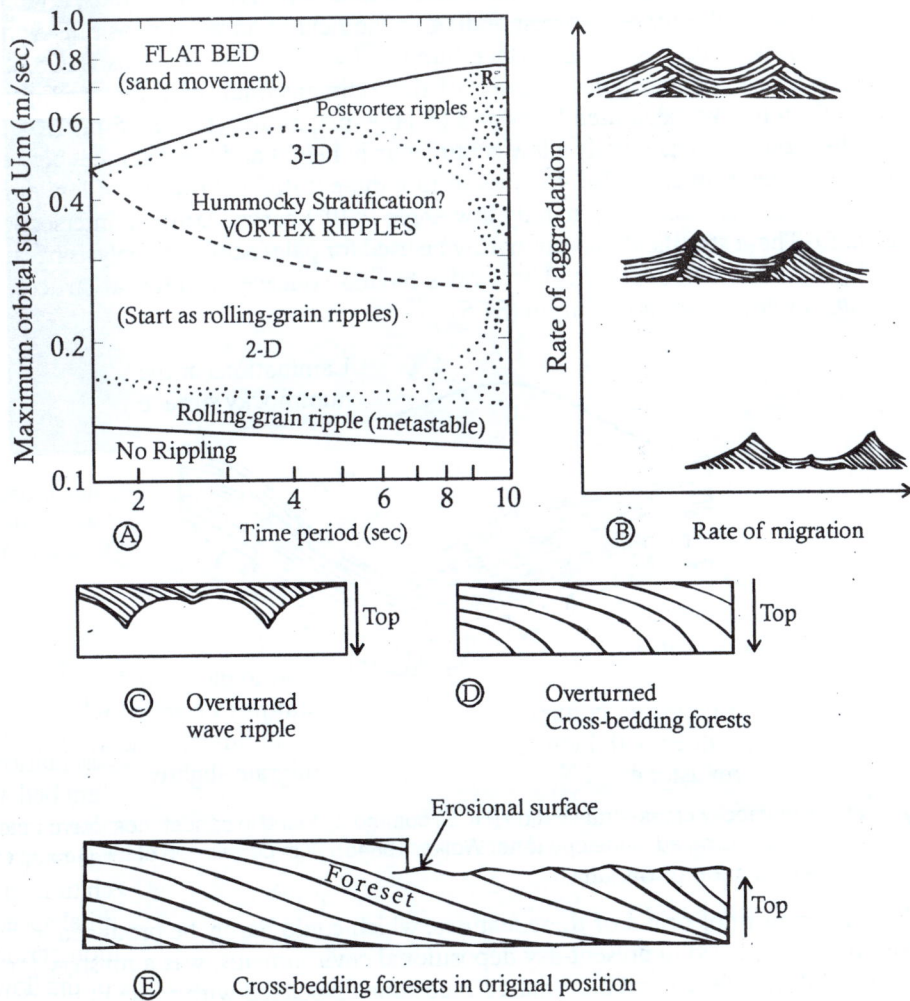

Fig. 5.24: (A) Physical conditions governing generation of wave ripples of various types (R indicates the fields of reversing-crest ripples). (B) Internal structures of wave ripples as functions of rates of migration and aggradation. (C-D) Internal structures of wave ripple and cross-bedding as indicators of depositional top and bottom of beds. (E) Cross-bedding foresets in original position. (A and B after Harms *et al.* 1982). With permission of SEPM (Society for Sedimentary Geology).

Wave ripples have been reported from various environments but they are most common in nearshore areas and in tidal flats. Symmetric and asymmetric wave ripples of small magnitude, often with tuning fork-like bifurcation in plan, are found on margins of water bodies not more than a few centimetres deep. Symmetrical wave ripples with unidirectional internal stratification have been reported from shallow-water, near shore areas. With sufficient sediment supply from suspension, these may develop into climbing ripples or ripple laminae in phase.

Features described in sedimentologial literature as *hummocky cross-stratification* have been assigned to oscillatory flows of moderate to long periods (5-10 seconds) and high orbital velocity (>0.5 m/sec) corresponding to the field of three-dimensional vortex repples (Harms *et al.* 1982). On surface hummocky cross-stratifications (HCS, also called hummocky megaripples) consist of three-dimensional, circular or elliptical mounds (hummocks) separated by swales of large magnitude (L = 1 - 5 m, H = 10s cm.). The mounds are generally a few centimeters in height and the distances between them may range from 2m to 5m. In cross-section these features show sets of low-angle, concave-up laminations and generally low-angle (<10°) curved laminae intersections (Fig. 5.25). These stratifications can hardly be used for palaeocurrent analysis because they dip in different directions. The swales having concave-up internal structures, constitute swaley cross-stratification (SCS).

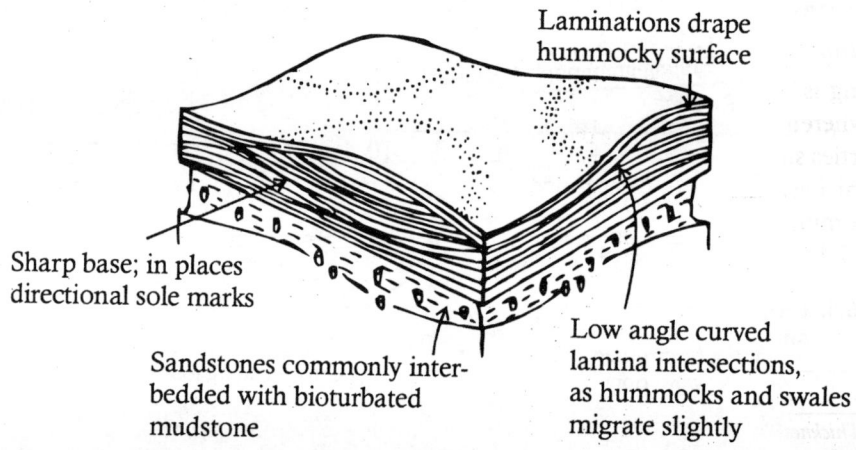

Fig. 5.25: Hummocky cross-stratification (HCS) commonly found in sandstones interbedded with bioturbated mudstone (after Walker 1984b). With permission of the Geological Association of Canada.

The mechanism of origin of these features, which could hardly be produced in the laboratory or observed in present-day depositional environments, was a mystery for a long time. Marine fauna and bioturbated mudstone associated with some hummocks suggest their marine origin. Inner-shelf storm surges at depths varying between fair-weather and storm wave bases (30-50m) were generally believed to be responsible for the origin of HCS.

More recent studies suggest that the hummocks and swales are the outcome of *combined flows* resulting from superimposition of unidirectional flows on oscillatory currents. During storm surges sand is carried into deeper sea in temporary suspension by high velocity unidirectional currents. At the same time the deposits are remodeled into hummocks and swales by oscillatory currents which reach below the ocean surface.

Reported originally from the Cretaceous shoreface deposits of the Rocky mountains, USA (Harms *et al.* 1975), hummocky cross stratifications have now been found in stratigraphic horizons of many different ages and locations (Dott and Bourgeois 1982).

Humocky megaripples (L = 2-5 m) with gently inclined internal laminations were reported from the inner Atlantic shelf of North America by Swift *et al.* (1983). Arrays of these scour pits and mounds were produced by intense storms. The structures were later modified by wind-driven flows. The processes involved are similar to those responsible for generation of megaripples in a tidal environment.

Bedforms produced by combined flows were reported by Banerjee (1996) from a Jurassic formation in Alberta, Canada. The size distribution of these bedforms was the result of a complex interplay of wave and current parameters together with climatic cycles of different periodicities. Swaley cross-stratifications were also reported from the Kansapathar Formation of the Chattisgarh Basin, India (Datta *et al.* 1999). These were produced in medium to coarse sand by oscillatory and combined flows. The relative frequency of trough cross strata and SCS within this formation provided clues to the relative proximity to shoreline.

Structures Related to Flat Beds

Bedding

Bedding is defined as a sequence of parallel layering within a body of sedimentary rock wherein individual layers differ from the overlying and underlying ones in physical properties such as colour, composition or texture. Beddings are classified according to their thickness. Sedimentary rocks often display a tendency to split along bedding planes. The commonly used terminologies for bedding and splitting properties are given in Table 5.4.

Table 5.4: Classification of bed thickness and splitting properties (modified after McKee and Weir 1953 ; and Potter, Maynard and Pryor 1980)

Thickness (mm)	Terminology for	
	Bedding	*Splitting property*
1200.0	Very thick-bedded	Massive
600.0	Thick-bedded	Blocky
50.0	Thin-bedded	Slabby
10.0	Very thin-bedded ⎫ Thickly laminated ⎬	Flaggy
5.0	Medium-laminated	Platy
1.0	Thinly laminated	Fissile
0.5	Very thinly laminated	Papery

Subaqueous horizontal bedding may be produced either during settling from suspension (without traction) or during transportation of sediments by running water. In the lower flow regime, plane beds may generate when the grain-size is large but the flow velocity is small. The upper flow regime plane beds are generated when the grain-size is small but the flow velocity is high. The upper flow regime plane beds may be identified by parallel arrangement of elongate mineral grains producing parting lineation (Fig. 5.26).

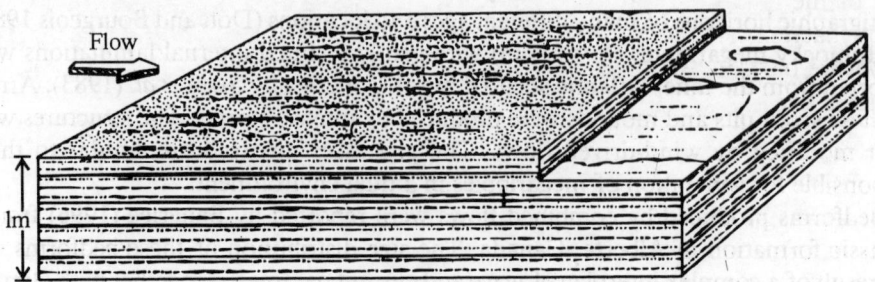

Fig. 5.26: Current lineation and parting lineation produced in the upper flow regime (after Harms *et al.* 1975). With permission of SEPM (Society for Sedimentary Geology).

Gradational, fining-upward sequences are common features in suspension deposits. Each layer of sediment within a *graded bed* contains an assortment of grains, both coarse and fine, but the mean size of the grains gradually decreases upward (Fig. 5.27). In the laboratory graded beds could be generated by allowing sediment laden flows to deposit their load from suspension under decelerating velocity condition (Kuenen and Migliorini 1950). Some of the graded beds in ancient marine and lacustrine sediments could have been produced by turbid flows which progressively deposited finer grains of sediments as flow velocity and carrying capacity of the current steadily decreased. Graded gravel beds are produced when slow moving unsorted debris flows (mud flows) deposit their sediment loads.

Grading may also be produced on the foreset of sand-dunes as a result of sorting during bed load transportation. As stoss faces of dunes erode, and the materials transported to the dune crest avalanche down the lee slopes, a vertical grading is produced due to accumulation of coarser grains towards the bottom. On rare occasions dispersive pressure caused by grain collision may force larger particles towards the surface of the slip face leading to what is termed *inverse grading* (Bagnold 1954; Jopling 1965). Some thinly bedded layers of marine origin are characterized by basal shell lags and a crude grading of carbonate clasts. These have been termed '*tempesites*' because of their inferred origin by storm surges. The term is also used for storm generated sand layers producing graded rhythmites (Reineck and Singh 1980).

Structures Preserved on Bed Surface

Shrinkage cracks, raindrop imprints, water-level (recession) marks, wrinkle marks, swash, and rill marks are some of the common features preserved on top of a bed surface. *Shrinkage cracks* may form either at the sediment-air interface by a desiccation process (desiccation cracks) or at the sediment-water interface by a synaeresis process. In plan desiccation cracks may look polygonal or non-polygonal (Fig. 5.28), where as *synaeresis cracks* are generally spindle shaped or sinuous. In cross-sectional view either may be U or V shaped. Desiccation cracks originate by algal or mud shrinkage (*mud cracks*) due to atmospheric drying. These are frequently used as evidence of subaerial exposure. Surface or substratal dewatering of submerged mud causes synaeresis cracks. Dewatering may be caused by the presence of concentrated brine, which sucks water out of the mud layer (Plummer and Gostin 1981). Distinction between desiccation and synaeresis cracks being difficult, a crack should be taken to be diagnostic of subaerial exposure only when it is associated with other shallow-water, periodically exposed features like raindrops, foam impressions or animal tracks.

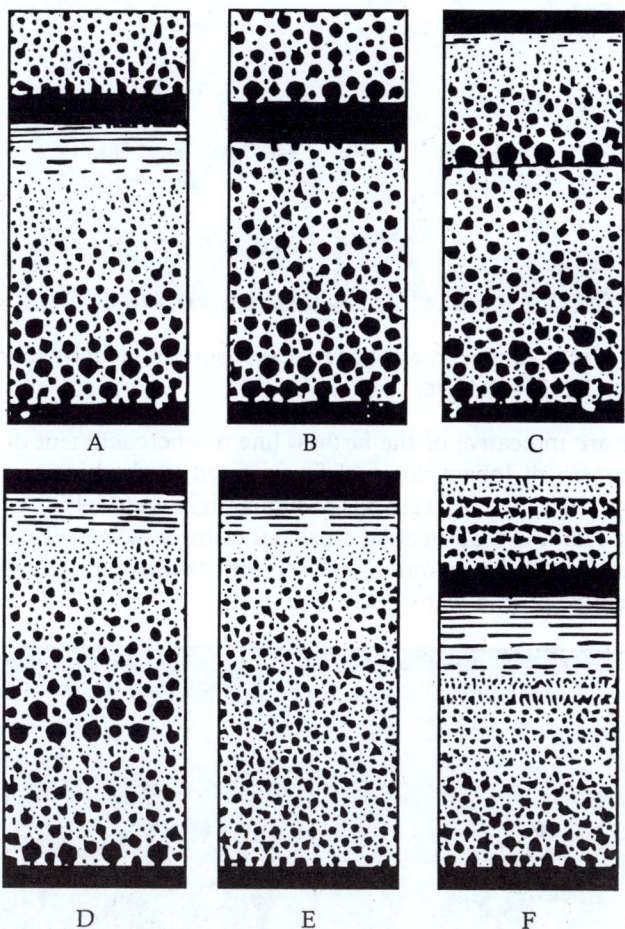

Fig. 5.27: Various types of graded bedding (after Kuenen 1953). Deep water lutites are shown in black.

Imprints of Intermittent Sbaerial Exposure: *Raindrop imprints* are preserved as circular, small, crater like depressions with raised rims. Other features produced by intermittent subaerial exposure include parallel markings on sloping banks (*water recession marks*), and *wrinkle marks* (*Runzelmarken* in German). The latter is produced by the action of wind on a thin film of water covering cohesive sediments. These are minute ripple-like features, a few millimetres in length and only a fraction of a millimetre in thickness.

Fig. 5.28: Mud cracks produced by subaerial exposure of modern flood-basin deposits in the Usri River, India.

Swash marks are indicative of the farthest line of encroachment of waves on land. These marks consist of lobate ridges of fine-grained sand which are convex towards the land. Broken shells, wood fragments or similar light materials often fringe the edges of these marks. Dendritic, bifurcating erosional features called *rill marks* are produced by erosion when water drains out of soft sediment surface at the falling water level. These are indicative of thin water cover (Fig. 5.29).

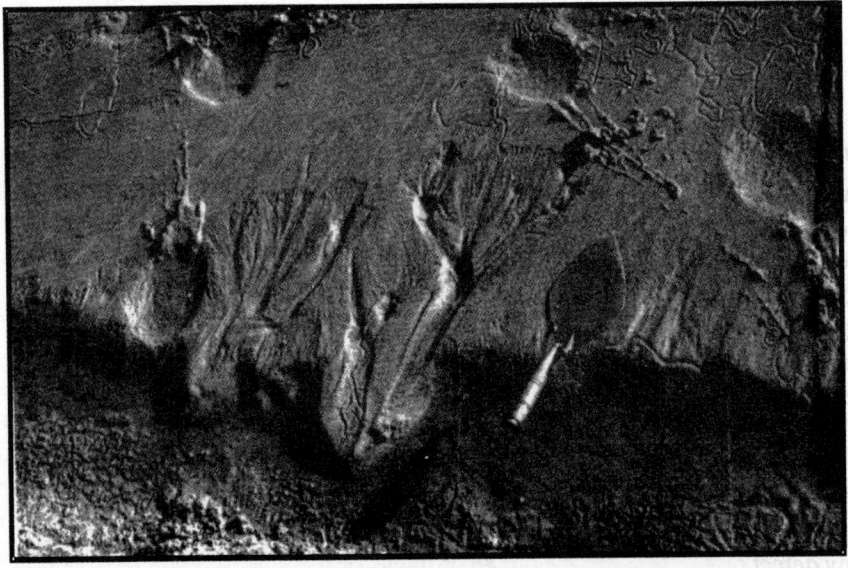

Fig. 5.29: Rill marks and worm trails in the Usri River point bar, India.

Current Crescent: Obstacles like shells, pebbles or wood fragments leave crescentic marks on soft sediment surface. These marks, called *current crescents*, are produced by continued excavation of sand on the upstream side of the obstacle due to generation of vortex. The eroded materials are deposited on the downstream side in the form of a flare (Sengupta, 1966, Johansson 1976, see Figs. 5.30, 5.31). The flare, preserved in geological record, may serve as an indicator of palaeocurrent because it always opens up towards the downcurrent direction. During the process of generation of a current crescent continuous excavation of bed material on the upstream side may cause the obstacle to tilt. Repetition of this process may lead to *imbrication*, a term coined to define overlapping shingling effect, generally of a series of flat pebbles (Fig. 5.32). Under suitable condition the scour width in current crescent preserved in geological record may serve as an indicator of flow velocity responsible for generation of the scour (Sengupta *et al.* 2005).

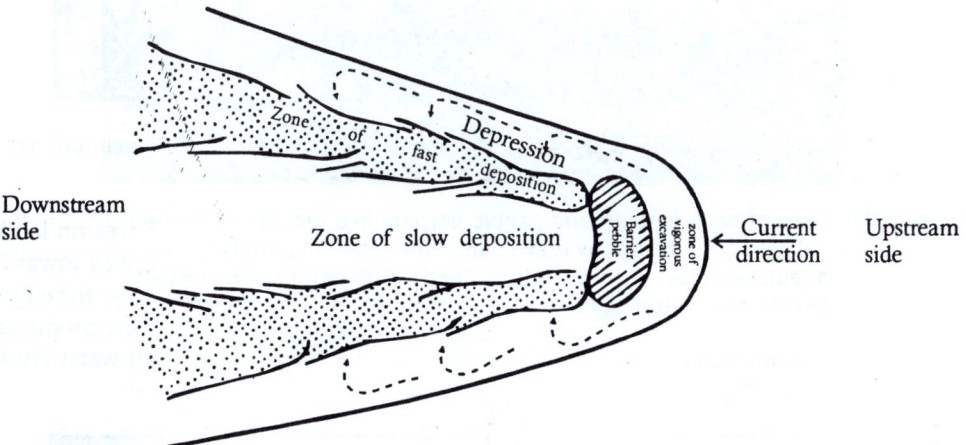

Fig. 5.30: Flow lines associated with a current crescent, plan view (after Sengupta 1966). With permission of SEPM (Society for Sedimentary Geology).

Structures Preserved on Undersurface of Beds

Scour marks of various shapes and sizes are produced when a current erodes an unconsolidated sediment surface. These are best preserved as moulds on the undersurface of overlying beds. Of these, the most common are the *flute marks* which occur as bulbous bodies, either singly or in rows, aligned parallel to the current direction (Fig. 5.33). Flutes flare and shoal out in the downcurrent direction to eventually merge to the sediment surface. Hence they may be used as indicator of palaeocurrent direction. Flute marks may range in size from a few millimetres to a few decimetres.

The most important prerequisite for generation of a flute is a vortex caused by a turbulent flow. Experiments have shown that flow generation accompanied by a strongly eddying current rotating in a nearly horizontal axis, transverse to the flow, is essential for generation of flutes (Allen 1973, 1975, see Fig. 5.34). Flute marks might generate at any defect on the bed surface but an obstacle, of the type necessary for production of a current crescent, is not essential in case of a flute.

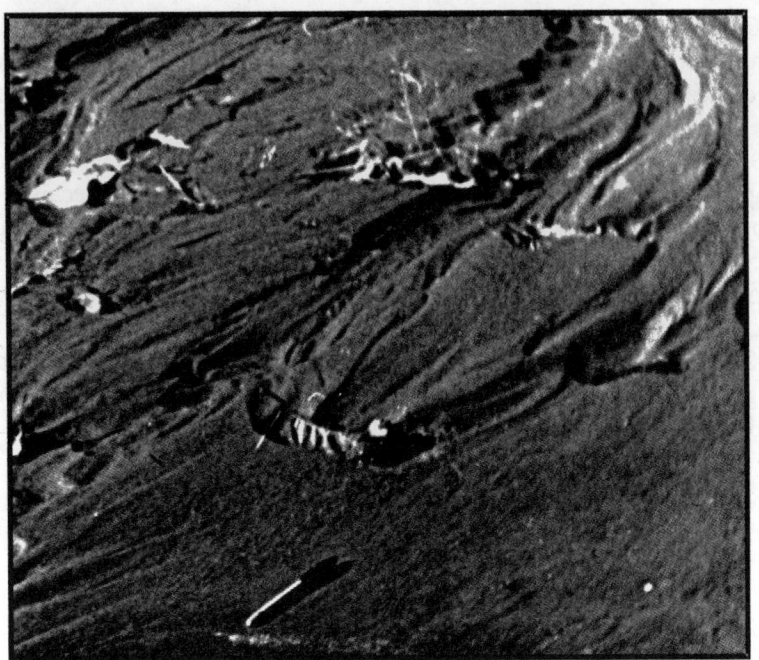

Fig. 5.31: Current crescents around pebble barriers in a recent stream bed. Most of the pebbles are covered by leaves and other plant materials. The arrow indicates direction of flow (from Sengupta 1966). With permission of SEPM (Society for Sedimentary Geology).

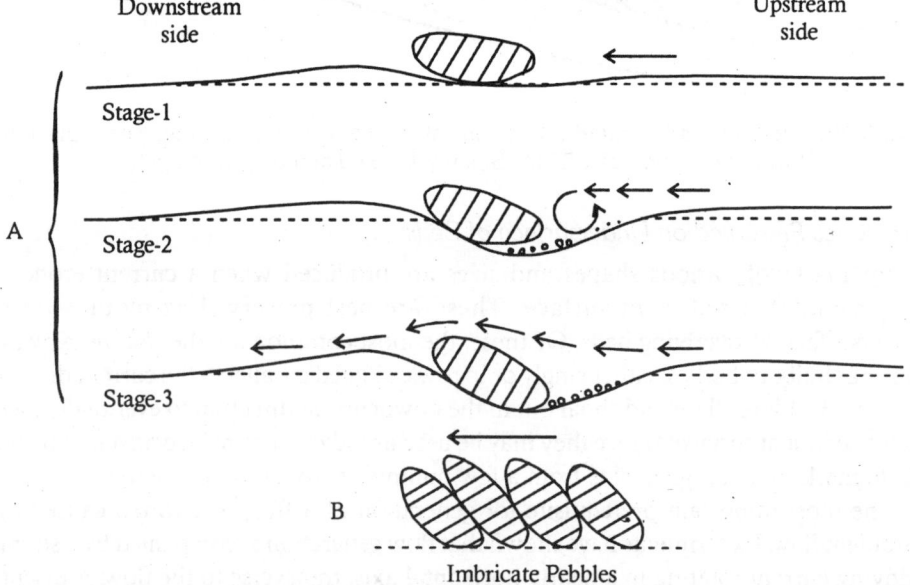

Fig. 5.32: Longitudinal cross-sections across Fig. 5.30 showing stages of development of current crescents and consequent increase of plunge of the barrier pebble leading to imbrication (based on Sengupta 1966).

Fig. 5.33: Isolated flute marks on undersurface of fine grained sandstone (Photo: courtesy of S.S. Das).

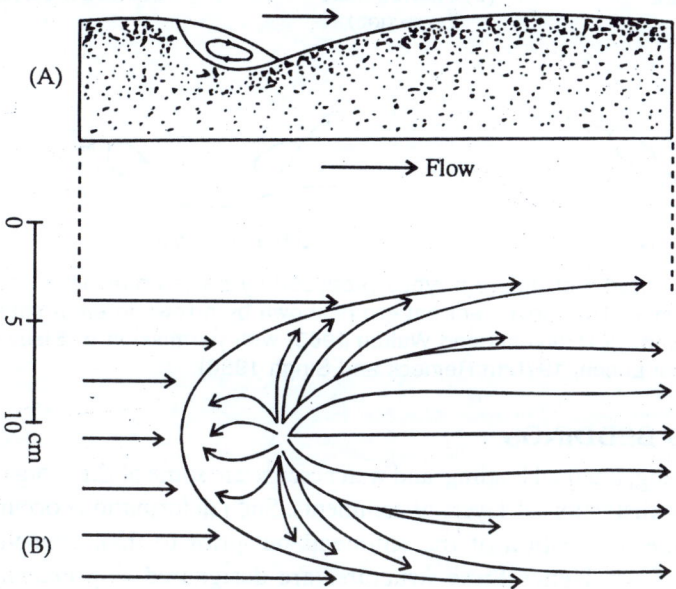

Fig. 5.34: Flow lines around flute mark. (A) Longitudinal profile showing separation of stream line. (B) Plan view showing skinfriction lines on surface of the flute and on the surrounding flat bed (based on Allen 1968a).

Tool Marks: Tool marks are produced on sediment surface by either stationary or moving bodies like a shell, a pebble, or a piece of wood. The tool may or may not be preserved in geological record. The more important of these marks are grooves, chevrons, prods and bounce marks. These are briefly described below after Dzulynski and Walton (1965), Ricci Lucchi (1970 cited by Reineck and Singh 1980).

Groove marks are long, linear depressions produced by objects rolled or dragged on soft sediment surface (Fig. 5.35A). Groove marks are often preserved on the undersurface of flysch sediments as sole marks.

Chevrons are continuous V-shaped marks (Vs pointing down-current) arranged in straight lines (Fig. 5.35B). Tool 'dancing'momentarily over cohesive sediment surface, before advancing forward, produce these marks.

Prod marks are asymmetrical elongate V-shaped depressions produced by objects hitting a sediment surface momentarily (Fig. 5.35C).

Objects approaching the sediment surface at a low angle produce *bounce marks*. Symmetrical depressions, tapering both in the up-and down-current directions, are produced in the process. The current soon lifts up the tool (Fig. 5.35D).

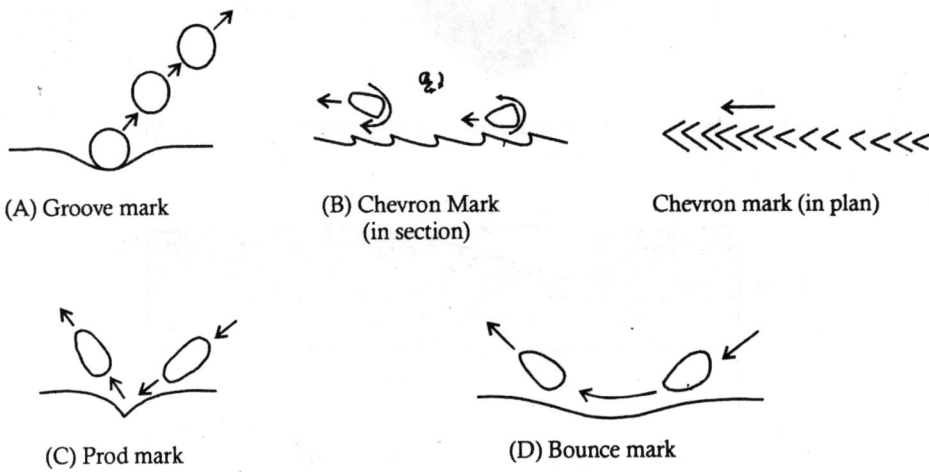

| (A) Groove mark | (B) Chevron Mark (in section) | Chevron mark (in plan) |

| (C) Prod mark | (D) Bounce mark |

Fig. 5.35: Tool marks of different types produced by bodies moving on soft sediment surface. The movement pattern is shown by arrows in each case. (A, B, C - based on Dzulynski and Walton 1965, with permission of Elsevier. D - after Ricci Lucchi 1970 in Reineck and Singh 1980).

DEFORMED BEDDINGS

Slumping, sliding, unequal loading, and water escape are some of the inorganic agencies responsible for distortion of layers of sediments. Such deformations occur at the time of, or soon after deposition of the sediments but prior to their consolidation into sedimentary rocks. Hence these structures are designated as *penecontemporaneous deformation structures* ('pene' means nearly). The process of ingestion of unconsolidated sediments by organic agencies leading to their deformation is called *bioturbation*. Penecontemporaneous structures have been grouped into the following types by Reineck and Singh (1982): load structures, ball-and pillow structures, convolute bedding, water-escape structures, dish structures and slump structures.

Sinking of sand layers into unconsolidated, hydroplastic mud may produce *sand lobes* varying in size from a few millimetres to a few decimetres. The underlying mud may be pushed upward in the form of tongues. At the sediment surface they may appear as bulges or knobby bodies. Casts of these load-generated features (*load casts*) may be preserved on the undersurface of beds as *sole marks*. When the sand layer is broken up into ellipsoidal, pillow-shaped masses and the underlying clay is pushed upwards into the sand in the form of tongues, the feature is designated as *ball-and-pillow*. Kuenen (1958) generated such features in the laboratory by subjecting a layer of sand overlying unconsolidated clay to vibrations. The sand layer in this experiment broke into discontinuous, kidney-shaped bodies, with the intermediate space occupied by clay (Fig. 5.36). Vertical movement of this type could have been responsible for most of the ball–and-pillow structures observed in nature. In some cases combined vertical and horizontal displacements due to slumping could have also been the cause.

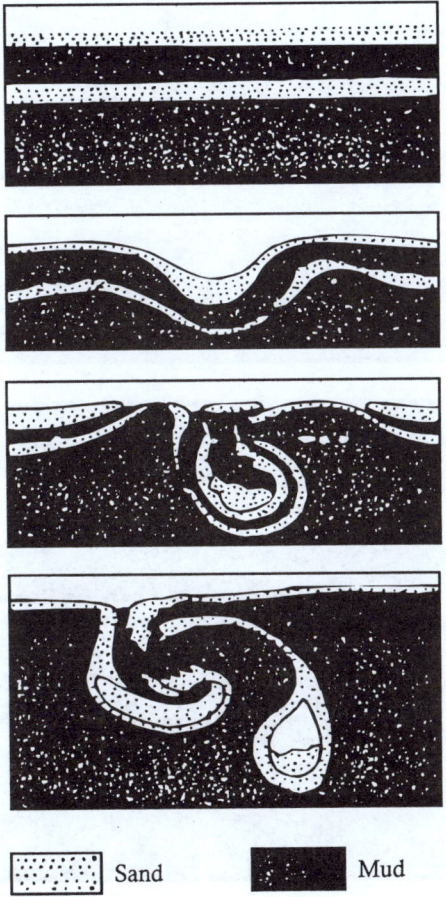

Sand   Mud

Fig. 5.36: Results of experiments on generation of pseudonodules by vibrating a layer of unconsolidated clay overlain by thin sheets of sand (based on Kuenen 1958).

Convolute Beddings

Convolute beddings are characterized by folds with sharp, irregular crests (often resembling flames) and broad, round troughs (Fig. 5.37). They are restricted to single sedimentation units, indicating that they are synsedimentary in origin. Sedimentary features of this type have been reported by many authors, both from ancient and modern sediments (see for example, Ghosh and Lahiri 1983).

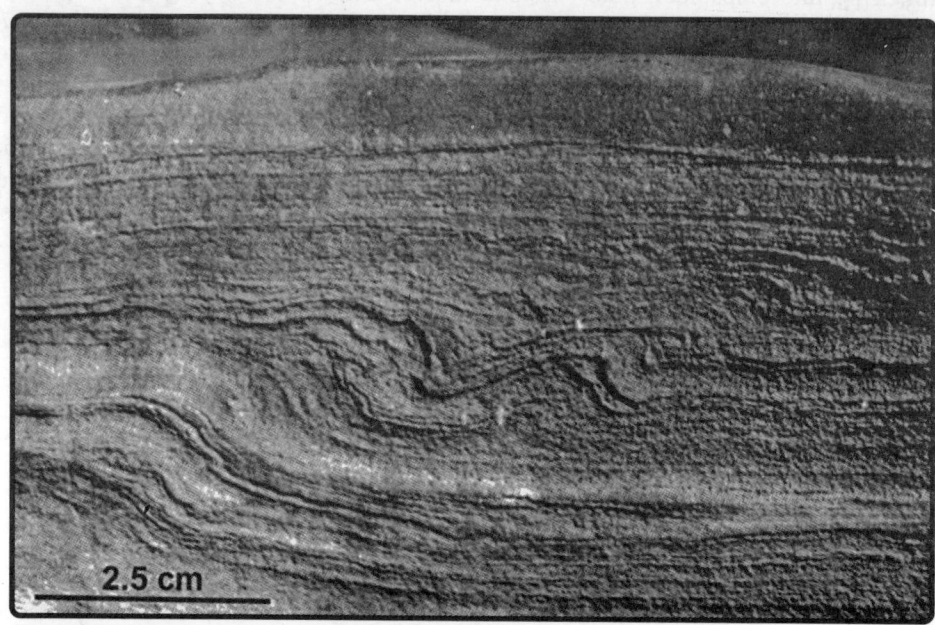

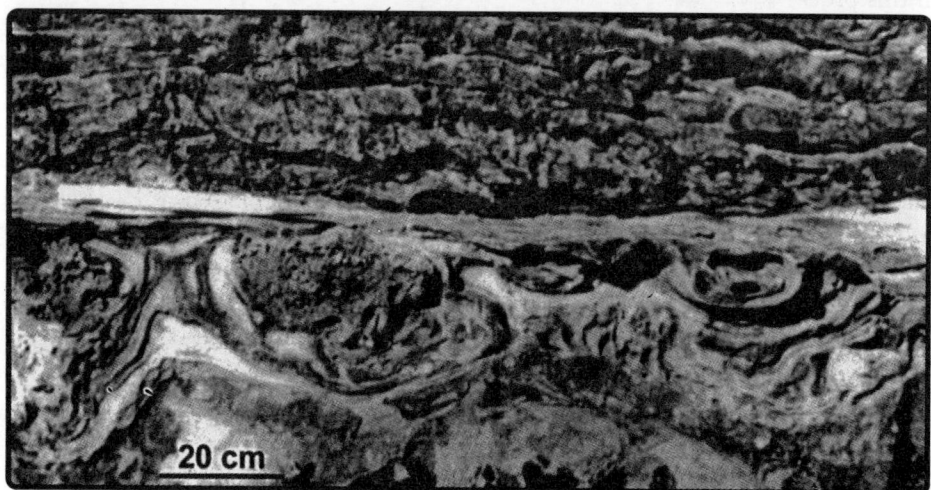

Fig. 5.37: Deformed sedimentary structures. (A) Contorted beds in (?) sandstone, Chattisgarh Basin, India. (Photo : courtesy of Sarbani Patranabis Deb). (B) Convolute laminations, Talchir Formation, West Bokaro Basin, India. (Photo: courtesy of H.N. Bhattacharyay).

Convolute laminations occurring at regular intervals were interpreted by Kuenen to have generated out of sand ripples whose crests were sucked up by water currents. The sucking was inferred to have been caused by differential pressure at the crests and troughs of ripples. Differential loading due to deposition of sand from above (McKee and

Goldberg, 1969), vertical expulsion of fluid leading to increase in pore pressure (Allen 1977), upward movement related to vertical expulsion of water or gas and vortex motion of turbulent cells (Chakrabarti 1977), are some of the other interpretations offered for these structures. The contortions running across planes of stratification (named '*interpenetrative convolutions*' by Allen 1977) look similar to the box-shaped forms produced at boundaries of liquids of varying densities when heated from below. This also led to the suggestion that some interpenetrative convolutions are generated by transfer of material by processes similar to convection current. Some researchers have also assigned convolutions to deposition of layers of sand over water-saturated mud or silt in the initial stage, followed by downward movement of sand and a compensatory upward flow of silt or clay (see Pettijohn 1975). Such movements, it has been suggested, could be caused by a sudden shock, as in an earthquake. In reality, however, evidence of downward movement is rarely found in natural convolutions. The process seems to involve only unidirectional upward movement of silt and clay caused by squeezing out of material from below.

When sand and mud are deposited together at a high rate, the sediment layers may be disturbed by escape of water unless the water can seep out of the system without disturbing the grains. The disturbances produced by water escape are of two kinds : (*i*) liquefaction caused by temporary suspension of sediment followed by settling consequent to collapse of the framework and (*ii*) fluidisation involving lifting of the grains by the drag exerted by the fluid. Four types of water escape structures may be produced by these processes (Lowe 1975): 1) *Hydroplastic or liquefaction layers* produced by internal re-arrangement of sediment layers. Not much physical movement is involved in this process. Some of the convolute beddings, deformed cross-bedding and pillar structures come under this group. 2) *Clastic sills and dykes* produced by intrusion of hydroplastic, liquefied or fluidized sediments into the adjacent bodies of sediments. 3) Various types of deformed bedding or cross-bedding produced by downslope movement of sediment layers. 4) *Dish structures* produced by particle rearrangement. These are generated when sediment water slurries are forced into a nearly horizontal passage until they reach points of vertical escape. Dish structures occur parallel to the sandy bedding planes as concave-upward laminae a few centimetres thick.

Slump Structures

Slump structures are caused by gravity flows of unconsolidated sediment layers. These may range from small-scale intraformational folds to large masses of slumps. Overturned or recumbent folds with associated thrust like structures are common within such slumps. Such features have been reported from oversteepened point bars, gullies of intertidal flats, glacio-lacustrine and fluvio-glacial sediments. In the latter case an overriding mass of ice may cause the sediments to slump. Subaqueous cross-beddings may also be folded with or without overturning. A combination of folding and faulting is common in eolian dune sediments. Avalanching of masses of cohesive sediments on the slip face of eolian sand dunes may also lead to slumping. In coastal dunes deformation is aided by the cohesive nature of sand. Graben-like features, a few centimetres in width, may be produced by slump-related gravity faults.

6

STRUCTURES OF CHEMICAL AND BIOLOGICAL ORIGIN

Not all primary structures are of mechanical origin. Sedimentary structures may be produced by chemical and biological actions also. Segregation and local concentration of minerals within host rocks, leading to generation of concretions, nodules and the like are caused by chemical action. Features like stylolites are produced by pressure solution. Stromatolites and trace fossils are indicative of biological action.

STRUCTURES OF CHEMICAL ORIGIN

Concretions and Nodules

Concretions are large spherical bodies whereas nodules are smaller. Both are produced by segregation and local concentration of a minor constituent within a host rock, for example, by accumulation of lime carbonate within shale and sandstone, silica within a carbonate rock, iron sulphide, phosphorous or barium within black shale. Manganese compounds may also lead to the formation of concretions. Segregation of the concretion forming mineral generally takes place around a nucleus, but its presence is not essential.

Concretionary bodies are classified according to their composition, morphological characters or by the time of their formation. Compositionally, they may be calcareous, siliceous, ferruginous or phosphatic. Morphologically, they are generally spherical, with concentric or radial internal structure but coarsely crystalline concretions with radial arrangement of acicular crystals are also known. These are named *spherulites* or *rosettes*. Spherical concretionary bodies are common in sandstones. Isotropic permeability of this host rock allows free flow of the minor constituent in all directions. Where permeability is unidirectional, as in shales, the concretions are generally flat and elongated. According to their time of formation the concretionary bodies are classified as syngenetic, penecontemporaneous or epigenetic. Manganese nodules on deep-sea floor are good examples of syngenetic concretions. Penecontemporaneous concretions form before compaction of the enclosing sediment. Epigenetic concretions generate after consolidation of the host rock. Following is a composition-wise description of some of the common concretionary structures.

Carbonate Concretions

The commonest calcareous concretions are *caliche* and *calcrete*. Caliche are thin, calcareous bodies occurring as crusts on soils which experience dry and wet climates alternately. Evaporation of groundwater drawn to the surface during the dry season allows precipitation of dissolved carbonate. Caliche may also occur on loess deposits of aeoline origin. Caliche may show crude concentric rings, but the primary sedimentary structures are often obliterated. In areas of low rainfall the carbonate dissolved in ground water may be initially precipitated in the soil as encrustation on roots of plants. These encrustations, called *rhizocretions*, develop into small calcareous nodules or *glaebules*, eventually coalescing to form calcretes.

Calcretes are composed of interlocking crystals of low-magnesium calcite and are much harder than caliche. They may have well developed macro and micro fabric. The cement of calcrete is constituted of sparry or acicular calcite grains. Calcretes occur in vadose zone as distinctive soil horizons (*pedogenic calcrete*), within or below the capillary rise zone. The latter are called *phreatic* or *groundwater calcretes*. The term calcrete is used for "whole family of near-surface, secondary carbonate accumulations, and prefixes such as pedogenic, vadose or phreatic should be applied when appropriate" (Tucker *et al.* 1990, p. 344). Well developed calcrete layers may display domal structures at the surface due to expansion of calcium carbonate. These are named *tepee*.

Variants of calcretes are known by different names. Concretionary carbonate bodies within marly host rocks are called *cornstone*. Spherical bodies, 1 cm to 10 m in diameter, are called *kugelsandstein* in Germany. These are produced by local precipitation of carbonates around nuclei in sandstones and shales. Disc shaped calcareous concretions occurring within silt layers in varve sequences are known as *imra stones* in Scandinavia.

Siliceous Concretions

Chart and *geode* are the two common varieties of siliceous concretions. Chart nodules are common in limestones. They are less frequent in siderite and shale. They vary in size from discs, a few centimetres in diameter, to elongate bodies several centimetres long. Their composition varies from monocrystalline quartz to chalcedony. They often show concentric internal features - a dark, black interior, surrounded by lighter coloured rings. Nodular charts and flints within limestones are possibly of epigenetic origin. Some of them might be the result of direct precipitation of masses of silica gel on the ocean floor.

Geodes are spherical or subspherical bodies of chalcedonic silica displaying central cavities. The cavities are filled, commonly by projecting crystals of quartz, and less commonly by crystals of aragonite, chalcopyrite or sphalerite. Presumably, the outer shell is produced by syngenetic deposition of colloidal masses of hydrated silica. This is followed by solution of the central portion of the geode and subsequent precipitation within the cavity. An alternative view assumes a calcareous concretion to be the precursor of a geode. Yet another view ascribes the origin of cavity to expansion due to generation of internal osmotic pressure within a layer of semi-permeable gelatinous silica membrane. Shrinkage cracks follow dehydration allowing passage of mineral bearing water into the cavity and deposition of a drusy lining over the primary chalcedony.

Ferruginous Nodules and Concretions

Local segregation of oxides and hydroxides of iron within host rocks may produce ferruginous nodules and concretions. These may also be produced by partial replacement of limestone pebbles by iron bearing minerals, or by oxidation of sideritic concretions or claystone pebbles. Partially filled hollows ('*box structures*') may develop when jointed siderite beds are oxidized to limonite. A variety of wired structures may result when joints and bedding planes of sandstones are filled up by segregated iron oxide minerals. Nodules and aggregates of pyrite and marcasite are also known. Brown to black nodules and layers of hematite (*ferricrete*) may develop on top of iron-rich host rocks in a manner similar to the generation of calcretes on carbonate-rich soils.

Nodules of Phosphates and Sulphates

Phosphatic nodules have been reported not only from phosphate-rich deposits but also from limestones and chalk. They occur on present day sea floor also. Phosphatic nodules are black or brown with dense, hard, shiny surface. Phosphatic spherulites, several centimetres in diameter, have been reported from the Cretaceous Montana Shale of Colorado. These are possibly the products of recrystallization of homogeneous phosphatic concretions (Pettijohn, 1975, p. 476).

Products of Pressure Solution: Stylolites

Serrated structures, called stylolites (*Druksuturen* in German meaning sutures generated under pressure) are often found in limestones and marbles (Fig. 6.1). These are produced by removal of rock material by interstratal solution circulating under pressure. Solubility of minerals like calcite, gypsum, and even quartz increase when subjected to pressure but dolomites resist pressure solution. Hence stylolites occur frequently in calcareous and gypsiferous rocks, but rarely in dolomites. They are also found, although infrequently, in quartzose sandstones.

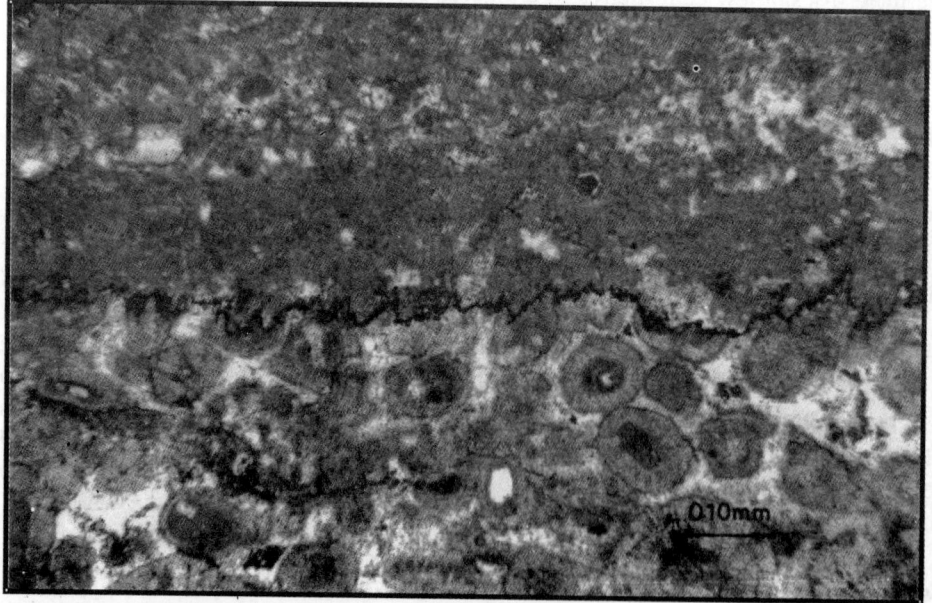

Fig. 6.1: Photomicrograph of a stylolite in oolitic limestone (Presidency College, Kolkata collection. Courtesy of Prabir Dasgupta).

Since stylolites are the effects of pressure solution a certain amount of porosity and permeability of the host rock is essential for their generation. This explains their absence in shale. The materials that go into solution at the stylolite surface eventually fill the available pore space of the host rock thereby reducing its porosity and permeability and hindering further growth of stylolites. Thus porosity and permeability of the host rock increase further from the plane of stylolitization. The insoluble residues are deposited at the stylolite surface as dark films. The tooth margins of stylolites are often marked by slickensides indicating movement of one part of the rock against another. A gross reduction of the bulk volume of the sediment, to the extent of even 25-30% is common in stylolites. The amount of reduction can be worked out from lateral displacement of the veins transecting the stylolite surface.

Stylolites transect intergranular cement in biomicrites and biospaites. Clearly therefore, stylolitization is a post-cementation phenomenon. The stages leading to generation of a stylolite, envisaged as early as 1926 by Stockdale, are as follows. (a) Production of an undulatory surface along the parting zone of a differentially soluble limestone. This is followed by (b) increase of pressure, apparently due to increase of overburden. The latter causes deepening of the interpenetrating parts, pressure being greatest at the crests and troughs of the undulations and least on the sloping sides. As a result the rocks on either side of the undulation dissolve, causing further increase of pressure. More modern views presume the process of stylolitization to begin in sediments which are loosely cemented. The process of stylolitization, by releasing calcium carbonate, leads the process of cementation to completion. Growth of a stylolite stops when permeability in the adjacent rock falls to a level that retards free flow of ions.

Stylolites vary in size from a few millimetres to several centimetres. The largest stylolite reported is from the Carboniferous Limestone of Yorkshire, where the amplitude is as big as one metre (Bathurst 1975). Microstylolites occurring in the carbonate rocks of the Late Precambrian Bhima Group of Karnataka, South India, were classified on their geometrical aspects by Malur and Nagendra (1988).

Occurrence and Implications of Structures of Chemical Origin

Innumerable reports on structures of chemical origin exist in geological literature. Perhaps the most spectacular of these are the very large size calcareous concretions in the Cretaceous Dakota Sandstone of Kansas. The diameter of some of these concretions go up to 9 m (Pettijohn, 1975). Calcretes have been reported, among other places, from the Old Red Sandstone Formation of Wales (Allen, 1965), and the Devonian Catskill Formation of Pennsylvania.

In the Arran Cornstone Formation of Scotland carbonate concretions form complete beds, bedding-concordant sheets, or bedding-discordant nodules and rods. These micritic concretions formed close enough to the surface, to be incorporated, after erosion and redeposition, as clasts into the overlying beds. Shrinkage played an important role in subsequent fracturing and reworking of the micrite (Tandon and Friend, 1989).

Calcretes might be useful in palaeo-environmental interpretations. In the Pennsylvanian shales, ancient marine, freshwater and brackish environments could be distinguished by $^{13}C/^{12}C$ ratio of siderite concretions (Weber, et al. 1964).

Calcic vertisols are a kind of calcic soil, displaying strongly vertical structures due to repeated desiccation of expandable clays. These require strongly seasonal condition for their deposition whereas tropical peats need continuity of precipitation. Systematic alteration of coal and other *paleosols* (see p. 189) with calcretes and calcic vertisols in cyclothems of the Sydney basin, Nova Scotia, Canada, therefore, was interpreted to imply strong variation in seasonality. In at least one of these cyclothems, calcic paleosols formed on an interfluve cut through a marine strata, suggesting that a more seasonal and probably drier climate prevailed during sea-level lowstand (Tandon and Gibling, 1994).

Occurrence of calcrete is particularly significant because of its palaeoclimatic implication, the essential condition for their generation being a rainy season alternating with arid or nearly arid periods of high evaporation. The calcretes found in the Late Triassic Maleri Formation of South India were inferred to be of similar origin (Robinson, 1964). Later studies indicated that these peloids of intraformational origin were derived by reworking of incipient pedogenic caliche profiles which developed in Maleri floodplains, abandoned channels and levées. Other lines of evidence like the predominance of smectites, poor floral content, paucity of evaporites within the Maleri point to low seasonal rainfall in a semi-arid environment (Sarkar, 1988).

Tandon and Narayan (1981) reported the occurrence of authigenic carbonates of three different varieties in the continental Siwalik Group of India. These are (1) calcrete conglomerate (2) 'case-hardened' conglomerate, and (3) cornstone. The calcrete conglomerate consists of *pisolites*, which are concentrically layered carbonate particles formed from carbonate-rich spring waters emerging on to the gravelly substrate of dry, abandoned channels. The '*case-hardened conglomerate*' with sparry calcite cement has limestone as its principal clast component. It has resulted from cementation of boulder conglomerate through continued dissolution and reprecipitation of calcite by meteoric water. *Cornstones* are nodular calcretes. These are associated with thinly bedded sandstones and red shales of levée and overbank origin respectively. This lithofacies is the result of concentration of carbonate through capillary action associated with pedogenic activity. The ooids in cornstones are micritic.

A combination of climatic conditions, carbonate availability, and sediment-starved condition induced by volcanism is inferred to have resulted to a stack of as many as fourteen calcrete profiles in the dinosaur bearing intertrappean regolith of Central India (Tandon, *et al.* 1998). Of these the calcretes having a distinctive sheet-type of morphology formed in soil vadose zone. Prominent shrinkage cracking is inferred to have favoured rapid formation of calcrete and associated rhizocretion.

Calcretes have been used successfully in sequence stratigraphic studies also. Nodular and associated groundwater calcretes in the coal bearing cyclothems of the Sydney basin of Australia mark the boundary between the underlying marine and overlying alluvial deposits. They are inferred to represent lowstand surfaces (see chapter 9 for explanation of the sequence stratigraphic terms). Microfacies of calcretes show evidence of replacement, displacement and shrinkage, suggesting that they were formed in relatively arid conditions. Thick alluvial sediments with red calcic *vertisols* overlying the calcretes are inferred to have formed within the transgressive systems tract under conditions of abundant sediment supply. The presence of calcareous palaeosols on lowstand surfaces suggests that the lowstands were times of relative climatic aridity (Tandon and Gibling, 1997).

BIOGENIC SEDIMENTARY STRUCTURES

Biogenic sedimentary structures are of the following three types (Frey, 1973).

1. Biostratification structures. This group includes stromatolites, byssal mats, biogenic graded bedding and related features.
2. Bioturbation structures. This group includes tracks, trails, grazing patterns, burrows, and other disruptive structures.
3. Biodeposition structures. This group includes fecal pellets, fecal castings, pseudotraces, coprolites and the like.

Stromatolites, discussed below, are the most important biostratification structures.

Stromatolites

Laminated structures, produced by trapping, binding and/or precipitation of fine-grained, calcareous detritus, by unicellular, filamentous organisms are called stromatolites. The more common stromatolites are the ones related to algal growth, but stromatolites are not fossil algae. Algal stromatolites occur as laminated, moundlike, columner or non-columner massses (Fig. 6.2 A & B).

Fig. 6.2 (A): Plan view of hemispheroidal stromatolites ('domed algal mats' or Cryptozoons) in Late Cambrian Hoyt Limestone exposed on a glaciated surface at Lester Park, USA. The stromatolites are the products of intertidal environment.

Some features of inorganic origin like *calcretes* and *geyserites* resemble stromatolites in outward morphology but their internal structures are different. Calcretes, for example, display radial, granular or blocky interior instead of laminations. Geyserites are opaline silica deposited non-biogenically near hot springs and geysers. In external morphology some geyserites may resemble algal stromatolites, but unlike stromatolites they show banded internal lamination and at times, even micro-cross–lamination. Moreover, unlike stromatolites they occur only around locales of water discharge.

Fig. 6.2 (B): Columnar stromatolites in vertical section in the Neoproterozoic Chattisgarh basin, India. (Photo: courtesy of Sarbani Patranabis Deb).

Stromatolites are found in limestones of all ages ranging from the Precambrian to Recent. They were most abundant during the Riphean when diversity of columnar stromatolites was also the highest (see Fig. 6.6 A & B).

Originally thought to be restricted to marine environments, stromatolites are now believed to form in any freestanding body of water allowing growth of microorganisms. A high rate of sedimentation is essential for preservation of stromatolitic structures but too high a rate of sedimentation retards algal growth. The rate of growth of the stromatolite binding organisms must also be higher than the rate at which these organisms are consumed by other biota like gastropods. In fact, decline of stromatolites during the Phanerozoic is generally attributed to the advent of grazing gastropods feeding on the algae. High salinity of restricted pools keeps the gastropods away allowing better growth of blue-green algae. The rate of algal growth is also high in the upper intertidal and supratidal areas where grazing invertebrates do not survive prolonged exposure. Tidal channels are also congenial to growth of algal mats.

Stromatolitic structures, both ancient and modern, have been studied extensively because of their importance in sedimentological and stratigraphic studies. A summary of the voluminous publications on stromalites appears in Walter (1976). Following is a brief review of the major ideas.

Morphology and Classification of Stromatolites

Stromatolites are organo-sedimentary structures, not fossil organisms. Although their existence depends on the sediment binding property of algae, stromatolites themselves lack the internal anatomical features of algae. The Linnean scheme of classification of algae cannot, therefore, be used, without reservation, for grouping stromatolites. The stromatolitic forms built by filamentous algae today are divided into two groups: *cryptozoon* and *collenia*. The former are laterally linked, vertically stakced, columnar hemispheroids attached to a surface. The latter are either discrete or laterally linked spheroidal or hemispheroidal bodies composed of concavo-convex laminae. Both the forms are similar to those found in geological record. The unattached, subspherical forms are called *oncolithi*. Hemispherical stromatolites lacking internal laminations are called *thrombolites* (Aitken, 1967).

Three of the major schemes of classification of stromatolites based on external morphology are discussed below.

Logan, Rezak and Ginsburg (1964) proposed a scheme of classification for stromatolites based purely on geometric criteria. Three basic combinations of hemispheroidal and spheroidal stromatolitic structures were proposed in this scheme.

1. Laterally linked hemispheroids (LLH) or linked domes. These are of two types. (a) Close lateral linkage (LLH-C), the space between structures being less than the diameter of the structures. (b) Spaced lateral linkage (LLH-S), the space between structures in this case is greater than the diameter of the structures. The scale of lateral linkage may vary enormously, from microundulations, a few millimetres in height, to large-scale laterally linked domes a few metres across. The latter are commonly associated with preexisting irregularities (Fig. 6.3 A).

2. Discrete, vertically stacked hemispheroids (SH) of two types: (a) those having a constant basal radius (SH-C), and (b) those having variable basal radius (SH-V). Their scale may vary from a few centimetres to metres across the base. The SH structures develop out of algal cappings over preexisting irregularities like mud cracks or old algal structures (Fig. 6.3 B).

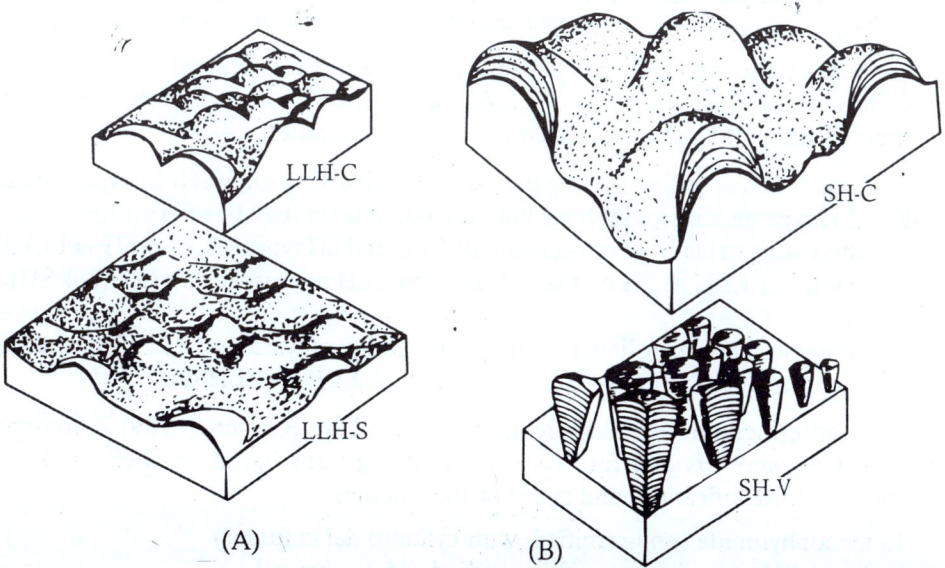

Fig. 6. 3: Stromatolities occurring as (A) laterally linked hemispheroids, and (B) discrete, vertically stacked hemispheroids. (after Logan *et al.* 1964). With permission of The University of Chicago Press.

3. Discrete spheroids (SS), ranging in size from fraction of a millimetre to about five centimetres or more. Three modes of occurrence are observed within these forms. (a) Mode I—inverted stacked hemispheroids. (b) Mode R—randomly stacked hemispheroids. (c) Mode C–concentrically arranged spheroids. Instead of coating a surface the algal mat in this case coats a shell or a lithic fragment lying on the bottom. The size may vary from fraction of a millimetre to about five centimetres or more in diameter. (Fig. 6.4)

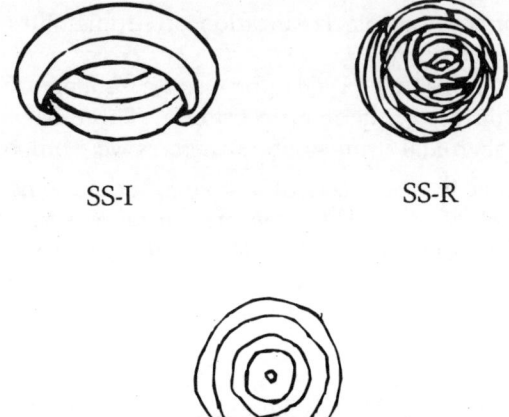

SS-I   SS-R

SS-C

Fig. 6.4: Stromatolites occurring as discrete spheroids. SS–I: inverted stacked hemispheroids, SS-R: randomly stacked hemispheroids, SS-C: concentrically arranged spheroids. (After Logan *et al.* 1964). With permission of The University of Chicago Press.

Stromatolite colonies found in nature today contain complex combinations of the above mentioned types. Logan *et al.* (1964) used the following arbitrary scheme for describing these combinations of macro and micro structures.

(a) Combinations of SH macrotype and LLH microstructure (SH/LLH) laminae.
(b) LLH type passing to SH type, both on macro scale (LLH → SH)
(c) Alternating structural arrangements of LLH and SH types (LLH → SH → LLH).
(d) Passage of LLH to SH on macroscale with LLH microstructure (LLH → SH)/ LLH).
(e) Compound stromatolites having combinations of various LLH, SH, and SS arrangements (LLH-C → SH-V → LLH-S → LLH-C/LLH-C).

In another scheme of classification proposed by Raaben (1969) columnar stromatolites were divided into four main supergroups based on their mode of ramification (diversification) and shape of the columns.

1. Conophytonida (non-ramified, with cylindroidal columns)
2. Kussiellida (passively ramifying, cylindroidal columns)
3. Tungussida (actively ramifying, columns widening upward, cup-like, with divergent axes)

4. Gymnosolenida (actively ramifying, pseudocylindroidal columns, axes running roughly parallel).

Conophytonida columns are circular or oval in transverse section and consist of parallel stems of varied size. They form lens-like or stratiform bioherms, 50m or more in thickness. The supergroup Kussiellida is characterized by subcylindroidal, frequently elongated columns, which are irregular in transverse section. In the upper portion these are usually subdivided into several minor columns running parallel to one another. Microstrata are dome shaped, varying in shape and height. Marginal parts protrude into the surrounding rocks. Cup-like columns, mostly irregular, characterize the supergroup Tungussida. Diameters of these columns rapidly increase upward. Structures actively ramify. Unlike Kussiellida the ramified bush expands upward. Axes of the columns are divergent. Gymnosolenida are actively ramifying stromatolites with columns noughly parallel and pseudocylindrical. When well developed they have a narrow base and wide top.

Hoffman (1976) proposed a morphological classification of stromatolites based on his study of the modern stromatolites in the Shark Bay of Western Australia. These stromatolites compare well with the Proterozoic stromatolites. He recognized three major mat types: colloform, smooth, and pustular. Four other types (gelatinous, tufted, blister and film) produce only stratiform sheets. An internal digitate fabric of multiconvex laminations produce a coarse fenestrate within the colloform mats. A fine fenestrate of convex laminations occurs within smooth stromatolitic mats. The pustular mats have irregular fenestrate fabric without laminations (Fig. 6.5).

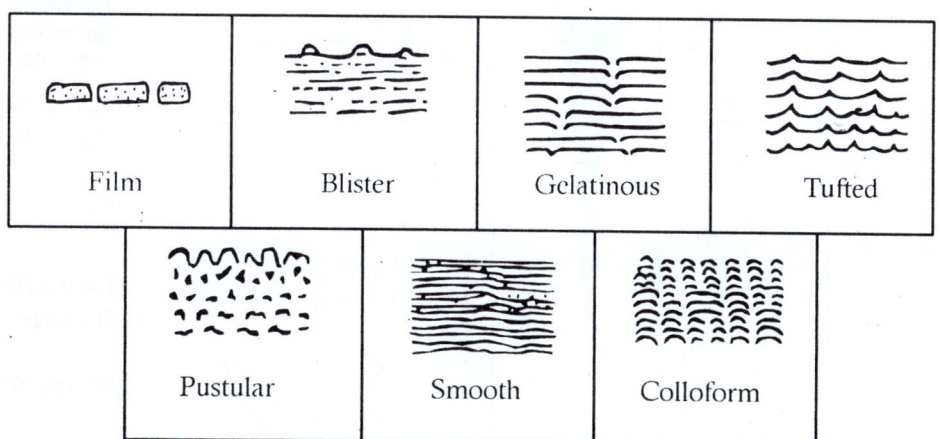

Fig. 6.5: Internal fabric of seven mat types of blue-green algae (based on Hoffman 1976) (With permission from Elsevier).

Hoffman noted that the algal mat types do not control the gross morphology of stromatolites. For example, smooth mats may produce structures ranging from stratiform sheets to high-relief columns. On the other hand, biologically distinct mat types may produce similar morphologies. The physical processes of environment ultimately control the distribution of mat types and also the stromatolite morphology. This aspect is discussed under environment.

Stromatolites in Sedimentological and Stratigraphic Studies

Stratigraphic Classification and Correlation

Stromatolites have been studied extensively because of their use in stratigraphy and sedimentology. They were particularly useful in the classification and correlation of the otherwise unfossiliferous Precambrian rocks (see Raaben, 1969 for a review of these studies). On the basis of the various assemblage zones of stromatolites the Proterozoic was divided into pre-Riphean (2700–1650 M. Y.), and Riphean. The Riphean again was classified into Lower (1650–1350 M.Y.), Middle (1350–950 M.Y.), Upper (950–670 M.Y.), and Vendian (670–570 M.Y.). This subdivision of the Riphean has been corroborated by radiometric age data (see Fig. 6.6). In the former USSR stromatolite zonation was used for stratigraphic correlation of extensive areas, and the Vendian was precisely established. Worldwide applicability of stratigraphic correlation on the basis of stromatolites has also been confirmed. In the western part of the Taoudenni Basin, Mauritania, West Africa, four stratigraphic units of Late Proterozoic age were distinguished and followed eastward as far as Algeria. Of these, the Atar or el Mreiti Group includes most of the stromatolite bearing carbonate beds. Two other groups also contain stromatolites, and one group is devoid of them (Bertrand-Sarfati and Trompette 1976).

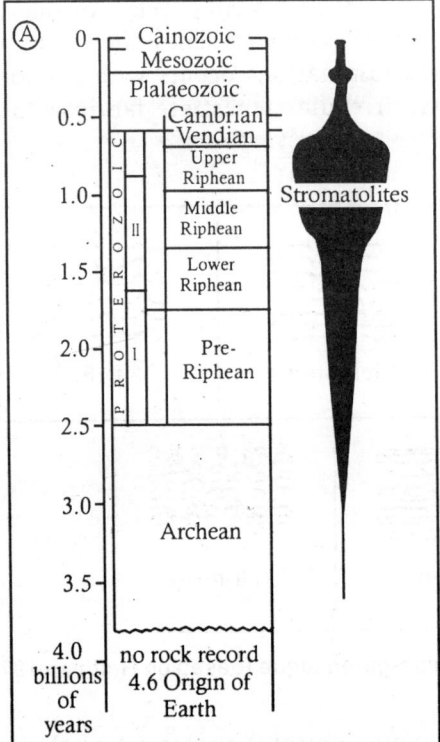
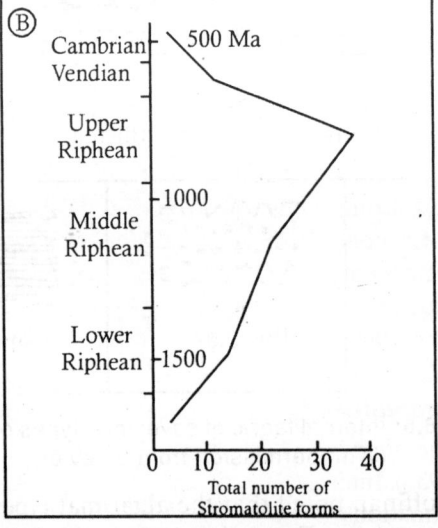

Fig. 6.6: (A) Abundance of stromatolites through time (Awramik, 1984). Reprinted with permission of John Wiley & Sons, Inc. (B) Number of stromatolite forms during Riphean and Cambrian (from Awramik 1971). Reprinted with permission from *Science*, v. 17, copyright (1971) AAAS.

Environmental Interpretation

Association of flat pebble conglomerate, oolites and cracked limestones with stromatolites indicates shallow water origin. Crinking of stromatolitic structures suggest drying, possibly in very shallow intertidal environment. Upward convexity of stromatolitic structures may serve as a clue to the depositional top of a sequence.

Morphology of stromatolitic forms can be used as indicators of depositional environment. For example, LLH-type stromatolites (Collenia) are found only in protected intertidal mudflats where wave action is not strong. The SH-Type stromatolites (Cryptozoons) occur in exposed intertidal mudflats with strong wave action because scouring action of waves prevents growth of algal mats between stromatolites. SS-type structures (Oncolites) are characteristic of low intertidal areas which are exposed to waves and agitated shallow water. The arrangement of laminae within these structures provides clues to the frequency of movement. Concentrically arranged spheroids are the result of frequent motion whereas randomly stacked hemispheroids are the result of less frequent motion (Logan, *et al.*, 1964).

Hoffman's studies in the Shark Bay enabled him to relate stromatolite morphology to the physical processes operating in different environments. His major findings, summarized below, provide clues to interpretation of ancient environments.

1. Stratiform cryptalgal sheets occur where wave and tidal scour are weak.
2. Discrete columnar structures occur in areas of strong wave and tidal scour. Their relief is proportional to the intensity of wave action.
3. Elongation of simple stromatolites is parallel to the direction of wave propagation and tidal scour, that is, generally perpendicular to the shoreline.
4. Elongate stromatolites occur in rows parallel to the shoreline.
5. Stromatolites tend to tilt seaward into the oncoming waves.

The pustular mats produce high relief circular columns at headlands, medium to low relief ridge-and-rill structures in bights, and flat pustular mats in protected embayments (Fig. 6.7).

Dimensions of these stromatolites also vary with environment. At rocky headlands, irregular columns of coloform mats, upto 1.0 m in height occur. These have discrete bladed branches with flattened tops. In the bights (a bend in a coast forming an open bay) stromatolites are more prolate and low (~0.5m) in relief. On the sublittoral floor of tidal pond in embayment stromatolites with digitate columnar internal structures occur beneath ovoid patches (<0.5m in diameter). The smooth mat stromatolitic structures are large (~0.8m) at rocky headlands. Near bights this relief is reduced (Hoffman, 1976).

Stromatolite Occurrences in India

One of the early reports in India of stromatolites of Proterozoic age was by Auden (1933). Interregional and intercontinental correlation in the Lower Himalayan carbonate formations and Vindhyans with the help of stromatolites was attempted by Valdiya (1969), who used the term *collenia* for these stromatolites. Riphean stromatolites were reported from the Jammu Limestones by Raha (1972). Geochronology of the Jammu Limestones, and also the stratigraphy of the Lesser Himalayas based on stromatolitic occurrences were discussed by Raha and Sastry (1973, 1976), and by Raha, *et al.* (1978).

Chaudhuri (1970) reported Precambrian stromatolites from the Pranhita-Godavari valley. Rudra and Maulik (1987) reported stromatolites of three morphological types - domal stromatolites, oncolites, and cryptalgal laminations, from freshwater Lower Jurassic Kota limestone of the Pranhita-Godavari Valley. Existence of these stromatolites within the lake margin sediments, and also in deeper water was taken to be an indication of repeated submergence and emergence of the lime-mud in the Kota Lake.

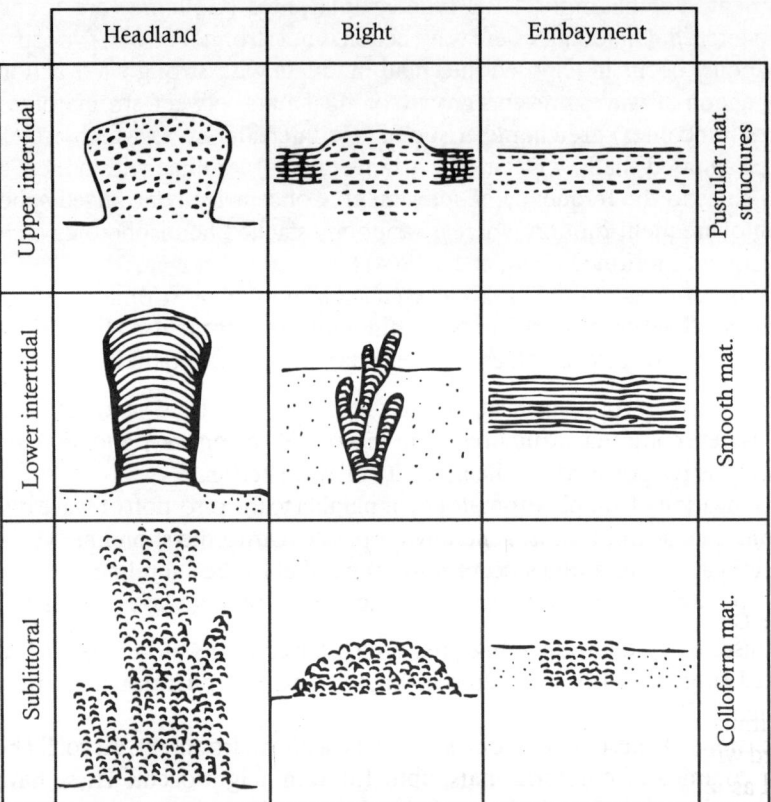

Fig. 6.7: Modification of stromatolite morphology with environment (after Hoffman 1976). Reprinted with permission from Elsevier.

Trace Fossils

Some organisms residing on sediment layers leave clues to their behaviour in the form of traces: trails, tracks, burrows and borings. These traces, preserved in rock record, are called trace fossils or *ichnofossils* (*lebensspuren* in German meaning life traces). They are quite distinct from body fossils. Trace fossils are common in sandstones and shales, but these have been reported from carbonate rocks also.

Classification of Trace Fossils

One of the primary methods of classification of trace fossils is based on the position of their occurrence with respect to the interface of sandstone and shale layers (Seilacher 1964). Traces preserved wholly within the bed are called 'full reliefs' and those occurring at the interface of two layers of rocks are called 'semireliefs.' Semireliefs again, may occur either at the upper or at the lower surface of a rock layer. Those occurring at the upper surface of a sandy bed are termed 'epirelief.' When a trace occurs on the lower surface (sole) of a rock layer it is called 'hyporelief.' Epirelief and hyporelief may be positive or negative depending on whether they bulge up or down respectively (Fig. 6.8).

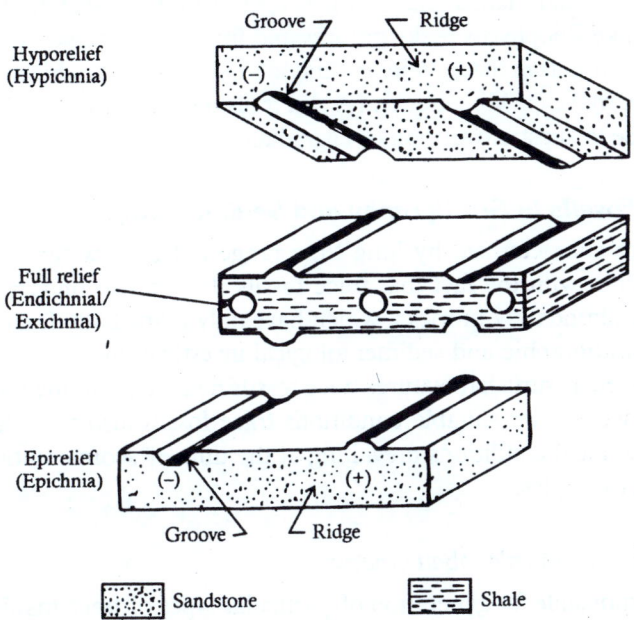

Fig. 6.8: Classification of trace fossils based on the mode of occurrence. (Based on Seilacher 1964, Lindholm, 1987. Terms within parentheses are after Martinsson 1970).

In a similar scheme of stratonomic classification Martinsson (1970) named the traces preserved wholly within a bed as *endichnia / exichnia* (burrows). Those on the top surface of a bed as *epichnia*, and those on the bottom surface as *hypichnia*. The positive reliefs were named 'ridges' and negative reliefs as 'grooves.'

Various morphological and behavioural classifications of trace fossils have been attempted from time to time. The following classification, proposed by Seilacher in 1953, is based on the inferred behaviour (life activity or habitat) of the animals concerned (see Frey and Seilacher 1980 for details).

Name	Habit	Mode of occurrence
Cubichnia	Resting	Traces of organisms resting on soft sediments.
Domichnia	Dwelling	Burrows and borings as dwelling places.
Fodinichnia	Feeding	Marks of organisms searching for food.
Pascichnia	Grazing	Marks of organisms exploiting nutrients.
Repichnia	Crawling	Tracks and trails of animals moving.
Fugichnia*	Crawling	Escape traces of animals buried suddenly.

(*Added to the list by Simpson 1975).

Animals that temporarily dig into the substrate to create isolated shallow depressions make resting traces. Rusophycus is a modern example of such an animal. Dwelling traces are made by suspension feeders and scavengers which provide relatively permanent homes. These traces are usually cylindrical. Skolithos is an example of such an animal. Grooves and furrows made by deposit feeders like Rhizocorallium may exhibit complex behavioural pattern and make radial and U-shaped patterns in search of food. Mobile deposit feeders like Zoophycos make grooves and furrows that make most economical exploitation of nutrients in the sediments. Animals like Cruziana, which travel from one place to another, make crawling trackways and trails. Escape structures (Fugichnia) are made in response to substrate erosion or aggradation.

Use of Trace Fossils in Stratigraphy and Sedimentology

Trace fossils are characterized by long time range but narrow facies range. These characters together with the fact that trace fossils show no secondary displacement and often occur abundantly in rocks, which have only a few body fossils, make them invaluable for stratigraphic and sedimentological investigation.

When preserved as surfacial features trace fossils help determining the depositional top of a sequence. Under suitable conditions trace fossils also provide clues to the physical and chemical conditions prevailing at the time of sedimentation. Following are some of the examples.

Depositional Facies and Palaeobathymetry

The animals responsible for generation of particular types of trace fossils are sensitive to environmental and bathymetric conditions. Following this clue Seilacher proposed that assemblages of trace fossils might be used for interpretation of depositional facies ('ichnofacies') and palaeobathymetry (Seilacher 1964, 1967). Following is a summary of Seilacher's major findings (Fig. 6.9).

1. In marine beach and intertidal zones, which are areas of rough water erosion and sedimentation, organisms like Skolithos and Arenicolites build deeply penetrating vertical tubes for self-protection. Both dwelling and feeding burrows dominate this zone.

2. Shallow shelves have Cruziana facies.

3. Relatively deep, tranquil water is characterized by Zoophycos facies. Zoophycos in general forms 3-D spiral structures. They branch out to build horizontal burrows. Such structures help them to collect food particles, which settle down from shallower levels and become part of the substrate.

4. In still deeper parts and within pelagic mud layers between turbidities, organisms like Helminthoida and Nereites splay out to collect the food trickling down from the water column above.

As an objection to this apparently idealized interpretation of palaeobathymetry, it was pointed out that ichno-assemblages are controlled, not only by water depth, but also by other physical parameters like the quality of light, salinity and oxygen content

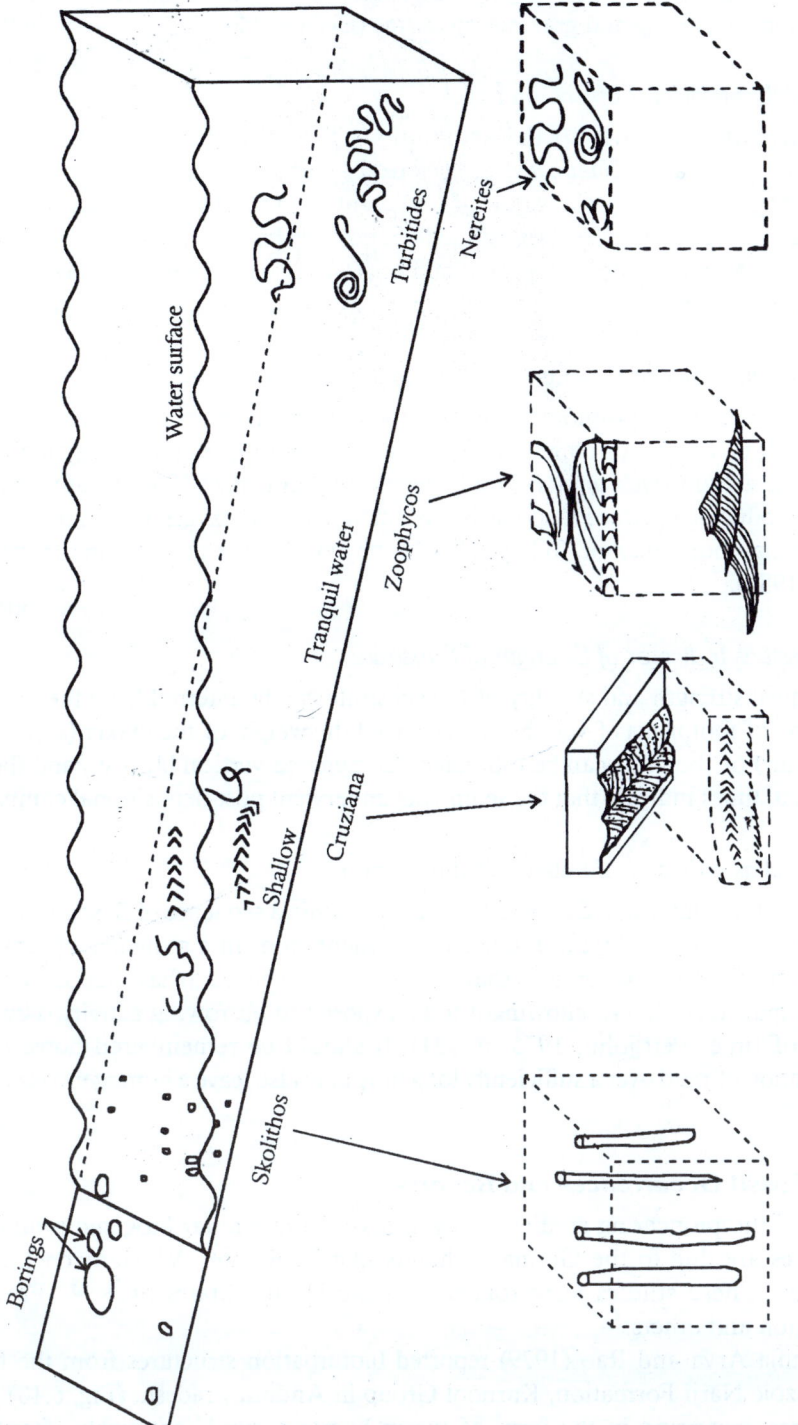

Fig. 6.9: Variaton of trace fossil assemblage with water depth. (Compiled from Seilacher 1967, Lindholm 1987).

of water. Direction of water current may also play an important role in controlling orientations of resting and grazing organisms (Ekdale 1988).

Ichnofossil as Indicator of Energy Level

Some organisms churn-up the sediments, thereby disturbing the layers. The phenomenon is called *bioturbation*. Excessive bioturbation suggests vigorous organic activity together with a slow rate of sedimentation. Conversely, total absence of bioturbation indicates lack of organic activity possibly because the prevailing condition was anoxic. An environment devoid of nutrients will also have the same effect. Vigorous mechanical agitation, as in a beach foreshore, might also retard organic growth.

Ichnofossil as Index of Toxicity

Burrowing activity is promoted in well-oxygenated environment. Increase of toxicity due to lack of oxygen or emission of H_2S may cause suppression of organic activity. Burrowing activity itself aerates freshly deposited sediments. Of all the animal species the one producing the trace Planolite is least tolerant of low oxygen level. Thalassinoides is somewhat more tolerant, followed by Zoophycos and then Chondrites (Savrda and Bottjer 1986).

Ichnofossil as Indicator of Strength of Substrate

Mechanical strength and stability of the substrate can be inferred from the depth and condition of footprints of vertebrates, provided the weight of the maker (say, a bird or a dinosaur) of the print can be estimated. Corrugated vertical burrows and flattened horizontal tubes indicate that the sediments underwent post-depositional compaction.

Ichnofossil as Indicator of Rate of Sedimentation

Seilacher (1962) interpreted absence of burrows within a sequence of flysch as indicative of very quick, essentially instantaneous, sedimentation. In a sequence of laminated, undisturbed sand and bioturbated shale, the sand is interpreted to have deposited quickly, and the mud very slowly, allowing it to be exposed to burrowing activity over a long period of time (Pettijohn, 1975, p. 131). It should be remembered however, that bioturbation of mud over a sufficiently long time may also leave a homogeneous residual layer.

Trace Fossil Occurrences and Reports

Much of the pioneering studies on trace fossils and similar biogenic sedimentary structures are due to the German scholars like R. Richter, W. Häntzchel, and A. Seilacher. These studies were followed in the United States by R.W. Frey, S.G. Pemberton and others.

In India Arya and Rao (1979) reported bioturbation structures from the Middle Proterozoic Narji Formation, Kurnool Group in Andhra Pradesh. (Fig. 6.10). These structures, occurring in the form of worm burrows, were tentatively classified as belonging to the Skolithos and Glossifungites assemblages.

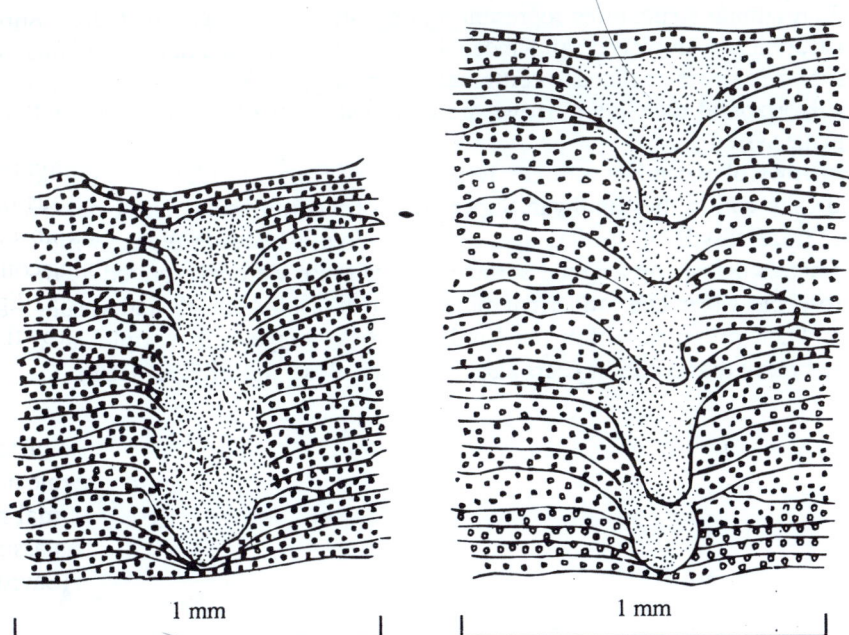

Fig. 6.10: Bioturbation structures from the Middle Proterozoic Narji Formation, Kurnool Group, India (from Arya and Rao 1979). Repreinted with permission from Elsevier.

Maulik and Chaudhuri (1983) reported three types of trace fossils from the Triassic red beds of the Gondwana sequence of the Pranhita Godavari Valley. These include (a) horizontal burrows (b) inclined passageways between successive galleries of horizontal burrows (c) vertical to slightly inclined burrows. The morphologies of these burrows are environment sensitive. The traces are vertical in the shifting conditions of the channel facies but they are horizontal in the stable flood plain areas.

Burrows and bioturbated sediments were reported from the siliciclastic tidal flat deposits of the of the Bhander Group of the Precambrian Vindhyan Supergroup near Maihar, Central India (Chakrabarti, 1990). The burrows vary from large-diameter, near-vertical stubby forms to microscopic thread-like features cutting across the sedimentary structures. Intense bioturbation of the sediments housing the burrows suggest that the trace makers were 'deposit feeders'.

Das and Rao (1992) reported numerous bioturbation structures from the Middle Proterozoic Chamuria Limestone belonging to the Chattisgarh Group, India. These burrows of microscopic dimensions were obtained from fine grained, pyritiferous limestone, deposited in dysaerobic lagoonal carbonate mud environment. Based on their shape, internal architecture, and their relationship to the host rock, these burrows are believed to belong to a number of ichnogenera: Skolithos, Monocraterion/ Rhizocorallium, and Polybessurus.

Sarkar and Chaudhuri (1992) reported several morphological varieties of trace fossils from the Middle and Late Triassic fluvial redbeds of the Pranhita-Godavari Valley. The distribution of burrows here was shown to be facies controlled. Skolithos, a type of vertical dwelling burrow, occurs both in channel and floodplain deposits. Taenidium, the horoizontal strucutures representing deposit-feeding organisms, are restricted to catastrophically emplaced sand sheets in channels and proximal floodplains. Vertical escape burrows are confined to continually accreting parallel laminated sands of channel bars. Horizontal dwelling burrows are confined to smaller sandsheets of floodplain drainage systems.

Burrows of wormlike animals were reported from the Mesoproterozoic Chorhat Sandstone within the Vindhyan Supergroup of Central India. These burrows, unlike other trace fossils of similar age, always follow a bedding plane, which is a few millimetres below the original surface. Hence these were interpreted as the work of microscopic, worm like "undermat miners" that excavated tunnels underneath microbial mats. Their presence within the Vindhyan rocks suggests that triploblastic animals existed more that a billion years ago (Seilacher, Bose and Pflüger, 1998).

7

SEDIMENTARY ENVIRONMENTS AND FACIES

INTRODUCTION: FACIES MODELLING

The term *facies*, derived from a Latin (also French) word, implies a facial expression characteristic of a particular condition. In modern sedimentology the term is used as an expression for 'the total field aspect of the rocks themselves ... The key to the interpretation of facies is to combine observations made on their spatial relations and internal characteristics (lithology and sedimentary structures) ...' (Middleton 1978a). The process of construction of a facies model for an environment involves identification of the diagnostic or common characters of the particular environment. This is done by studying a large number of modern and ancient situations and then formulating general principles about that environment after elimination of the local details. The purpose is to obtain 'a norm for comparison, a guide for future observation and a predictor for new geological situations' (Walker 1978, 1984a). With this end in view, considerable effort has been made during the last few decades to collect information about the modern environments of sedimentation. The following pages summarize the present-day knowledge of the more important environments, marine, non, marine and mixed. A genetic explanation for an ancient phenomenon should come by way of analogy with these modern environments.

MARINE ENVIRONMENTS

Traditionally the ocean-floor is divided into three zones according to depth: *nentic* (200m and above), *bathyal* (200-2000m) and *abyssal* (below 2000m). The deep-sea trenches are located in the *hadal* zone at depths exceeding 6000m. The more modern schemes use ocean-floor gradient as the principal criterion for classification (Fig. 7.1).

The subdivisions thus obtained are: *continental shelf* (average slope 0.1°), *continental slope* (4°), *continental rise* (0.05°-0.6°) and *abyssal plain* (0.05°). The facies associations in these areas are briefly discussed below after Walker (1984 b & c)..

Continental Shelf

The continental shelf is that area of the sea-floor lying between the shoreline and the shelf break. The sediments here are either siliciclastic or carbonate dominant. The sediment distribution in siliciclastic shelves is controlled mainly by three types of currents: meteorological (storm) currents, tidal currents and intruding ocean currents.

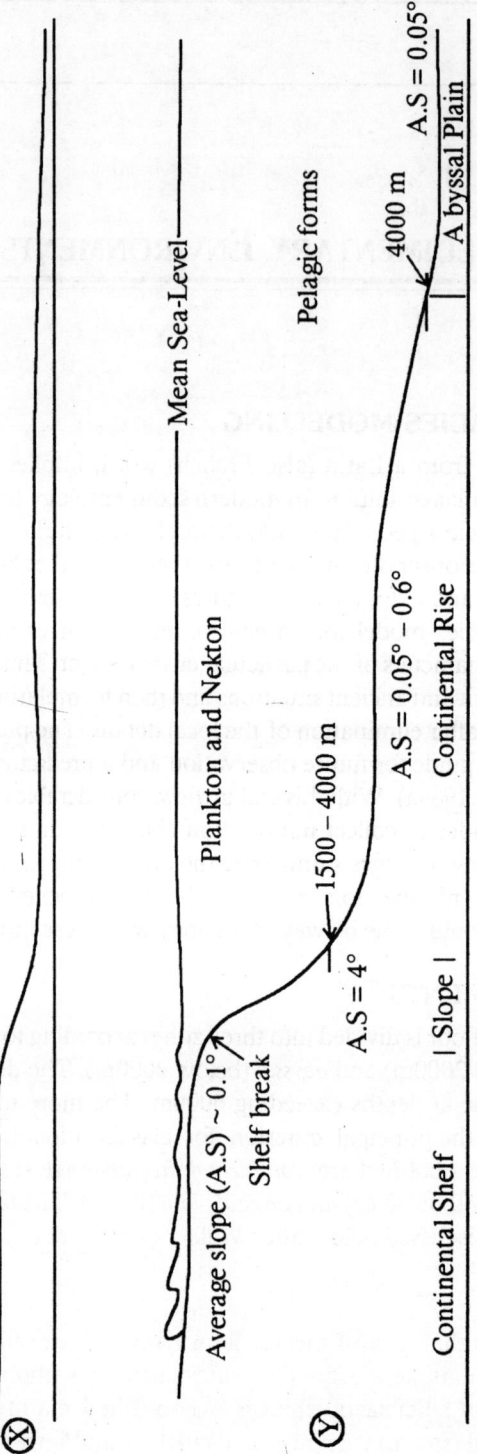

Fig. 7.1: Morphological classification of ocean floor according to gradient (based on Cook et al. 1982). (X) vertical exaggeration 1X (Y) vertical exaggeration 20X approximately. A. S. = average slope.

Siliciclastic Sediments on Shelf

Strom-dominated Shelf. About 80% of the continental shelves of the world today are frequented by storm waves. These waves agitate the ocean-floor above the storm wave base, 'liquefying' the sediments and generating suspension currents of high density (*turbidity currents*) in the process. Deposition from these suspension currents below the wave base takes place in the form of *turbidities*. Hummocky cross-stratifications are found below the storm wave base in ancient deposits. The features which dominate modern shelves are sharp-based, cross-bedded, linear sand ridges a few kilometres wide, several kilometres long and 3-12 m thick, running subparallel to the shorelines. Ancient counterparts of these sand waves are lacking.

Tide-dominated Shelf. Seventeen per cent of the world's continental shelves today are dominated by tidal waves. In the North Sea, adjacent to the coasts of England and the Netherlands, tidal waves have generated large-scale subaqueous sand waves over extensive areas ranging between 5000 and 15,000 km^2 (Fig. 7.2). The crests of these sand waves, which are about 35 m high, lie below 10 to 15 m of water, with their steeper sides dipping at low angles. Smaller sets of cross-bedding, dipping up or down the master bedding surfaces, constitute the internal structures of these sand ridges. Similar features have been reported from the Atlantic coast of North America also.

Shelves Dominated by Oceanic Currents. The continental shelf south-east of Africa, which is under he influence of a major oceanic current called the *Agulthus Current*, is an example of such a shelf. Extensive elongate strips of sand waves, measuring 20 km x 10 km x 17 m approximately, occur on this shelf at a depth of about 40 m below the Indian Ocean surface. These sand waves have been generated by reworking of the sediments already present on the shelf. No comparable ancient example is known.

Carbonate Sediments on Shelf

Study of limestone facies in a large number of geologically ancient basins, followed by their interpretation, using Holocene sediments as keys, led Wilson (1975) to formulate a generalized model of carbonate sedimentation on shelf. He recognized (*i*) an inner, shallow, low energy zone (*undathem or shelf*), (*ii*) an outer, deeper, low energy zone (*fondothem or basin*), and (*iii*) a shallow, high-energy zone (*clinothem or shelf margin*) separating these two. Wilson further classified the three zones into nine facies belts covering the whole gamut of carbonate sediments on shelf, ranging from deep marine basin to evaporites and sabkhas inland. Following is a brief account of these nine standard facies belts summarized from Wilson 1975 (Fig. 7.3).

Belt 1A: This belt is divided into two groups: (a) *geosynclinal troughs* filled with rapidly deposited limestone turbidities. (b) slow sedimentation with intermittent debris flows ('*leptogeosynclinal* deposits'). Sediments in these two belts are as follows.

(a) 'Allodapic' limestones—basinal sequence of breccias, exotic boulders, lime sands derived from shelf and slope carbonates. Their thickness is often great. Sedimentary textures and structures of terrigenous flysch are common.

(b) Leptogeosynclinal troughs do not contain extensive allodapic limestones. These are only intermittent sites of sediments from outside the basin. Sediment is chiefly pelagic, mostly siliceous sediment accumulated in water deep enough for extensive solution of calcium carbonate. The rock types include radiolarites, red

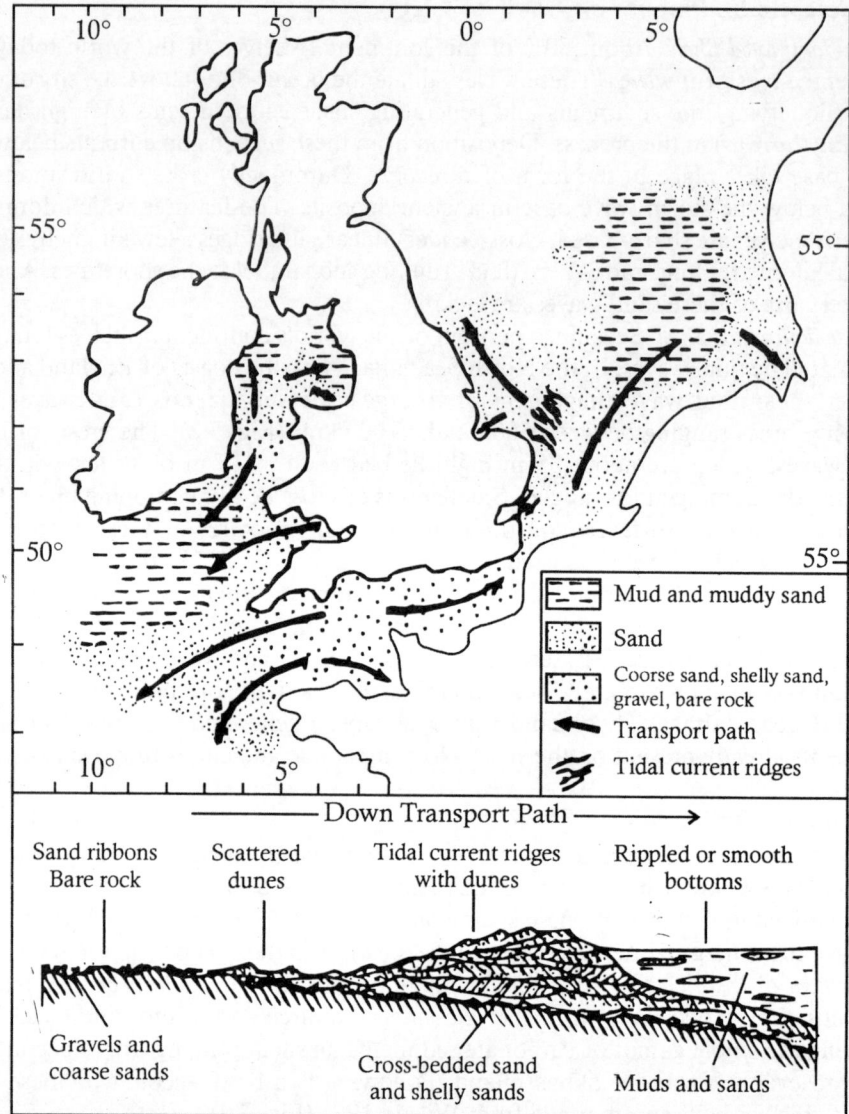

Fig. 7.2: Sedimentary structures in tide dominated shelf off the coast of Britain. The profile shows the structures typically encountered along a transport path (after Allen 1970, with permission).

biomicrite and red nodular limestone, light-coloured pelagic lime mudstone, dark coloured basinal micrite, pelagic micropeloids, bioclastic grainstones and packstones etc. These rocks are generally cherty, the silica being derived from solution of the opaline silica of organisms like microplanktons. The strata are generally thin bedded. Crumplings and contortions due to channeling down of breccia are noticed. Syngenetic conglomerates and breccias formed by solution and growth of concretions at sea floor may be seen. The biota of geosynclinal strata is chiefly pelagic.

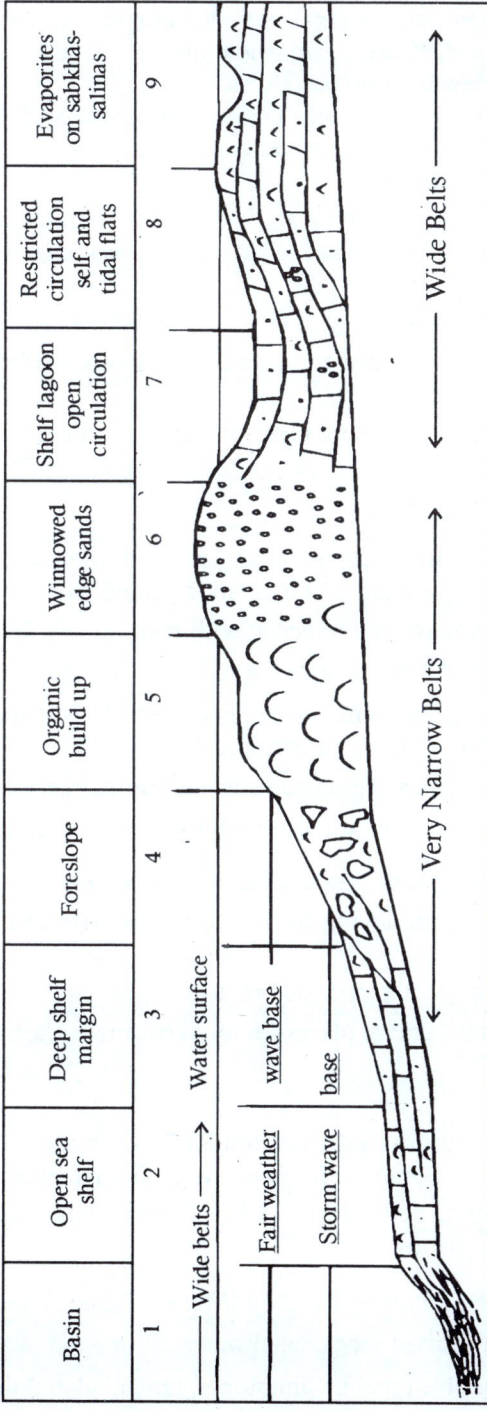

Basin	Open sea shelf	Deep shelf margin	Foreslope	Organic build up	Winnowed edge sands	Shelf lagoon open circulation	Restricted circulation self and tidal flats	Evaporites on sabkhas-salinas
1	2	3	4	5	6	7	8	9

Fig. 7.3: Carbonate facies belts on shelf after Wilson 1975 (wave bases inserted from McLane 1995). With permission of Springer Science and Business Media.

Belt 1B: *Cratonic Basin Carbonates (Starved, Mostly Euxinic) (Fondothem)*

Location: Deep intracratonic and marginal cratonic (miogeosynclinal) basins removed from carbonate producing shelf areas. Contribution from land is slight, often wind blown. A starved and somewhat deep basin results.

Physico-chemical conditions: generally euxinic, stagnant and reducing, below oxygenation level and below wave base. Water depth: thirty to several hundred metres. Bottom water hypersaline and dense.

Rock type: dark thin beds of limestone with dark shale or silt and some thin-bedded anhydrite. Terrigenous clastics like quartz silt and shale are admixed with carbonates. Chert is common.

Colour: Dark brown or black.

Sedimentary structures: Planar lamination, cross-lamination, small-scale rhythmic bedding.

Biota: Nektonic-pelagic fauna abundant. Graptolites, planktonic bivalves, ammonites, sponge spicules.

Belt 2: Shelf Facies (*Deep Undathem*)

Physico-chemical condition: water depth 10-100s metre deep. Normal salinity. Good current circulation. Shelves are wide. Sedimentation - uniform.

Rock types: Fossiliferous limestone interbeded with marl.

Colour: Grey, green, red, and brown.

Sedimentary structure: thin to medium, wavy to nodular beds. Ball and flow structures. Thoroughly burrowed. Mud-mounds, pinnacle reefs.

Terrigenous clastics: quartz silt, siltstone and shale occur interbeded with limestone.

Biota: diverse shelly fauna. Brachiopods, corals, cephalopods and echinoderms present.

Belt 3: Basin Margin or Deep Shelf Margin Facies (*Clinothem*)

Location: toe of a carbonate-producing shelf, below wave base and barely at oxygen level.

Water depth: as in Belt 2 or perhaps 200-300 m deep.

Rock type: Fine grained limestone, at places cherty. Terrigenous clastics: rarely present. Chert common.

Colour: Dark to light.

Sedimentary structures: laminated lime mudstones, rhythmic beds. Massive unlaminated lime mudstone. Some graded beds. Megaslump features.

Biota: mixture of older forms derived from the shelf, benthonic organisms living on the slope and some pelagic forms.

Belt 4: Foreslope facies of Carbonate Platform (Marine Talus, *Clinothem*)

Location: above the lower limit of oxygenated water.

Rock type and texture: variable types of limestones, lime mud and sand, boundstone, sedimentary breccia. Lime silts and bioclastic wackestone-packstone. Reworked material with locally derived organic debris, reef rudstone.

Colour: dark to light.

Sedimentary structures: megaslumps in thinly bedded strata. Large exotic blocks. Syngenetic slumps, pull-aparts, breccias, clastic injection dykes.

Terrigenous clastics: shale, silt, fine sand drifted from above mix with carbonate or fill cavities.

Biota: mostly bioclastic debris from upslope and colonies of encrusting organisms.

Belt 5: Organic Reef of Platform Margin

Location and physical conditions: three types of profiles are known (*i*) downslope accumulation of carbonate mud and organic detritus (*ii*) ramps of knoll reefs with bioclastic sands (*iii*) frame-constructed reef rims.

Rock type: massive limestone and dolomite in places consisting solely of organisms. Bioclastic debris. Patches of organic boundstone with interstices filled with lime mudstone. Grainstone and packstone in upslope accumulations.

Colour: light.

Sedimentary structures: massive organic framework.

Terrigenous clastics: essentially absent.

Biota: colonies of sessile frame-building organisms may dominate.

Belt 6: Winnowed Platform Edge Sands

Location and physical condition: sholas, beaches, offshore or tidal bars, aeoline dune islands. Marginal sands range from well above sea level to 5 to 10m deep. Clean sand is winnowed and deposited by waves, tidal, or longshore currents. Environment is well oxygenated but not hospitable to marine life.

Rock type and texture: calcareous or dolomitic lime sand. Rounded, well-sorted grainstones, often oolitic or rounded bioclasts.

Colour: light.

Sedimentary structures: medium to small scale festoon cross-bedding. Eolian dunes with large cross-bedding. Old soil horizons and root casts.

Terrigenous clastics: quartz sand and calcarenites.

Biota: worn coquinas of benthonic animals living on reef and foreslope. Large bivalves and gastropods common. Fragmented remains of large algae and some foraminifera.

Belt 7: Open Marine Platform Facies (*Shallow Undathem*)

Location: in straits, open lagoons, and bays behind outer platform edge.

Physical condition: Shallow water, a few metres to tens of metres deep. Salinity varies from normal marine to somewhat higher. Circulation is moderate. Water condition suitable for growth of organisms but stenohaline forms are excluded.

Rock types and texture: variable limestone and lenses of land derived clastics. Grainstone to mudstone. Lenses of lime sands of shelly and angular fragments. Beds of bioclastic wackestones, biostromes.

Colour: light and dark.

Sedimentary structures: medium to platy bedded burrowing and pelleting of sediment common. Ball and flow compaction structures, wavy bedding.

Terrigenous clastics: when present, generally in well-segregated beds intercalated with limestones.

Biota: mollusks, sponges, arthropods, foraminifera and algae common. Patch reefs present. Fine sediments occur trapped in marine grasses. Organisms (e.g. brachiopods, echinoderms, cephalopods) requiring normal marine salality are present but may be rarer than in open sea.

Belt. 8: Facies of Restricted Circulation on Marine Platform

Location: restricted lagoons behind or between barrier reefs, behind coastal spits or within atolls. Generally shallow, hypersaline water, both reducing and oxygenated conditions. Diverse intertidal environments.

Rock types and texture: mostly lime mud with dolomite. Grainstones rare. Channels contain lithoclastic grainy sediment, pelleted mudstones and wackestones.

Colour: light.

Sedimentary structures: laminated mudstone, fenestral (birdeye) fabric, algal stromatolite, graded bedding, dolomite and caliche crusts. Cross-bedding in channel sands.

Terrigenous clastics: rare except wind-blown material in well-segregated layers.

Biota: limited fauna and flora. Gastropods, algae, foraminifera (miliolids), ostracods abundant locally.

Belt 9: Platform Evaporite Facies

Location: Supratidal and inland pond environment of restricted marine platform in evaporative climate—sabkhas*, salinas, salt flats. Sporadic marine flooding.

Rock types: nodular and wavy anhydrite or gypsum interlaminated with dolomite. Commonly associated with redbeds.

Colour: variable. Red, yellow, brown.

Sedimentary structure: laminate, planar, mud cracks, stromatolites. Gypsum rosettes and selenite blades. Nodular and flaser structures. Caliche crusts.

Terrigenous clastics: redbeds and windblown sediments.

Biota: No indigenous fauna excepting blue-green algal stromatolites and brine shrimps.

---

* Sabkhas are depressions formed in extremely arid conditions (for example deserts) by removal of loose particles by wind. Sabkhas go down to the groundwater level or to the zone where concentration of salts drawn by capillary action takes place. In marginal marine conditions, in arid climate, the tidal flats subjected to periodic tidal influx may lead to hypersaline pools. Such conditions have been termed platform evporite facies by Wilson (1975). The typical deposits within these pools are laminated dolomite together with nodular anhydrite, gypsum and halite. Some wind-blown terrigenous materials may also be present. Burrowing organisms are absent due to hypersaline conditions. The most well known coastal sabkha today is in the coastal areas of the Persian Gulf.

Continental Slope

The slope bordering the continental shelves are marked by networks of submarine canyons and gullies. Water-wet terrigenous sediments are carried *en masse* by gravity into the deep sea through these canyons. The phenomenon is called '*mass-transport*'. Three major types of mass-transport are recognized—rockfalls, rockslides and sedimentary gravity flows. Rockfalls cause talus accumulation at the toe of slopes. Rockslides can be either translational (glide) or rotational (slump). Several types of sedimentary gravity flows, from debris or mud flows to turbidity current flows, are known. In debris flows the clasts are supported by the cohesive strength of the mud matrix. The water-wet mud behaves like a plastic body. In a turbidity current the clasts are supported by fluid turbulence and the whole mass behaves as a fluid. This also includes liquefied flows in which the sediment is supported by upward displacement of fluid (*dilatance*), and fluidized flow in which the sediment is supported by upward motion of escaping pore fluid. Intermediate between plastic and fluid flows are grain flows in which the sediment is supported by dispersive pressure (see Fig. 7.4).

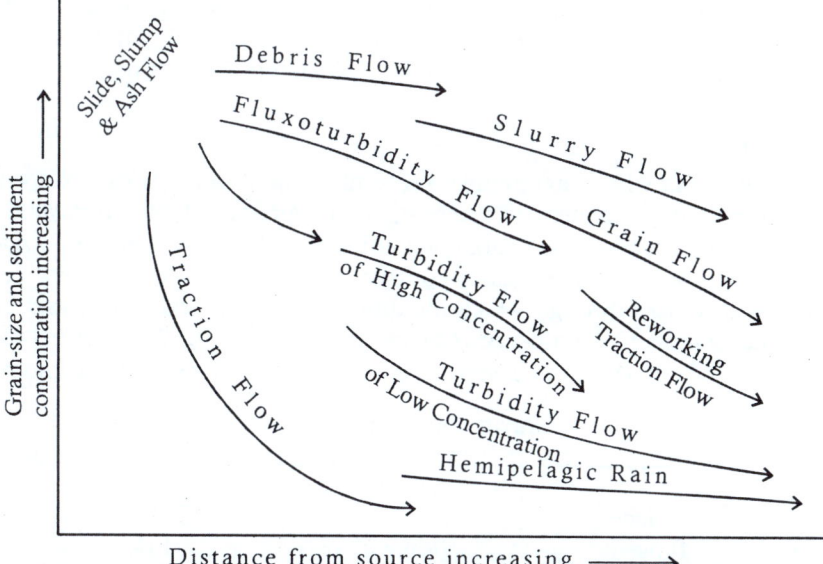

Fig. 7.4: Inferred flow processes in turbidity and related flows (based on Howell and McLean 1976, Howell and Normark 1982,) With permission of American Association of Petroleum Geologists.

Continental Rise and Deep-sea Fans

Terrigenous sediments transported to the ocean floor through feeder channels and canyons splay out to produce *deep-sea fans*. Gravitational processes effect transportation of these sediments. These fans constitute the major part of the continental rise. Of the many different flow processes through the feeder channels the following three are more important.

Debris Flows

Debris flows are produced when a mixture of clay and water acquires a density higher than water and transports large clasts of sand and even gravel. The large-size grains are deposited when the transporting medium looses its density and becomes increasingly fluid. Deposition takes place *en masse*.

Carbonate Debris Flows

The debris flows and turbidites are not necessarily composed of silicic materials only. Carbonate sediments also contribute to turbidites. Corals and algal fragments, obviously of shallow water origin, are sometimes transported to much deeper water by turbidity currents and deposited there. Spectacular carbonate slump masses, called *olistoliths*, have been reported from Nevada in USA and also from Italy. Carbonate debris flows consist of clasts of varying sizes embedded in lime mud. Enormously thick carbonate debris flow deposits have been reported from the Cambro-Ordovician rocks of Newfoundland. Carbonate turbidites are known in the Alps and Apennines. Carbonate turbidity grain flows and debris flows of recent origin occur in the Bahamas in USA.

Grain Flows

Grain flows take place when interaction between the grains is strong enough to produce dispersive pressure forces which keep the grains apart. Scattered grains floating in a sandy matrix are deposited out of this process. Sole marks are common.

Fluidised Flows

Fluidised flows result when intergranular pressure maintains the grains in suspension. Transportation occurs with increase of interstitial fluid pressure. Deposition takes place when the difference between interstitial pressure and hydrostatic pressure decreases.

All the above mentioned processes contribute substantially to turbidity current deposits. The distances through which the different types of sediment travel on the ocean floor depend on their grain-size and concentration (see Fig. 7.5). The net volume of these mass flow deposits can be enormous - of the order of many cubic kilometres.

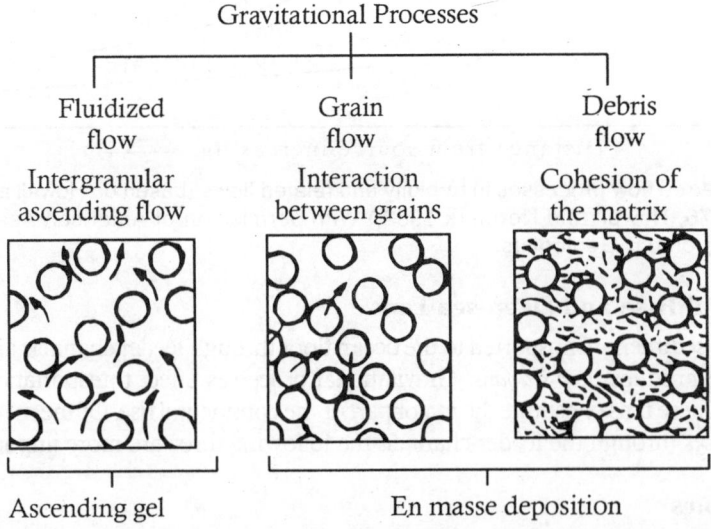

Fig. 7.5: Mechanism of cohesion and deposition of sediment in aquatic environment by gravitational processes (based on Middleton and Hampton 1976).

These can be transported with astonishing velocity. For example, a maximum velocity of 20 m/sec was estimated for the turbidity current produced by the Grand Banks Earthquake of 1929 (Dunbar and Rodgers 1957).

Depositional Sequence Within Deep-sea Fans

Bouma Sequence

The sequence of sedimentary structures produced within a submarine fan by the deposits of mass flow is as follows (see Fig. 7.6):

Top   e.   pelitic division

           d.   upper division of parallel lamination

           c.   current ripple lamination

           b.   lower division of parallel lamination

Bottom   a.   graded or massive beds

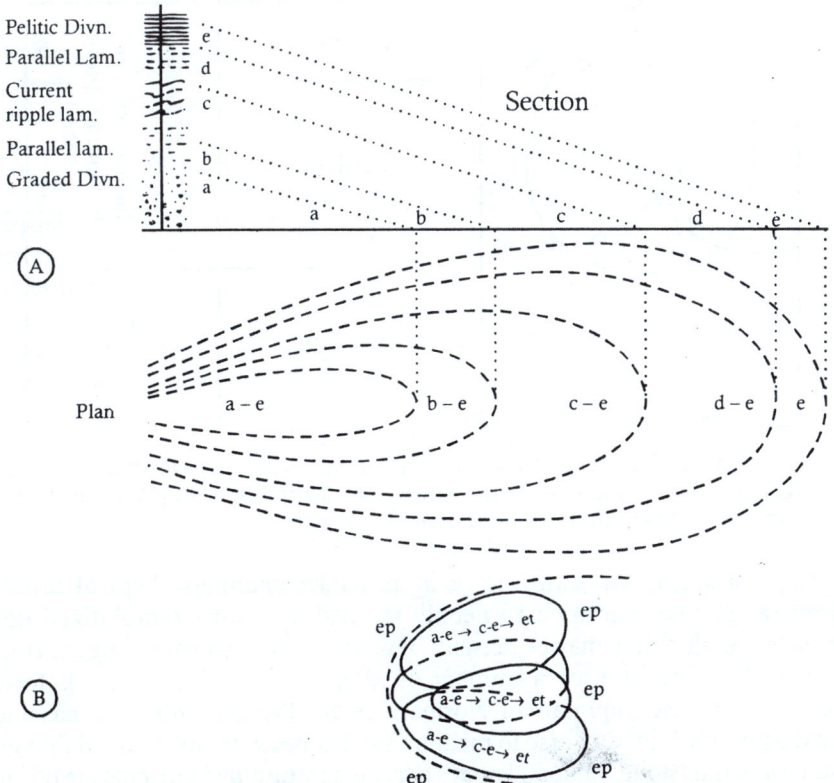

Fig. 7.6: (A) Schematic representation of an ideal Bouma sequence in plan and section. (B) Hypothetical pattern of filling of a basin by turbidites. Note that each cone does not fill the entire basin (modified after Bouma, by Howell and Normark 1982). With permission of American Association of Petroleum Geologists.

This sequence, idealized by Bouma (1962), has been named after him. The complete Bouma sequence (T_{a-e}) is rarely found in geological record but even a partially preserved Bouma sequence may help in the identification of an ancient fan deposit.

Marphology of Deep-sea Fans

The following generalized facies model for deep-sea fans was formulated by Mutti and Ricci-Lucci (1975) on the basis of studies by Normark on modern deep-sea fans and by Mutti and associates on the ancient fans of the Apennines, Italy. An idealized fan, according to this model, consists of the following components (Fig. 7.7).

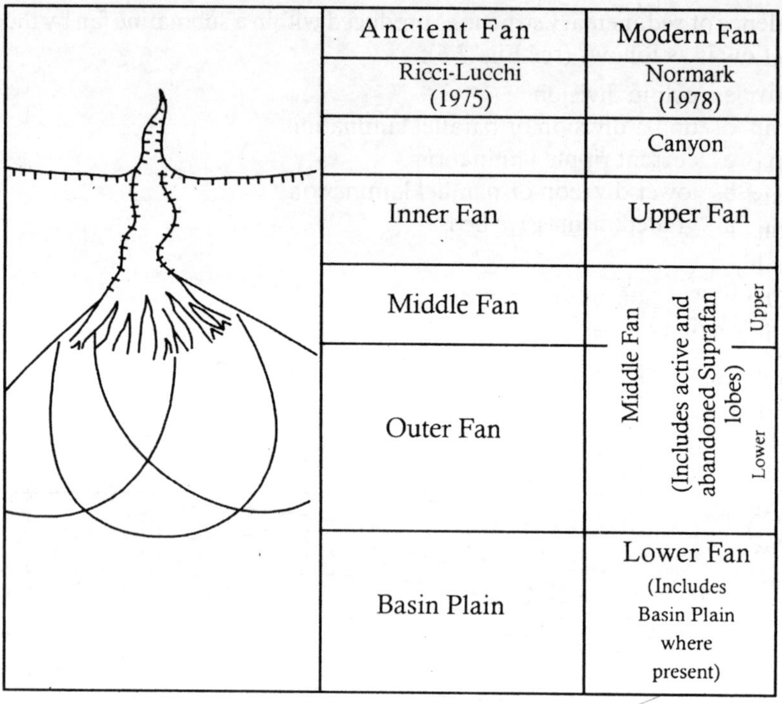

Fig. 7.7: A simple classification of submarine fans proposed by Normark (1978), Ricci-Lucchi (1975) (modified after Howell and Normark 1982). With permission of American Association of Petroleum Geologists.

An inner (or upper) fan with one or more feeder channels. Typical lithofacies: conglomerates, coarse-grained to pebbly sandstones. Some remobilized deposits suggesting mass slumping may be present. Basal scouring and channeling are common.

A middle fan consisting of a number of *shallow*, branching channels feeding a depositional lobe (the 'suprafan' of Normark 1978). Typical lithofacies: basal gravels and mud chips overlain by coarse to medium sandstones with interbedded fine-grained sandstone and mudstone. The sandstones deposied by turbidity currents are of uniform thickness for a long distance and exhibit a classical Bouma sequence.

An outer (lower) fan, which begins beyond the limits of the distributary channels, contains tabular mudstones with interbedded, graded, thin sandstones. The upper part of the Bouma sequence is preserved in these deposits. Sole marks are common. The depositional lobes, which are topographically smooth, grade laterally into a basin plain.

The basin plain (lower fan of Normark 1978) consists essentially of pelagic and hemipelagic mudstones with a few graded sandstones. The deposits are transitional with the outer fan fringe.

Abyssal Plain Deposits

The abyssal depths of the ocean are dark and cold. Blue-grey to lead-coloured mud, transported from land to the ocean-bottom by suspension currents, floor the bathyal zone of the ocean, but, the abyssal zone is covered with red-coloured mud. This red mud, having the chemical composition of an average igneous rock, is believed to have been blow.a into the deep sea from volcanic vents located in the adjoining land areas. Prolonged oxidation of fine-grained terrestrial particles slowly settling through oxygenated water columns might also be responsible for the red colouration.

In relatively deep water, upto a depth of about 3000 m, pteropod ooze, cocoliths and foraminifera predominate. The carbonate precipitation in this zone is essentially in the form of aragonite (orthorhombic $CaCO_3$). Between 3000 and 5000 m globigerina ooze, consisting mostly of low-magnesian calcite (hexagonal $CaCO_3$), dominates the ocean floor. In the deeper parts of the ocean, where the water is cold, pH is low and pressure is high, carbonate remains mostly in solution. Increase of partial pressure of CO_2 (pCO_2) due to dissolved CO_2 at great depth also allows ocean water to hold carbonate in solution. The deposits of the abyssal ocean floor contain only about 20% carbonate, whereas the deposits in the shallower parts of the ocean contain 90%. The depth above which calcite-rich sediments accumulate is called the *calcite compensation depth* (CCD). Below this depth only calcite-free sediments are found because the rate of calcite supply is balanced by calcite dissolution at CCD. Deepwater carbonate, wherever found, has been either transported from the shallower parts of the ocean by suboceanic density currents, or dropped onto the ocean floor from the pelagic zone where floaters and swimmers abound. Cocolith ooze, formed by the latter process, is an important constituent of the fine-grained limestone called chalk.

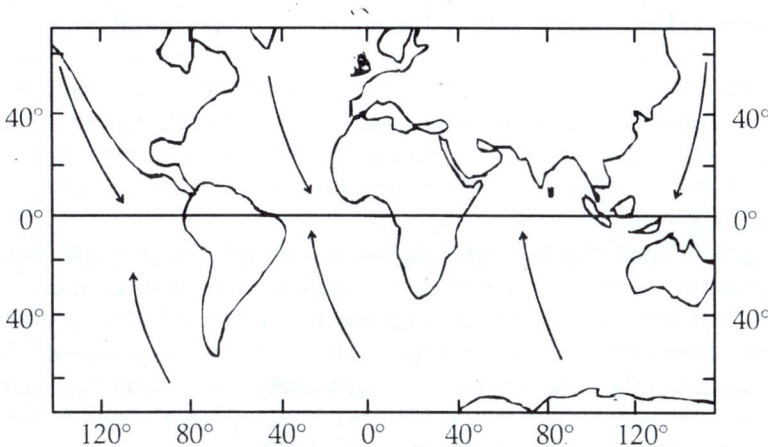

Fig. 7.8: Approximate direction of increase in the depth of carbonate compensation (CCD) in the major oceans of the world.

Calcite compensation depth varies from ocean to ocean but its average depth is about 4500 m. In the Pacific Ocean CCD occurs at a relatively shallow depth (4200-4500 m). In most of the North Atlantic and in some parts of the South Atlantic also (between latitudes 40 ° N and 40° S), it is at 5000 m, or still deeper. In the equatorial

Pacific CCD is about 5000 m. Increase in the amount of biological productivity leading to abundant supply of calcareous material accounts for this increase in CCD near the equator. In the Indian Ocean CCD is at about 5000 m near the equator but its depth decreases to about 2000 m at 60° latitude (Kennett 1982, Tucker *et al.* 1990). (Fig. 7.8).

The biochemical precipitates below the carbonate compensation depth are of siliceous nature. Local concentrations of ferromanganese concretions ('*manganese nodules*') are found in some pockets of the deep-sea floor. These are ascribed to exhalations from suboceanic volcanic or hydrothermal vents but the precise mode of their origin is not known.

Marine Sediments in Stratigraphic Record

The typical facies associations found in marine shelf, continental slope and continental rise are briefly described below.

Shelf Sediments in Stratigraphic Record

Sedimentation on continental shelves can take place under three conditions: transgressive, prograding and aggradational (Galloway and Hobday 1983). Transgressive sequence may be either tide-dominated, or storm-dominated. Both of these produce fining-upward sequences. The transgressive sequences start with lag gravels at the bottom and end with bioturbated silts and muds on the top, passing successively through coarse and fine-grained, cross-bedded sands (Fig. 7.9A). The storm-dominated transgressive sequences start with lag gravels at the bottom and end in silty and muddy sediments on the top, passing successively through silts and muds (with lenticles of sand), multistorey sets of hummocky cross-stratification and thinly graded, parallel laminated sands (Fig. 7.9 C). The hummocky cross-stratifications are storm generated. The thinner sets are deposited as the depth of water increases due to transgression. Laminated mud, rich in organic carbon, glauconitic and phosphatic muds may be laid down on top of the succession of a deep shelf deposit. Many of the outer shelf mudstones are phosphatic with a very high organic carbon content. Some of these have been correlated with the phases of worldwide eustatic transgressions by Vail *et al.* (1977).

The prograding shelf sequences typically coarsen upward. Two prograding sequences, one dominated by storms and the other, a mixed energy shelf succession, have been identified. The storm-dominated prograding successin is almost the reverse of the storm-dominated transgressive sequence (Fig. 7.9 B). A progressive seaward fining of sediments, caused by the storm-induced sorting processes, is represented as a coarsening-upward vertical sequence. The typical sequence produced in a mixed energy (storm and tidal) prograding shelf is represented in Fig. 7.9 D.

Aggradational shelf sediments are generally fine grained. Silt and sand are deposited by traction currents. Mudstone is deposited when the rate of sediment supply is slow due to gradual rise in sea-level or alternatively, due to subsidence of the land area. Periodic transgression and progradation may cause sandy sediments to alternate with thick sequences of mud. In ancient geological record mixed energy aggradational sandy shelf successions are represented by thickly cross-bedded quartz arenites.

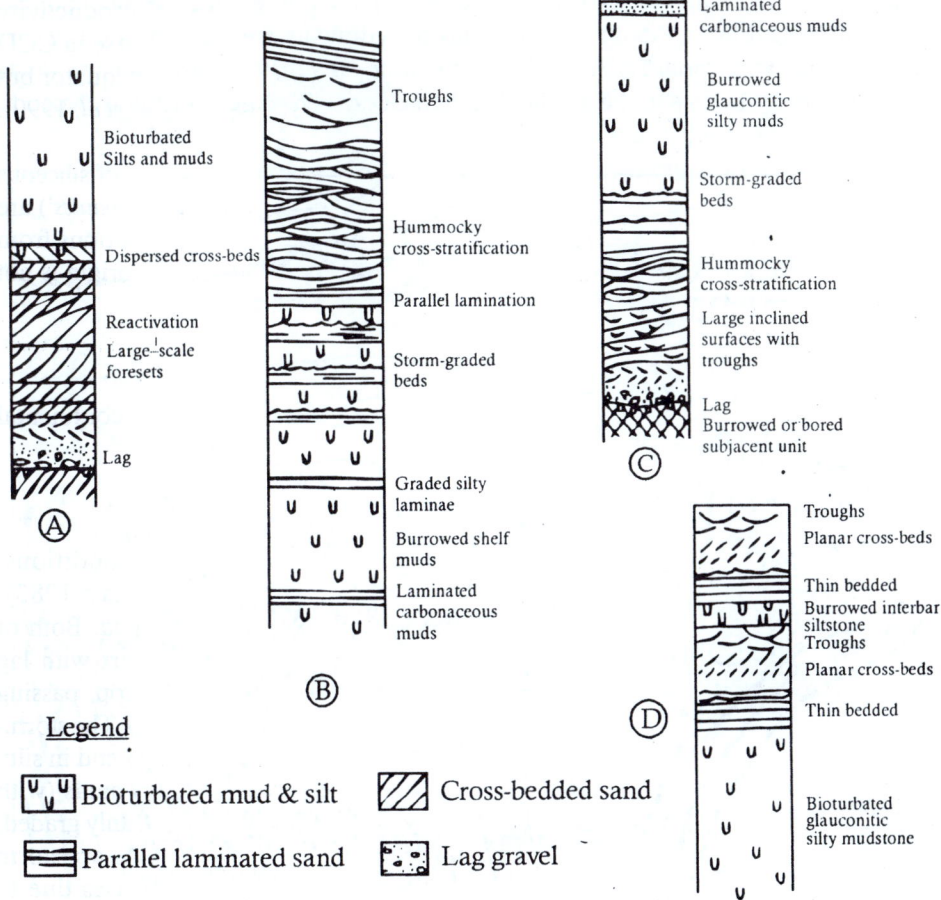

Fig. 7.9: Idealised depositional sequences in (A) transgressive, tide-dominated shelf, (B) prograding, storm-dominated shelf, (C) transgressive, storm-dominated shelf and (D) prograding shelf dominated by storm and tidal actions (after Galloway and Hobday 1983). With permission of Springer-Verlag, Heidelberg.

Continental Slope and Fan Deposits is Stratigraphic Record

The deposits of the continental slope and talus include debris sheets, rudite sheets and sedimentary breccias of very large dimensions. The latter are called *olistostromes* (Greek: *Olistomai*-to slide, *Stroma*-layer). Derived from shallow water siliciclastic and carbonate deposits, the slope breccias may be of diverse composition, varying from well-bedded calcareous sands and calcarenites to rounded blocks of reefal limestones resembling bioherms. Downslope they may grade into deep-water *pelagic* or *hemipelagic* sediments. The *pelagic* sediments are open sea deposits containing calcareous ooze, skeletal remains of foraminifera, pteropods and coccolithophorids. The terrigenous component is small (< 20%) in these sediments but very fine-grained pyroclastics (*tephra*) are common. Siliceous (radiolaria, diatoms) ooze and skeletal remains are common below the Carbonate Compensation Depth. *Hemipelagic* sediments are admixtures of fine-grained pelagic and coarser grained terrigenous components. Deep-sea fans occur on the ocean floor.

The largest submarine fan known today covers an area of about 3,000,000 km² on the Bay of Bengal floor. About 2 billion tons of sediments are washed into this fan annually through the '*Swatch of No Ground*' submarine canyon, to be distributed through a channel-levee system up to the northern Indian Ocean. The biggest of these channels is about 27 km wide and 180 m deep (Fig. 7.10).

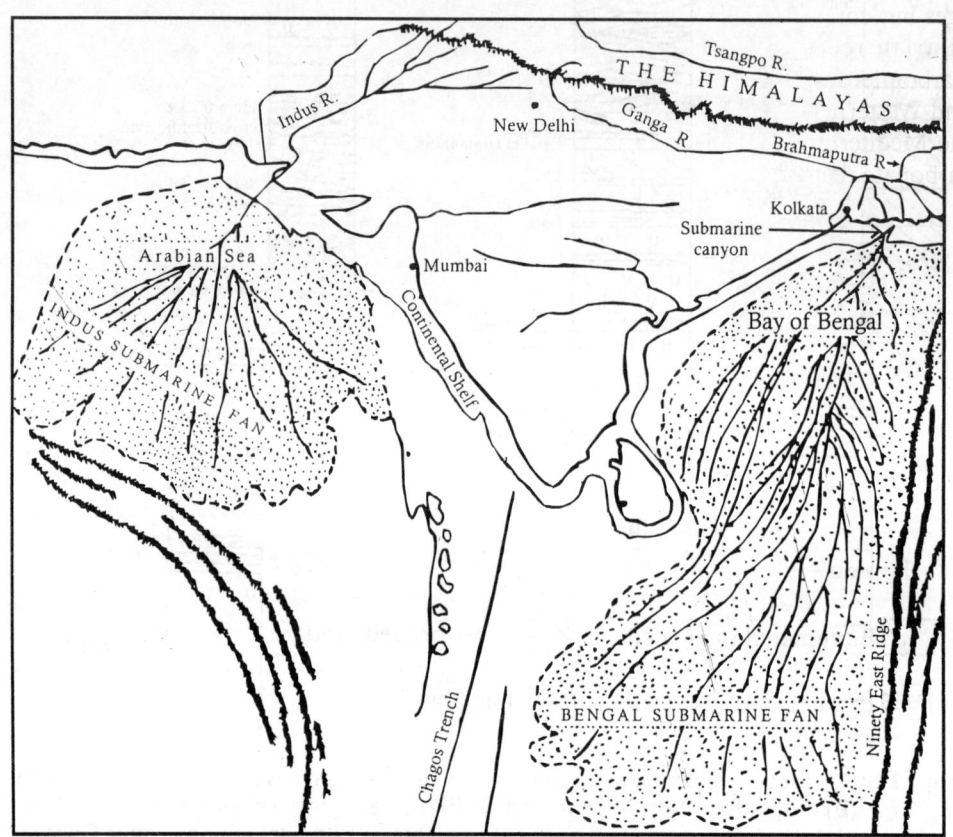

Fig. 7.10: Sketch map showing extension of the Bengal and Indus Submarine Fans. Based on Map of the Indian Ocean Floor by National Geographic Society (1967).

A submarine fan migrates through space and time to produce a vertical sequence which is preserved in the geological record. A progressive migration from the proximal to the distal areas (from the inner to the outer fan) may be caused either by uplift of the source area, or by an increase in the rate of sedimentation. Lowering of the sea-level will also have the same effect. In the vertical succession produced from this process the coarse-grained inner fan deposits overlie the finer-grained outer fan sediments (Fig. 7.11 A). When the rate of sediment influx decreases, or the sea-level is raised, a retrogradation of the submarine fan takes place. In the vertical sequence the coarser inner fan deposits are successively overlain by the progressively finer middle fan and fan fringe sediments (see Fig. 7.11 B).

Many examples of marine shelf deposits exist in geological literature. The Lower Palaeozoic sections of the upper Mississipi valley and the Great Lakes of North America are often cited as typical examples. In both the areas shelf sediments were deposited on the southern extension of the Canadian Shield when the conditions were very stable (Pettijohn 1975). The Jura Quartzite of Scotland and the Gog Quartzite (Lower Cambrian) of the Southern Rocky Mountains are ancient examples of tide-generated sand bodies (Walker 1984b). Reefoidal complexes are recorded from all geological ages but coral reef builders evolved on a large scale from the Silurian onwards. The Silurian reefs of Gotland, Sweden, the Devonian reefs of Alberta, Canada, Carboniferous reef complexes of North England, the Permian reefs of New Mexico and Western Texas, Triassic reefs of the Dolomite Alps, Itlay, and the Tertiary reefs of the Mediterranean and the Persian Gulf are some of the most famous examples of carbonate reefs in the geological record.

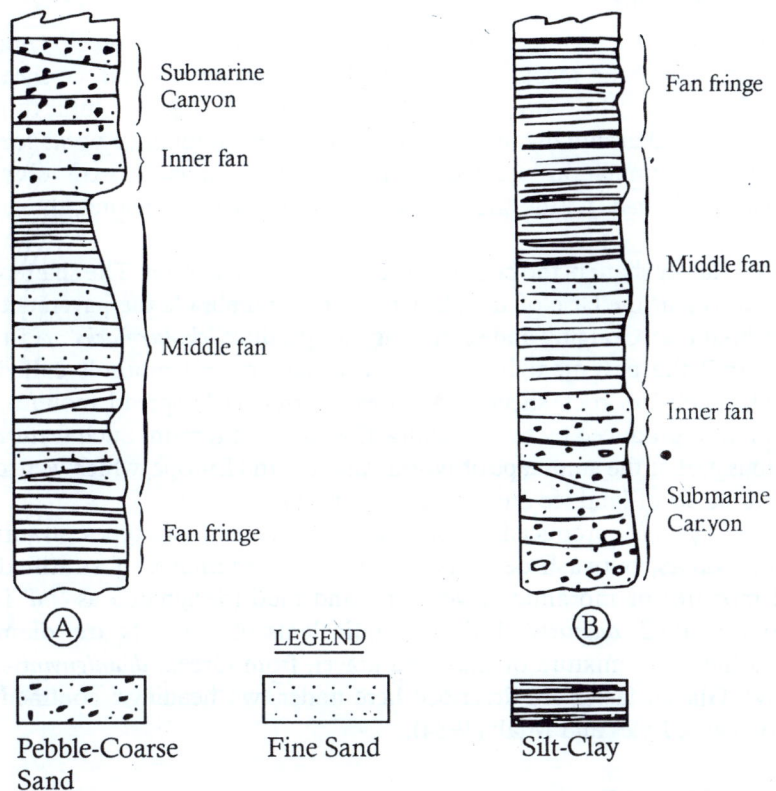

Fig. 7.11: Examples of vertical sequences produced by migration of submarine fans : (A) a progradational sequence, (B) a retrogradational sequence (based on Howell and Normark 1982). With permission of American Association of Petroleum Geologists.

In India the shale-quartzite-arenite sequence of the Kaimur Formation in Rajasthan is believed to have been deposited during the Proterozoic by an epiric sea on a very gently dipping, storm-dominated continental slope. Many sedimentary structures, including hummocky cross-stratifications, have been reported from these sediments (Bose, Chaudhuri and Seth 1988). The Cretaceous Nimar Formation of M.P., India, was also interpreted as the product of sedimentation on a storm-cum-fair weather-dominated continental shelf (Bose and Das 1986).

Extensive marine sediments have been reported from the Himalayas by many authors. In one of the more recent studies 'a complex jumble of volcanic and sedimentary rocks' occurring between Lamayuru Formation (Triassic to Cretaceous) in Ladak, and the Triassic flysch in southern Tibet, have been described as '*ophiolitic melange*' (Thakur and Virdi 1979; Virdi 1986). The Lamayuru Formation in Ladak contains exotic masses of limestones which are believed to have accumulated at the toe of an ancient continental slope by slumping. Olistoliths and grain-flow deposits have also been reported here (Virdi 1986).

NON-MARINE ENVIRONMENTS

Glacial Environment

Glaciers are masses of moving ice produced by recrystallization of precipitated snow. Two types of glaciers are recognized: valley glacier, and continental ice sheet (Figs. 7.12 A, B). On crossing the snow line a valley glacier starts losing its volume by melting and evaporation and eventually reaches a zone where the amount of wastage exceeds the amount of accumulation in the upper reaches. The annual loss of volume of a glacier is called *ablation*. As the snout of a glacier vanishes by melting, the glacier is said to recede.

A moving glacier shatters the bedrock by exerting pressure on it. The shattered rock fragments, frozen at the base of the glacier, scratch and abrade the pavement below, producing striations. Gouging and scratching also produce '*chattermarks*'. As a glacier slides downhill the pavement is sculptured to asymmetric mounds called *roches moutonnée* (French for 'rock sheep'). A series of irregularly spread mounds, called '*drumlins*', is also produced by smoothening of a glaciated terrain. Such features form an ubiquitous part of the landscape of North America and Europe, which were covered by extensive sheets of land ice during the Quaternary.

The deposits ('*drifts*') produced out of glacial action are of two types: stratified and unstratified. *Varves*, outwash deposits and *eskers* are examples of stratified drifts. Unsorted mixtures of morainic gravel, sand and mud (designated as '*till*' in older literature) are called '*diamicts*'. Lithified equivalents of diamicts are 'diamictites' (meaning a thorough mixture of mud and gravel, from Greek: *diamignymi* - to mix thoroughly). Glacial facies are described here under two headings: continental and marine, following Eyles and Miall (1984).

Continental Glacial Facies

These include grounded ice and also, glacio-fluvial, glacio-lacustrine and periglacial facies. Ground ice grows into ice sheets of continental proportions. Two types of facies associations are possible within ground ice sheets, depending on whether the ice margin is inactive, that is, frozen to the substrate, or actively sliding over the bed. In the case of an inactive margin, the frozen basal debris is draped by the diamicts of melt-out englacial debris. These may be resedimented to fill basins elsewhere. These basin fills are eventually exposed as 'hummocks' after melting of the surrounding ice cover. Typical vertical profiles of these hummocks show unstratified, *in situ* aggregates of meltwater diamicts at the bottom, overlain by graded and stratified diamicts which are the products of re-sedimentation.

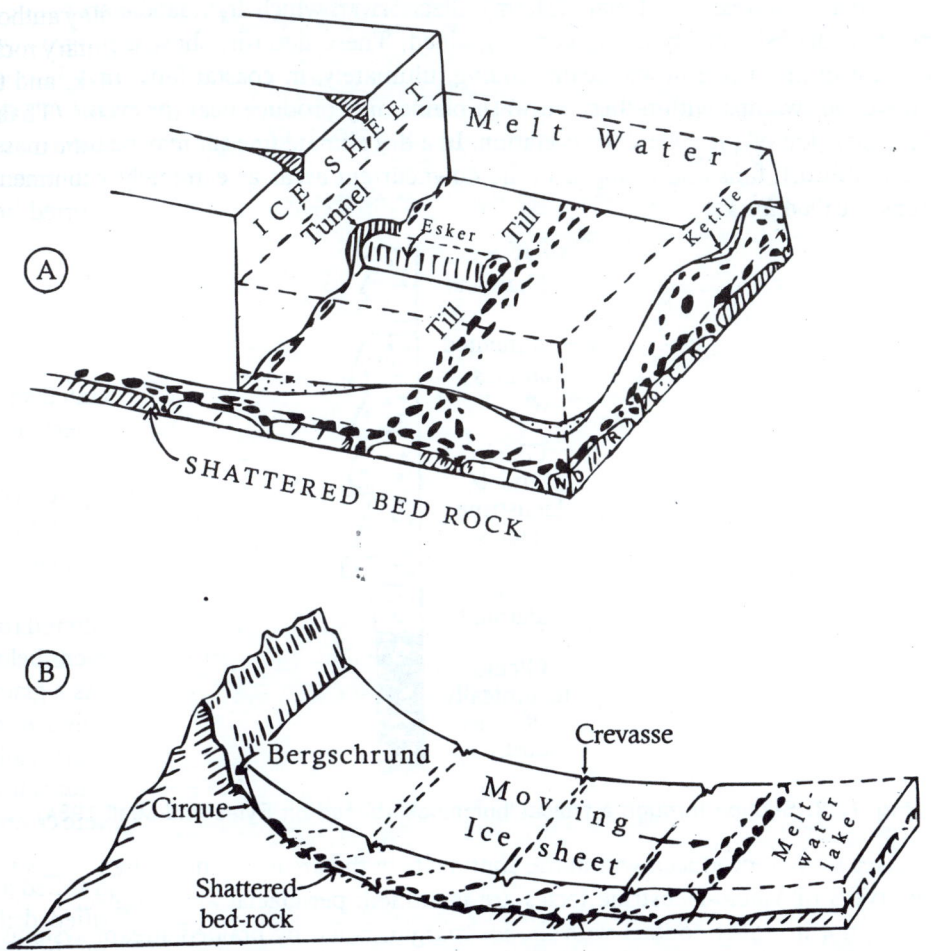

Fig. 7.12: Schematic representation of the deposits associated with (A) land ice sheet and (B) valley glacier.

When the glacier continues to slide over the bed, being continuously lubricated by the basal meltwater, faceted clasts result from intense abrasion (see Figs. 3.1 and 3.2). These clasts, with a strong preferential alignment to the direction of ice movement, are incorporated into the basal diamicts. They may be overlain by channel fills produced from subglacial stream drainage. *Eskers*, the deposits of the subglacial streams, may be of various types and forms, depending on the site of deposition and the nature of the conduit through which the glacial stream flows out. In channels bordered by high walls, antidune structures may develop due to high-flow velocities, whereas within tunnels, parallel or cross-bedded structures dominate. These structures are preserved in vertically stacked cycles. Deltaic sedimentation at the mouth of the tunnel produces subaqueous fans. The proximal parts of these delta are marked by cobbles and pebbles. The grain-size decreases downstream (Banerjee and McDonald 1975).

A glacio-fluvial facies is produced from the deposits of the meltwater draining the outwash plains. These deposits, which may extend to hundreds of kilometres, are called *sandar* (plural for *sandur*, an Icelandic term). Glacial rivers which drain sandar are braided and their deposits are typically coarse grained. These deposits show a progressive decrease in grain-size downstream, ending, ultimately, in coastal fans. In a humid climate the swamps within these outwash plains may produce peat (or even coal) by decomposition of the subarctic vegetation. In a dry climate fine silt may be blown out of the outwash fans and dropped in the downcurrent areas as extremely will-sorted deposits called *loess*.

Diamicts

Resedimented
diamicts

Glacio
fluvial/
lacustrine
facies

In situ
diamicts,

Glacio-
tectonically
deformed
substrates

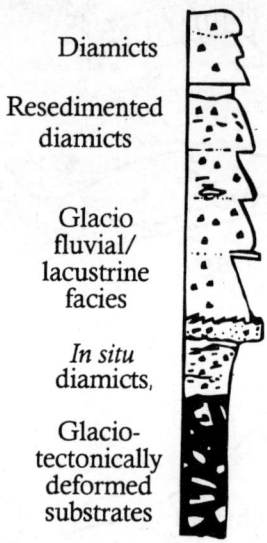

Fig. 7.13: Section through a typical 'hummock' (based on Eyles and Miall 1984).

Lakes may be produced within the continents in the areas evacuated by ice sheets. Two types of glacio-lacustrine facies are identified: periglacial and proglacial. The former, not in direct contact with the ice margin, is fed by braided stream systems. *Varves* with a sandy or silty lithofacies deposited during the meltwater (summer) season, followed by 'winter clays' are common in periglacial lakes. Varves, traditionally believed to be seasonal deposits ('annual couplets'), have also been interpreted as graded units of a Bouma sequence deposited by turbidity currents within periglacial lakes (Lambert and Hsü 1979). The silt unit ('summer layer') of a varve may be graded by density underflows with pelagic materials incorporated from suspension (Banerjee 1973). A diamict lithofacies forms a minor constituent of the periglacial sequence.

Periglacial lakes, which are in direct contact with the ice sheets, are fed by meltwater conduits, subaqueous fans and by ice rafting. The accumulation of suspended sediments and stratified rafted debris (diamictites in these lakes) are overlain by cross-bedded, deltaic silty sand. Dropstones are common.

The Talchir (Lower Permian) deposits in India may be cited as examples of continental glacial facies. The sequence at Dhakoi, for example, starts with a conglomeratic facies at the base and is followed upward successively by massive sandstone and a fine sand to shale facies (Fig. 7.14). The conglomerates were deposited as longitudinal bars within proglacial outwash plains during retreat of the Talchir ice lobe. Blocks of diamictites enclosed within these conglomerates represent ice-rafted morainic materials. The overlying sandy facies, which is often cross-bedded, represents channel sands and migrating bars of braided streams. The finer clastics were deposited in protected or abandoned parts of glacial outwash plains. A fining-upward cycle was produced by shifting of the braided channel bars within the glacial outwash plains (Casshyap and Tewari 1982).

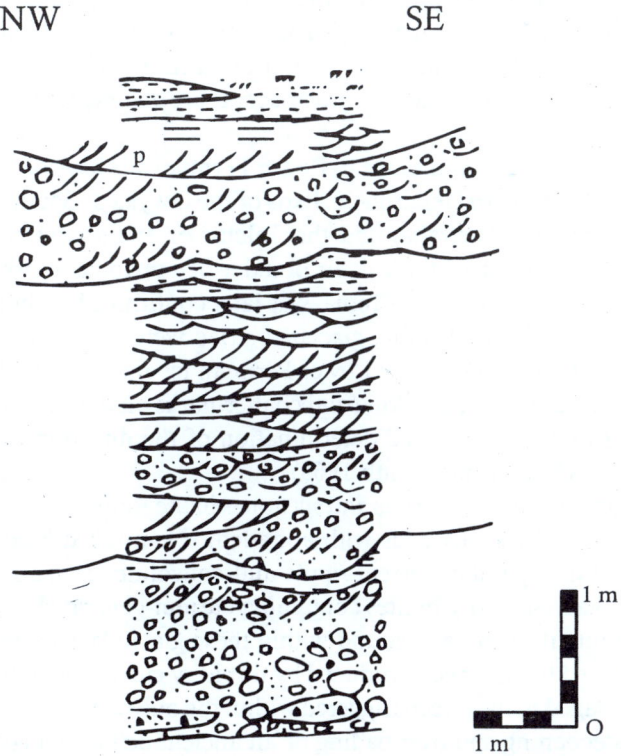

Fig. 7.14: A typical Talchir sequence exposed at Dhakoi, India (after Casshyap and Tewari 1982). With permission of SEPM (Society for Sedimentary Geology).

Marine Glacial Facies

Two different types of glacio-marine sediments, proximal and distal, are recognized. In the proximal areas dense, sediment-laden meltwater produces bottom currents and submarine fans. Diamicts with ice-rafted clasts are common deposits at the fan apex. Downstream, the deposits are much finer, cross-bedded sand forming the mian fan body with mud deposited from suspension. Pebbly diamicts may be incorporated into this deposit by gravity flows. In the steeper channel banks horizontally stratified and inversely graded gravels are deposited by gravity flows.

A generalized model for marine glacial facies was proposed by Edwards (1986), based on Pleistocene examples. Four depositional areas within marine glacial facies were recognized : isostatic, submarine retreat, marginal and proglacial. These deposits, Edward assumed, are controlled by the relative effects of rise and fall in the sea-level due to eustatic and isostatic causes.

Recognition of Ancient Glacial Environment

Geologically young glaciated terrains may be recognized by the presence of such geomorphic features as *roches moutonnées*, eskers and drumlins. But these surfacial features are rarely preserved in older geological records. An association of striated pavements, chattermarks, varves and diamictites with typically pentagonal, flat-iron-shaped, faceted pebbles indicates glacial activity.

Glacially transported quartz grains are characterized by conchoidal breakage patterns, semi-parallel steps, arc-shaped microchattermarks, parallel striations, imbricated breakage blocks and irregular indentations. These surface textures, identifiable under transmission electron microscope, are diagnostic of glacial transport (Krinsley and Donahue 1968).

Tillite beds of Upper Carboniferous to Lower Permian age have been reported from all continents of the Southern Hemisphere. Parts of the Dwyka of South Africa, Tubarao of Brazil, Kamilario of Australia, and the Talchir of India may be mentioned as examples. The glacial origin of the Talchir Boulder Bed of India, proposed by pioneer workers of the Geological Survey of India, was later confirmed by a host of geologists. At places such as Irai, central India, glaciated striae were incised on the underlying Precambrian limestone pavement by Lower Permian ice movement, allowing faithful preservation. The direction of plucking and chattermarks on the underlying limestone, elongate troughs and striae enabled determination of the direction of ice movement (towards NE and NNE) at Irai (Smith 1963).

The Talchir glacio-lacustrine shales and siltstones in the Raniganj Coalfield in eastern India are punctuated by sandy beds with a graded, sole-marked base and laminated top. Slump structures, parallel current-ripple and convolute laminations, mark these laminated beds. Varve-like rhythmites containing current-generated structures such as current-ripple lamination are also found within these glacio-lacustrine sediments. All these are interpreted to have been produced by turbidity current activity. These density currents are envisaged to have been triggered within the outwash glacial lakes by slumps caused by oversteepening and overloading of an ancient delta front (Banerjee 1966).

Eolian Environment

Eolian activity is strong in an arid climate - in sand deserts located in rain-shadow zones within continents, and in glacial outwash plains. The climatic conditions in these areas do not permit growth of vegetation and wind blows unabated for long distances. Eolian activity in coastal dunes is independent of climate. Vegetation plays an important role in stabilizing these dunes. Disintegration and exfoliation of rocks in a desert environment is caused mainly by exposure to sunrays. Sediment transportation is essentially by suspension and saltation. The impact of the saltating grains may cause the finer particles to creep over the surface without losing contact with the ground (*surface creep*). The impact of the sand grains in motion (*deflation*) is another phenomenon responsible for erosion in deserts.

Bedforms of various dimensions are produced by the action of wind: ripples (wavelength $\lambda = 0.01–10$ m), dunes ($\lambda = 10–500$ m) and *draas* ($\lambda = 0.5–5.0$ km). Each of these may be further subdivided into transverse and longitudinal elements leading to a total of eight subgroups, all apparently independent (Wilson 1972). A more comprehensive scheme of classification of eolian bedforms is given in Table 7.1 (after McKee 1979). Wilson's draas are recognized in this scheme as the result of combination of two or more bedforms of the same type. Sand seas of still larger dimensions (~10,000 km^2) are named *ergs*. These are produced by redistribution of sand initially deposited elsewhere.

Table 7.1: Morphology and classification of eolian bedforms (after McKee 1979)

Morphology	Name	Association
Sheet–like	Sheet sands	
Thin elongate strips	Streaks	Compond—two or more of the same type combined by overlap or superimposition
Circular to elliptical mound, dome–shaped	Dome	(Wilson's draa).
Crescent in plan	Barchan	
Connected crescents	Barchanoid	
		Complex–two different basic types occurring together, either superimposed (Wilson's draa),
Asymmetrical ridge	Transverse (reversing)	or adjacent to each other
Symmetrical ridge	Linear (seif)	
Central peak with arms	Star (pyramidal)	
U-shaped	Parabolic	

The external morphology and internal structures of the more important eolian bedforms are shown in Fig. 7.15 (A and B). With increasing complexity of wind regimes the dune form changes from simple transverse to complex star dunes and draas (see Fig. 7.16). In the sand roses shown in this diagram the arm length of a rose is proportional to the amount of sand that can be moved by the wind from a given direction towards the centre of the rose. (Stereonet plots of wind movement prepared from the sand roses will give a diametrically opposite picture). With a decrease in Resultant Drift Potential (RDP) and the RDP/DP ratio, the pattern of sand accumulation changes from lateral migration to vertical accretion. The Drift Potential (DP) is a measure of the total sand-moving capability of the wind expressed in vector units (v.u.) without regard to wind direction at the locality. RDP is a measure of the resultant or net sand-moving capability of wind at a locality in vector units (Fryberger 1979). Thus RDP/DP measures wind directional variability. By plotting RDP/DP against Equivalent Sand Thickness (EST), Wasson and Hyde (1983) could discriminate the four major dune types: linear, barchan, transverse and star (Fig. 7.17).

Eolian Facies Models

Two different models, one applicable to transverse bedform migration and the other to longitudinal bedform migration, have been proposed (Brookfield 1984). In the transverse model, unidirectional wind flows, sand patches and barchan dunes are initiated with the onset of aridity. With increasing aridity, these change over to transverse dunes and compound transverse draas. These forms decrease in size with decreasing aridity and sand supply. In the longitudinal model the initial sand patches and barchans are followed subsequently by linear dunes, linear complex draas and star draas. These longitudinal patterns are rarely found in ancient eolian dunes.

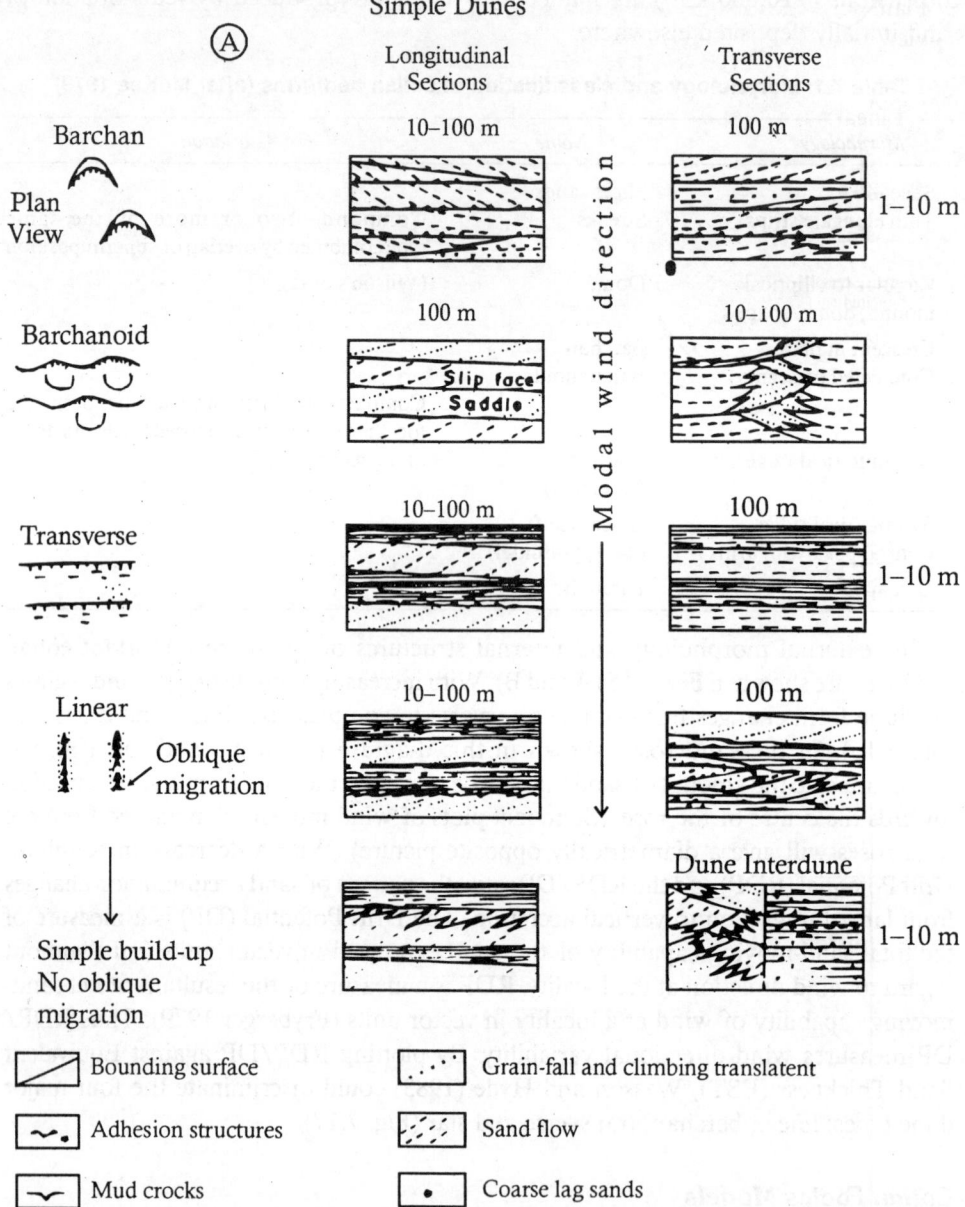

Fig. 7.15: Schematic representation of stratifications found in (A) simple and (B) complex eolian dunes of different types. The longitudinal sections are parallel to the resultant wind direction, the transverse sections are perpendicular to it (redrawn from Brookfield 1984). With permission of Geological Association of Canada.

(B) COMPLEX AND COMPOUND DUNES

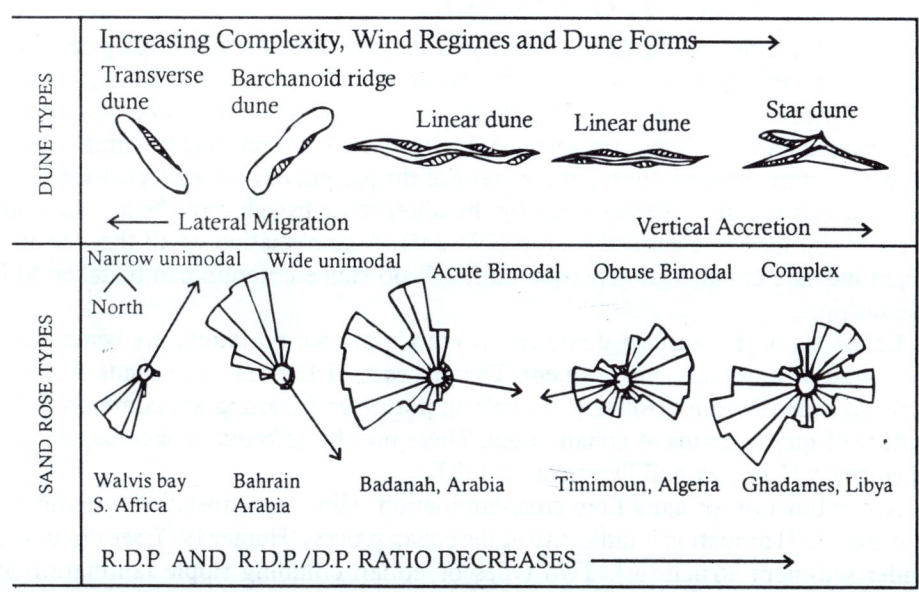

Barchanoid draa

Plan view 10-100 m

Wind modes

100-1000 m

100-1000 m — Interdune Saddle

Slip face Interdune

Linear draa 10-100 m

Migration
Transverse

Longitudinal (|| to migr direction)

100-1000 m

100-1000 m — Interdraa

Draa

Interdraa

STAR DRAA 10-100 m

B

1000-5000 m

1000-5000 m — Draa

Interdraa

Draa

Interdraa

A   B

	Bounding surfaces		Grain-fall and climbing translatent
~	Adhesion structures		Sand flow
⌄	Mud cracks	.	Coarse lag sands

Fig. 7.15: (B)

Increasing Complexity, Wind Regimes and Dune Forms ⟶

DUNE TYPES

Transverse dune   Barchanoid ridge dune   Linear dune   Linear dune   Star dune

⟵ Lateral Migration   Vertical Accretion ⟶

SAND ROSE TYPES

Narrow unimodal   Wide unimodal   Acute Bimodal   Obtuse Bimodal   Complex

North

Walvis bay S. Africa   Bahrain Arabia   Badanah, Arabia   Timimoun, Algeria   Ghadames, Libya

R.D.P AND R.D.P./D.P. RATIO DECREASES ⟶

Fig. 7.16: Relationship between eolian dune forms and wind regimes, DP = Drift Potential, RDP = Resultant Drift Potential, RDP/DP is a measure of lateral migration. The proportion of vertical accretion increases with decrease of RDP and RDP/DP (after Ahlbrandt and Fryberger 1982). With permission of Amcrican Association of Petroleum Geologists.

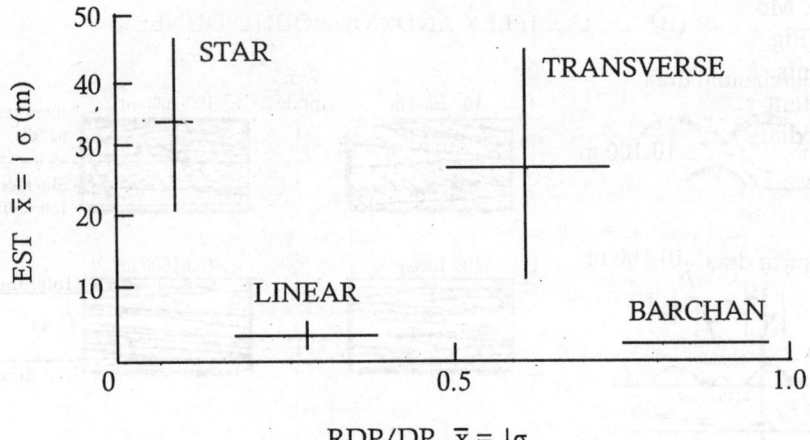

Fig. 7.17: Plots of equivalent sand thickness (EST) vs. wind direction variability (RDP/DP) as a tool for discriminating different types of eolian dunes (after Wasson and Hyde 1983).

Vertical sections of eolian deposits worked out in the Permian Thornhill and Dumfries basins of Scotland show barchan dunes marking the onset of a desert. These are overlain by large barchanoid draas corresponding to the main eolian phase when arid climate predominated. The overlying small dunes mark the dwindling eolian phase, which were superimposed by alluvial fans during semi-arid climate (Brookfield 1984).

Recognition of Ancient Eolian Deposits

Forms and magnitudes of dunes of eolian and subaqueous origin overlap to a large extent. These cannot be used as criteria for recognition of ancient eolian dunes, therefore. The presence of lag gravels on avalanche toes, convex forests, arcing either up or down, deformation and reactivation surfaces, and progressive flattening and thinning of cross-bedding surfaces towards the top mark many of the present-day eolian deposits. Modern eolian dunes are also characterized by the absence of trough cross-beds. The upper parts of slip faces are generally steep (29°–34°). A combination of all these features might indicate eolian sedimentation although no single criterion can be taken to be diagnostic.

Eolian sands, because of their selective mode of transportation, are better sorted than sand of any other environment. The presence of fines in these sands skews the grain-size distribution positively. Rounding, pitting and frosting are common on the surface of quartz grains of eolian origin. These may be detected by scanning electron microscopy (Krinsley and Doornkamp 1973).

A combination of sand-flow cross-lamination, climbing translation stratification and grain-fall lamination is indicative of the eolian process (Hunter 1977; see discussions under sediment structures). Two types of eolian climbing ripple lamination are recognized; translatent laminae and ripple-form laminae. In the former only the bounding surfaces between the ripples can be seen but in the latter the forests are also visible. Grain-fall lamination is produced by deposition from suspension in the zone of flow separation. Sand-flow lamination is caused by downslope avalanching on the lee face. Most of these laminations can be identified on the eroded surface of barchan dunes (Fig. 7.18). The degree of porosity may also help in the identification of eolian sediments. Sand-flow cross-strata are loosely packed and highly porous. Climbing translatent strata are tightly packed and less porous. The grain-fall laminae are intermediate between the two in porosity and grain packing (Hunter 1981).

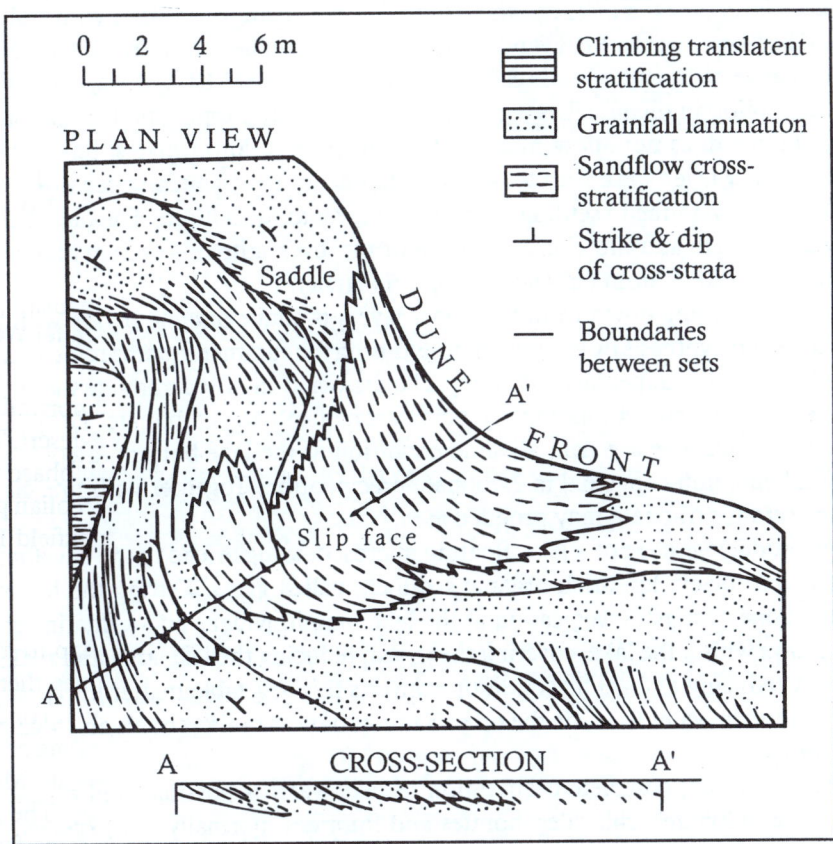

Fig. 7.18: Plan and sectional views of a barchanoid dune showing different types of stratification and cross-stratification (modified after Hunter 1977). With permission of International Association of Sedimentologists.

The Permian Toroweap Sandstone of Arizona, the Mesozoic Navajo Sandstone of Utah and Arizona, the Aztec Sandstone of Nevada, and the Lower Bunter Sandstone of England are some of the well-studied eolian deposits. The Late Proterozoic Venkatpur Sandstone of southern India is interpreted to be of eolian origin on the basis of (*i*) grain-flow strata, (*ii*) inverse graded translatent strata, (*iii*) adhesion laminae and (*iv*) grain-fall strata. Nine facies, representing both eolian and related aqueous environments, have been recognized within what is believed to be a well-developed erg. Cosets of large-scale cross-strata are present within this sand sea which terminates at an alternate sequence of eolian and marine sediments. The horizontally stratified units within the Venkatpur sandstone have been variously interpreted to represent interdune, inland sabkha, sand sheet and coastal sand flat deposits (Chakraborty 1991).

Lacustrine Environment

Lakes occupy natural depressions surrounded on all sides by barriers or rims. These rims may be of tectonic or volcanic origin. Damming of river valleys, solution, glacial excavation and meteoric impact may also generate lakes.

Organic activity is an important part of lake sedimentation. The upper, well-oxygenated layer of water in lakes, where sunlight can penetrate, is replete with biological activity. The lower layers are cold and depleted of oxygen. The zone of rapid temperature variation between the two layers is called the *thermocline* (Fig. 7.19 A). A salinity layering may also develop in inland lakes connected to the sea. Ice capping on the lake surface in cold climate does not allow mixing of the upper and lower layers, but in spring, when the upper layer warms up to 4° C (the temperature at which water is heaviest), the lake 'turns over' due to settling of the heavy layer to the bottom. The phenomenon is repeated again in autumn. During periods of turnover, nutrients carried to the upper layer from the lake bottom enhance biological activity.

Organisms settling down from the upper layers produce organic debris within the lake mud. Photosynthesis by phytoplanktons during the summer months removes much of the CO_2 from the upper layer of water. The dissociation of bicarbonate that follows increases alkalinity, thereby aiding in precipitation of calcium carbonate. The presence of carbonate indicates a warm, alkaline condition at the lake-bottom, a cold, acidic condition being non-congenial to carbonate precipitation. Together with fine clastics, the lake-bottom carbonate may produce *marl*.

In the productive stage of a lake, floating microscopic algae lead to the development of a brown to black, lipid-rich organic sediment called *gyttja*. During the low-water stage, the zone of aquatic vegetation at the lake margin widens and a floating mat of sedges grows within the lake, covering the gyttja, reducing the lake area and making it smaller. A reducing and acidic condition converts this vegetable matter into *peat* (Fig. 7.19 C). A rise in water level, on the other hand, produces oolitic shoals and algal mats (Fig. 7.19 B).

Unlike sea-water, complete evaporation of lake-water is possible, and a variety of salts such as sulphates, chlorides, borates and fluorides, normally not expected from evaporation of sea-water, precipitate within the lake sediments. Authigenic silicates such as feldspars and zeolites may also develop. Reduction of iron at the lake-bottom may cause precipitation of pyrite (FeS_2). Reducing and restricted basin conditions are needed for such precipitation.

Reducing (Euxinic) Basin

Hydrologically closed lakes, as also all other basins cut off from open sea by barriers or sills, enjoy reducing environment. The term euxinic, derived from the Latin name of the Black Sea (*Pontus Euxinus*), implies 'cut-off or silled' condition. Being cut off from the Mediterranean by a barrier the Black Sea enjoys an euxinic condition (Fig. 7.20, X and Y). Other well-known areas enjoying euxinic conditions today include the Drams Fjord of Norway, Saanich Inlet of British Columbia and the Baltic Sea.

To be euxinic a lake should be located in a climatic zone wherein seasonal 'turnover' of upwelling is not possible. The Black Sea is euxinic because it is located in a humid

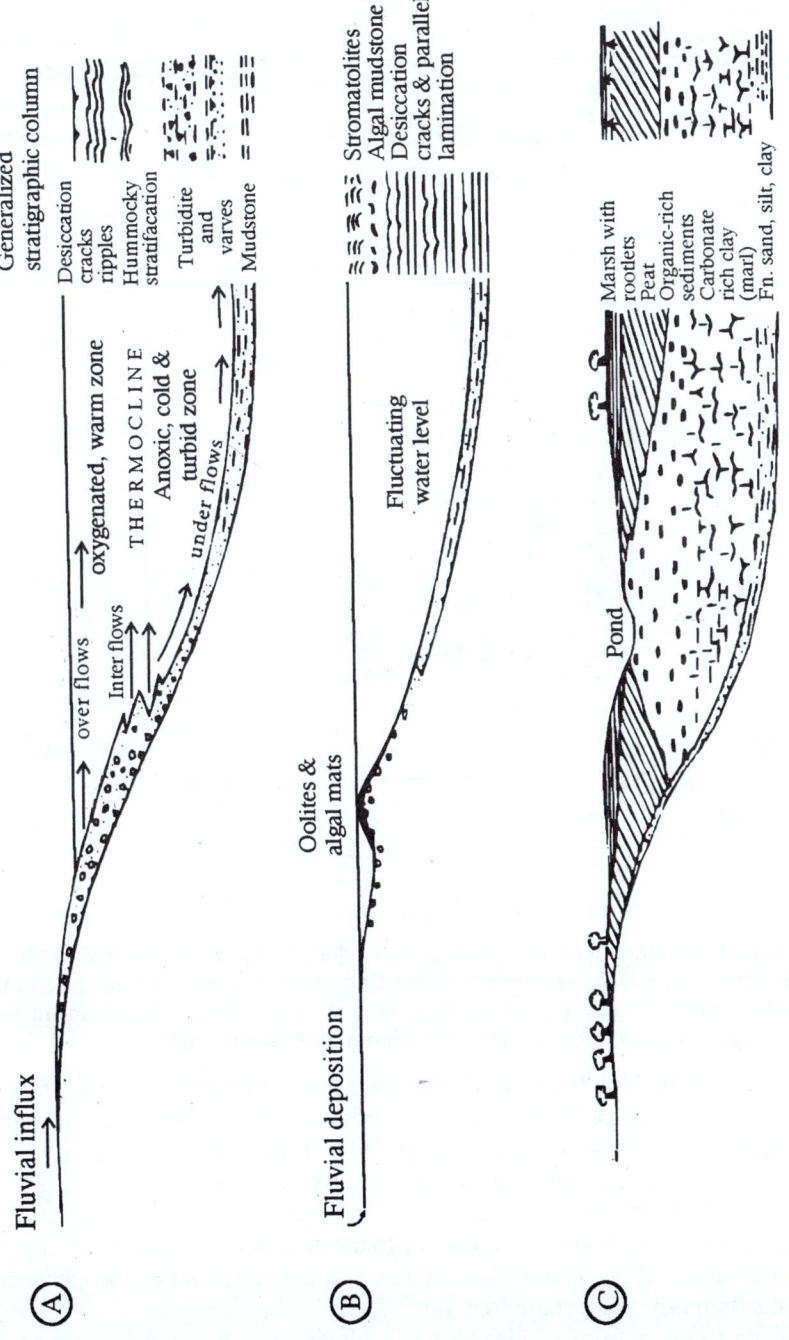

Fig. 7.19: Schematic representation of lacustrine sedimentation : (A) with high sediment input during low water level, (B) during high water level and (C) during low sediment input (after Galloway and Hobday 1983; Fouch and Dean 1982).

area. The Mediterranean and the Red Sea are not euxinic because they are located in arid regions where evaporation exceeds inflow of water and, consequently, the bottom waters are well oxygenated (Demaison and Moore 1980).

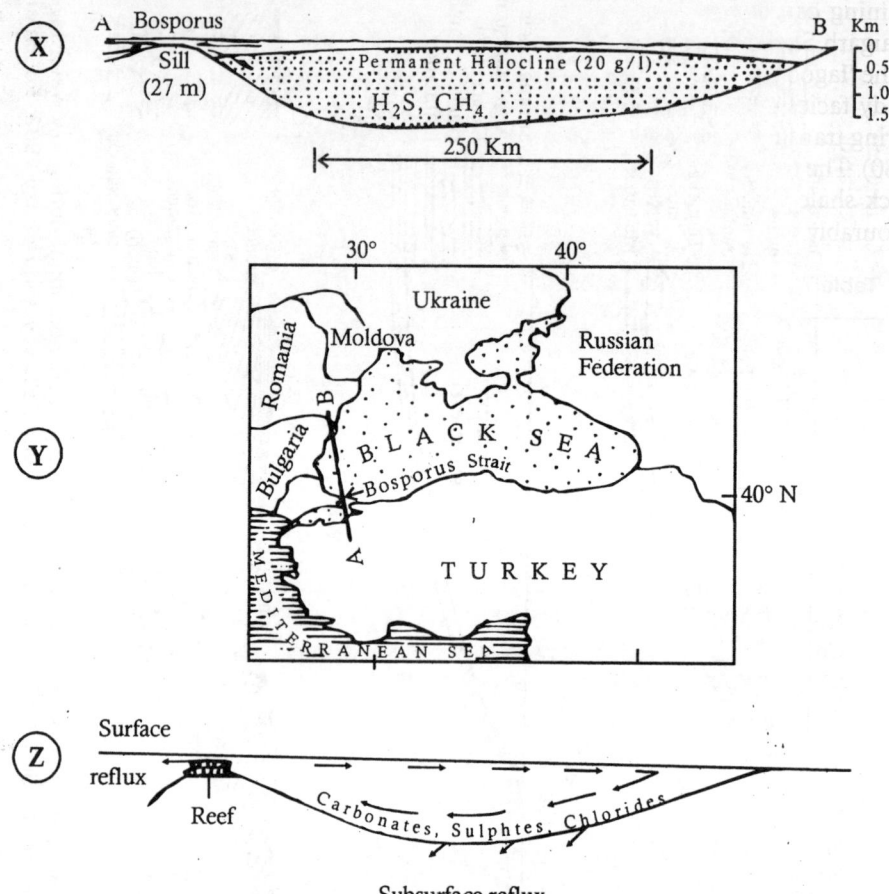

Fig. 7.20: (X) A-B section across the Black Sea (Y) sketch map of the Black Sea showing a modern euxinic environment (after Demaison and Moore 1980); (Z) Model cross–section showing locales of deposition of carbonates, sulphates and chlorides in a typical euxinic basin (after Krumbein and Sloss 1963).

The conditions prevailing below an active wave base in restricted basins are acidic as well as reducing. Temperature increases towards the lake-bottom. Organic matters, settling down to anoxic bottom conditions produce a black ooze (*sapropel*) which is rich in ferrous sulphide. The anaerobic organisms thriving in lake-bottoms absorb trace elements such as molybdenum, copper, vanadium and uranium. The colour of the shale deposited at the lake-bottom is black due to the presence of dispersed manganese granules (Goldhaber 1978). These black shales contain much larger proportions of trace elements than any other shale (see Table 7.2).

Black shales are known from a wide range of geological horizons but they are perhaps most common in the Palaeozoic and the older rocks. The world's well-known euxinic deposits include the Precambrian pyritiferous and carbonaceous shales of the Iron

Belt, Michigan, the Kupferschiefer of the West European Permian, the Permian Black Cherts and the Phosphoria Formation of the Rocky Mountains. In the eastern part of the Precambrian Vindhyan Basin in India, an isopach map shows a stratigraphic basining caused by a local sagging of the basement. Pyritiferous black shales (the Bijaigarh Shales) found within this depression are interpreted to have been deposited in the 'lagoonal pond' which enjoyed euxinic conditions during the Precambrian. The sandy facies overlying and underlying the shale were deposited at the lagoon fringe during transgressive phases of the Vindhyan sea (Banerjee and Sengupta 1963; Singh 1980). The trace element proportion of the Bijaigarh shale is less than that of an average black shale but the concentrations of Ni, Co, Cu and Mo in these shales compare favourably with those of oxygen-free sediments (see Table 7.2).

Table 7.2: Trace elements in shales and black shales (average estimates in ppm)

	Shale*	Black Shale**	Bijaigarh*** Shale	Bijaigarh*** Sandstone
Ba	800	300	85	79
Sr	299	200	26	39
Sc	10	10	4	15
W	–	–	6	2
Li	–	–	3	–
Cr	160	100	–	–
V	130	150	12	77
Ni	21	50	7	1
Co	12	10	12	9
Cu	38	70	62	29
Y	33	30	31	–
Mo	0.74	10	16	9
Tn	–	–	34	22
Ga	40	20	17	24
Pb	20	20	7	17
Sn	–	–	17	–
Zr	200	70	47	–

* Analysis of average shale (Green 1959, cited by Vine and Tourtelot 1970).
** Analysis of average black shale (Vine and Tourtelot 1970).
*** Analysis of average black shale and sandstone, 'Bijaigarh Sub-Stage', Vindhyan 'System', (Precambrian) (Sinha and Singh 1964).

Recognition of Lacustrine Facies

Morphological and sedimentological characters of large lakes and shallow ocean basins overlap in many respects. Rivers debouching into large lakes produce fan-shaped deltas similar to those bordering seas. In cross-section these show simple, steep forests overlain by flat topsets (typical Gilbert-type delta). The finer sediments deposited at the lake centre below an active wave base are generally thin bedded and parallel laminated. Varves may develop as seasonal deposits within lakes located in a temperate climate. Storm surges produce hummocky cross-stratifications that are smaller in dimensions than those formed under marine conditions.

There is no single criterion which is distinctive of a lacustrine deposit. Identification of an ancient lake is possible only by taking several of the following criteria together

(Pickard and High 1972). Lakes have overall circular or elliptical shapes and are surrounded by rims or barriers ('positive structural elements'). They are of limited size compared to shallow seas. In plan lake deposits show concentric patterns with gradual increase of sediment thickness towards the centre. Narrow shore and nearshore deposits surround these sediments. Mean grain-size of the sediments decreases while the proportion of organic content increases towards the lake centre. Non-marine invertebrates (bivalves and gastropods), plants, algae and stromatolites are common.

Other significant characters of lake deposits are the presence of chemical and biochemical precipitates in a terrestrial setting such as evaporites, chert-banded sedimentary iron ores, freshwater limestones and marls. Finely laminated, graded turbidites at the centre; large-scale delta forests, flaser beddings in the fringe areas and occasional hummocky cross-stratification within the deposits may also indicate a lacustrine environment. As the coarse, gravelly and sandy sediments of the lake fringe approach the finer sediments at the lake-bottom during progradation, a coarsening-upward sequence is produced. Such a sequence in a non-marine environment, together with some of the other features described above, point to a lacustrine environment.

Fluvial Environment

Rivers originate in elevated areas and flow through flat plains before debauching into the sea or large lakes. The pattern of a typical drainage network is schematically represented in Fig. 7.21. The bigger loop in this diagram is called the contributing net. Many tributaries, big and small, contribute to the water discharge of the principal river within this net. The smaller loop is the distributing net. Distributaries flowing out of the principal stream within this loop carry water into the ocean or large lake. While doing so, they often cut through the silt laid down by the stream leaving triangular or delta shaped bodies.

On leaving the mountain gorge a river deposits most of its course sediment load near the mountain front which is often faulted. Sparse vegetation and copious rainfall cause quick denudation and redeposition of the sediments in the form of *alluvial fans*. Humid, tropical climate is not congenial to the development of alluvial fans. Such a climate causes chemical rather than mechanical weathering and also produces dense vegetation, which protects the slopes. In glaciated valleys paraglacial fans develop where tributaries join the main glacier. Two types of alluvial fans are recognized: dry or mudflow fan and wet fan. The former results from ephemeral streams and the latter are produced by perennial flows (Schumm 1977).

On entering the plains a river can take four different patterns. In plan view these are: straight, braided, anastomosing and meandering. Of these, the braided and the meandering types are more common. Braiding is facilitated by rapid fluctuation in water discharge, high gradient, a large proportion of coarse sediment load and readily erodable banks (Rust 1978). In a braided stream mid channel bars develop due to deposition of coarse-grained sediments which the stream is unable to transport. These deposits further decrease the competence of the stream by increasing roughness of the channel. The enhanced rate of sediment accumulation, which follows, leads to the development of alluvial islands. The channel, constantly subdividing and rejoining, in an attempt to bypass these islands, produce an inter-twinning pattern resembling a braid (Fig. 7.21 B). Attempts have been made from time to time to define braiding patterns quantitatively. In one of the more recent studies a braid channel is defined as the sum of mid-channel lengths of all channels in a reach divided by the mid-line length of the widest channel (Friend and Sinha 1993).

The transverse cross-section of a stream channel is dependent on its sediment load and water discharge. The cross-section, which is parabolic near the head of a stream, changes to a rectangular shape downstream (Fig. 7.21 C). This change is effected by an increase in shear stress downstream due to an increase in water discharge. The banks, unable to adjust themselves to the increasing shear stress, cave in, thereby modifying the channel cross-section (Leopold, Wolman, and Miller 1964).

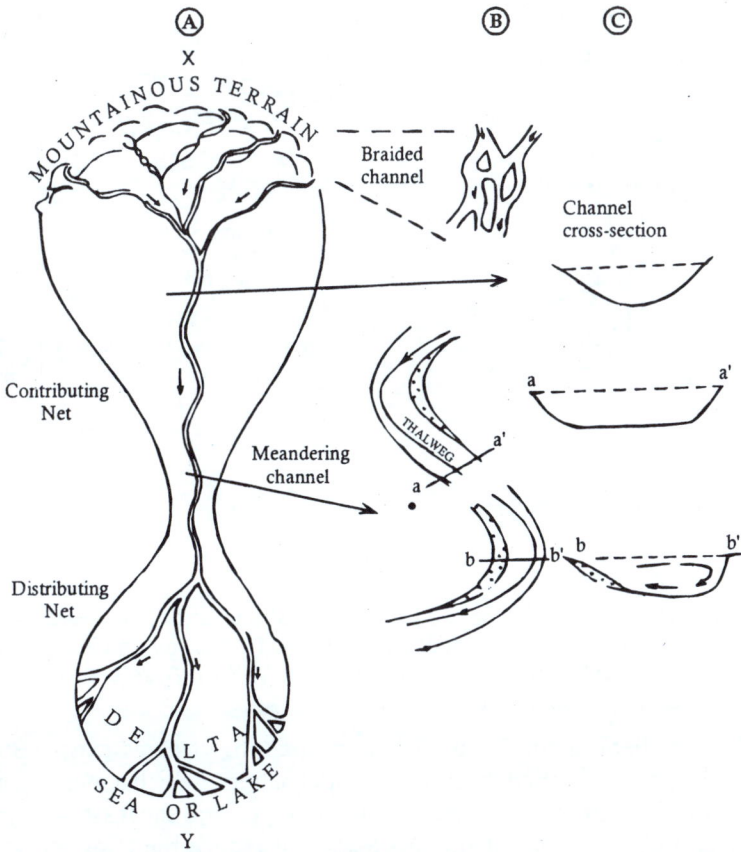

Fig. 7.21: Schematic representation of fluvial patterns and cross-sections of fluvial channels (based on Allen 1965a; Leopold, Wolman and Miller 1964).

As a stream adjusts its course with respect to its water discharge and sediment load, its longitudinal profile changes gradually until it assumes a smooth (graded), slightly concave-up pattern (Fig. 7.22 A). During the process of adjustment an overloaded stream deposits a large part of its load and also straightens its course by downcutting. The limit of downcutting, called the *base level of erosion*, is the sea level. When a stream reaches this base level and cannot cut further, it meanders in an attempt to decrease its gradient by increasing its effective length.

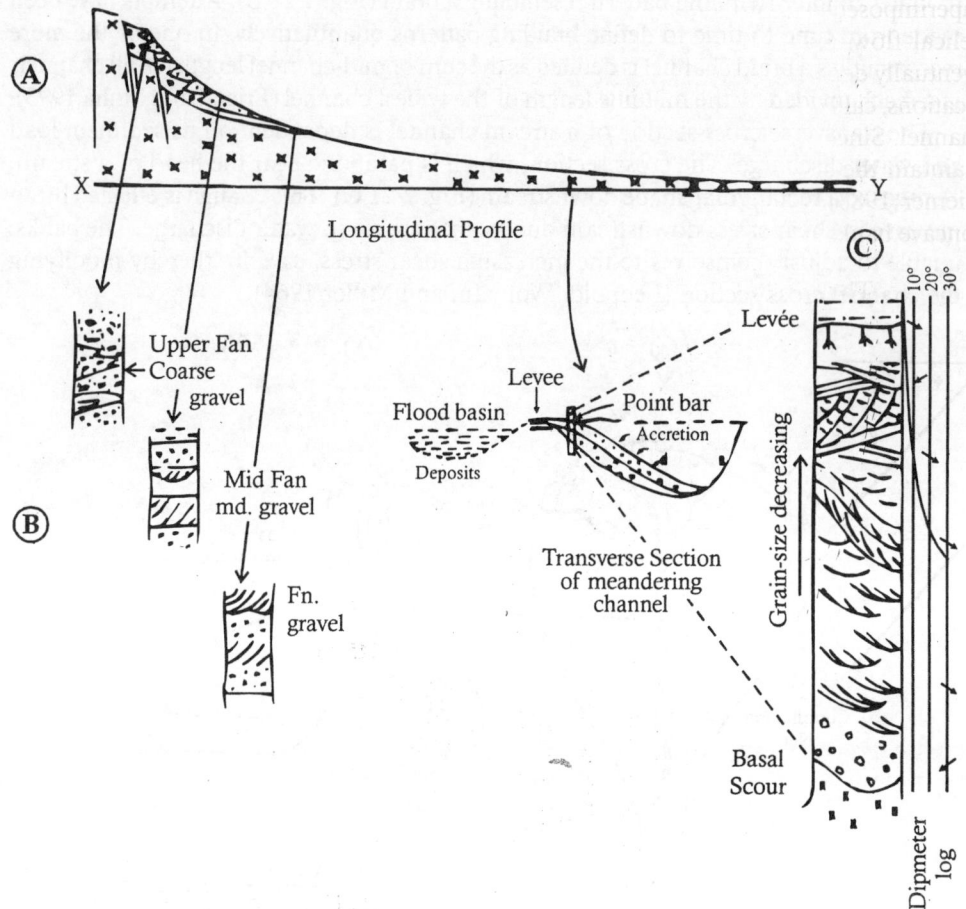

Fig. 7.22: Typical longitudinal profile of a meandering stream showing the vertical sequence
of sediments at different locations (based on Galloway and Hobday 1983) with
permission of Springer Verlag, Heidelberg, X-Y of (A) refers to Fig. 7.21. Note
'red pattern' in dipmeter log in C.

Several other theories for the origin of meandering streams exist. Leopold and
Langbein (1966) demonstrated that the sinusoidal pattern, which a meandering stream
describes, is the form in which a river is required to perform the least work in turning.
It is the most probable result of the processes that reduce the total energy loss to a
minimum. Non-erodable clay banks and bed loads of fine-grained material facilitate
meandering. Hence meandering rivers have gentler gradients than braided streams.
They are also characterized by finer sediment loads and more steady water discharge,
but exceptions are not unknown.

Within a meandering channel the upper layer of water moves faster than the lower
layer, which is in constant friction with the bed. The fast moving upper layer of water,
impinging on the concave bank due to centrifugal action, starts eroding it. As the surface
water plunges downward on the concave bank, some water from the lower level emerges
at the surface on the opposite bank to maintain continuity. This circulatory motion,

superimposed on the downstream motion of water in the stream channel, generates a helical flow. Sediments caught up in this flow are transported downstream to be eventually deposited by lateral accretion at the inner convex bends of the stream. These locations, called *point bars*, are the main areas of deposition within a meandering stream channel. Since the point bars are deposited on alternate banks of the stream they help maintain the sinusoidal pattern characteristic of a meandering channel (Bridge and Diemer 1983). The deepest channel (called *thalweg*) also almost touches the inner concave banks on alternate sides (Fig. 7.23).

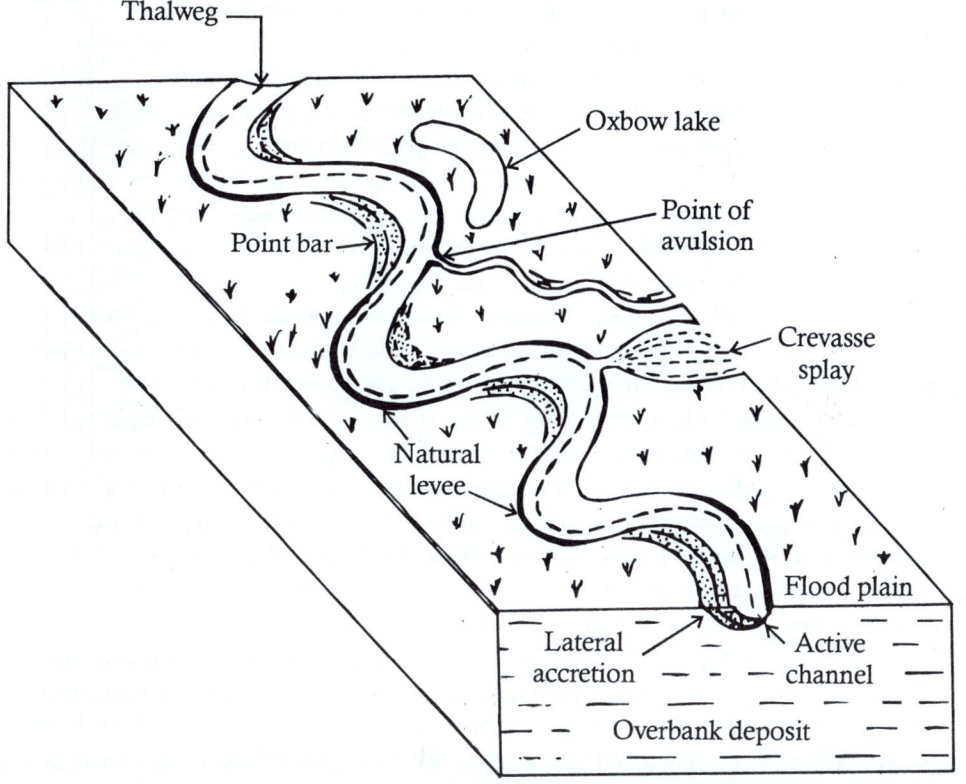

Fig. 7.23: Three dimensional view of a meandering river channel showing the morphological units.

Large transverse ridges (curvilinear ridge-and-swale), perpendicular to the stream flow direction, were mapped by Sundborg (1956, p. 207, 271) in the Klarälven River of Sweden. These bars, which resemble the geometry of transverse dunes, and have smaller ripples superimposed on them, might be considered as incipient point bars. Detailed studies on the Usri River point bars in India showed aggregates of strips of sediments, which have been added laterally. In transverse profiles these show up as series of steps. At the same time the longitudinal profiles of these point bars resemble large, asymmetrical dunes with avalanche faces pointing downstream (Purkait, 2000). As in the dune like bedforms generated in laboratory channels, these show accumulation of coarser grains at the toes suggesting that the sorting processes within large point bars are similar to those of large dunes (Sengupta *et al.* 1999).

The sediments at the bed of most streams progressively fine downstream and also develop lognormality. This means that the size distributions depict normality when plotted on logarithmic (phi) scale (Sengupta *et al.* 1991). Grain-size distributions within the point bars of the Usri River were found to conform to the log-normal law. After a certain distance of transportation the distribution becomes log-hyperbolic. Within each point bar there is a general tendency for the grain-size to decrease downstream. However, this pattern is not consistent along the entire course of the river (Purkait 2002). Controlled experiments in laboratory channels show that the phenomenon of grain sorting is related to the process of bedform migration. The coarser grains, entrapped at the toes of bedforms, are gradually removed from the flow, while the relatively finer ones are carried forward. Concentration of grains belonging to narrow size range in the stream leads to lognormal grain-size distribution (Sengupta *et al.* 1999).

Fining-upward sequences develop within the point bars because the coarse-grained materials, which are the first to be deposited, are followed upward by the finer sediments (Fig. 7.22C). Lateral accretion may produce the '*ridge-and-swale*' topography at the surface (Fig. 7.23). Within the channel the depth varies along its length producing a series of alternating shallows ('*riffles*') and deeps ('*pools*'). The pools are located nearer the concave erosional banks. The riffles are situated in between the consecutive pools. These features, generated by autogenic erosional-depositional processes are sometimes correlated with the mechanism of evolution of meanders (Clifford 1993).

A channel is said to be meandering when its sinuosity exceeds the arbitrarily set critical value of 1.5. Sinuosity in this case is defined as curvilinear channel length divided by linear valley length of the stream channel. Many natural streams describe very irregular sinuous patterns. Simulation studies indicate that sinuosity of more regular patterns should be around 3.14 (Stolum 1996). LeRoux (1992), and Ghosh (2000) estimated sinuosity to range between 1.36 and 5.24 for ideally disposed symmetrical meander loops. They have also demonstrated that consistency of current (or palaeocurrent) directions in a stream is inversely proportional to the sinuosity value.

From time to time a meandering channel may abandon a portion of its loop by locally increasing its slope. *Chute cut-off* is produced when a new meandering channel is excavated on the bar enclosed by the loop. When the abandoned loop is left as an *oxbow lake*, and the steam continues to flow through the shortest possible channel, it is called a *neck cut-off* (Fig 7.24). The deposits formed within the chute and the neck cut-off are important features of an alluvial flood basin. These prismatic, U-shaped or triangular (in cross-section) deposits are essentially coarse-grained except in the uppermost parts. They reveal a discordant relationship with the older alluvium on the concave side. On the inner, convex margin they grade laterally to the point bar sediments.

Natural *levees* and *crevasse splays* are the other two important features associated with a meandering channel. Levees are natural embankments bordering the concave banks with slopes towards the flood basins. They are constituted of alternating coarse and fine sediments deposited by sediment-laden water overflowing the channel. During high floods the levees are breached at places by crevasse. Coarse sediments, which find easy routes from the main stream channel to the flood basin through the crevasse, constitute the crevasse-splay deposits. These deposits are lobate or tongue-shaped in plan. Finer grained sediments accumulated by the process of aggradation occupy the flood basins (Fig. 7.23).

When a stream looses its capacity to carry its sediment load, it may shift its course to a newer channel, which is more suitable for sediment transportation. The phenomenon, called *avulsion*, is not to be confused with channel shifts associated with chute cut-off, which is on a much smaller scale. The loss of sediment carrying capacity of a stream, which induces it to shift to a newer channel, may be due to one or more of the following reasons. (1) Decrease of the gradient of the stream. (2) Sudden increase in gradient elsewhere within the same alluvial plain. (3) Sudden increases in the sediment load and/or decrease in water discharge. In some cases channel avulsion has been demonstrated to originate as a crevasse. Continued avulsion develops an anastomosing system of channels within an expanded crevasse splay. Out of these a single channel incises the older splay deposits and initiates the growth of a new alluvial ridge (Fig. 7.23).

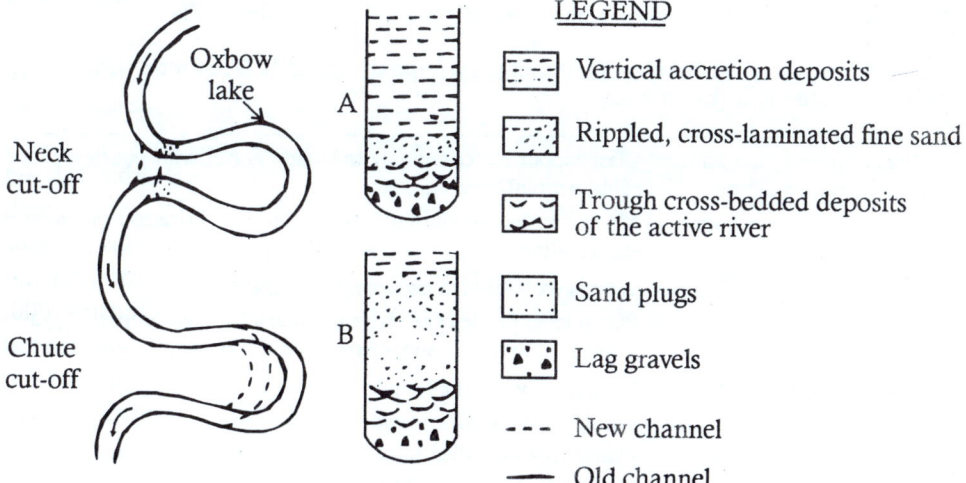

LEGEND

☲ Vertical accretion deposits

▨ Rippled, cross-laminated fine sand

▨ Trough cross-bedded deposits of the active river

▨ Sand plugs

▨ Lag gravels

- - - New channel

—— Old channel

Fig. 7.24: Neck and chute cut-offs in meandering channels. Vertical sections of the deposits produced in neck and chute cut-offs are shown in A and B respectively. In chute cut-off the channel is abandoned gradually, but in neck cut-off the meander loop is suddenly abandoned and sealed off by deposition of (stippled) sand plugs leaving the main channel as an ox-bow lake. Hence in the latter case the cross-laminated sandy deposits are thin but the vertical accretion deposits are thick (based on Walker and Cant 1984). With permission of Geological Association of Canada.

Anastomosing river systems consist of an interconnected network of low-gradient channels of similar order. The individual channels, which may be either straight or sinuous, are deep and narrow with stable banks composed mostly of silt and clay. Large, stable vegetated islands with natural levees and wetlands separate these channels. Unlike the braided system, where the channel is confined to a single pair of floodplain on either side, each island of the anastomosing system may act as a floodplain. Noncohesive fine silt-clay constituting these islands accounts for their stability.

Anastomosing channels are characterized by relatively low discharge and the islands separating them remain exposed subaerially in all seasons (Smith and Smith 1980). The present day braided, meandering and anastomosing river channels can be discriminated in slope vs. discharge diagrams (Leopold and Wolman 1957). For obvious reasons such techniques are not applicable to ancient streams.

Fluvial Facies Models

Two distinct facies models are possible within fluvial systems: a gravelly facies associated with alluvial fans and braided rivers and an essentially sandy facies associated with meandering systems. The characters of these two facies assemblages are summarized below.

Gravelly Alluvial Facies

The facies characterizing fans and braid-plain deposits are presented in Table 7.3. The sediments containing at least 50% gravel have been termed coarse grained (prefixed 'G') in this scheme. Only small quantities of sand are present in this facies.

Table 7.3: Facies typical of fans and braid-plain deposits (Miall 1977; as modified by Miall 1978a and Rust 1978a).

Major Facies	Gm :	Clast-supported, commonly imbricate gravel with poorly defined subhorizontal bedding.
	Gms:	Muddy matrix-supported gravel without imbrication or internal stratification
	Gt :	Trough cross-bedded clast-supported gravel
	Gp :	Planar cross-bedded gravel, transitional from clast-supported gravel through sand matrix-supported gravel to sand (Sp)
Minor Facies	Sh :	Horizontally stratified sand
	St :	Trough cross-stratified sand
	Sp :	Planar cross-stratified sand
	Fm :	Massive fine sandy mud or mud
	Fl :	Laminated or cross-laminated very fine sand, silt or mud
	P :	Pedogenic concretionary carbonate

Alluvial Fan Facies: Debris flows are characteristic of 'dry' alluvial fans. Downstream decrease in grain-size and radiating sediment transportation patterns are common in these flows. An accumulation of a threshold amount of loose muddy detritus is essential for such flows. The flows are initiated when an abundant water supply is available. In areas of steady rainfall the threshold amount of debris cannot accumulate because of the continuous rather than abrupt process of erosion prevailing there (Schumm 1973). Exceptions to this rule, however, are not unknown. Debris flow deposits are common in both semi-arid and paraglacial environments (Rust and Koster 1984).

Downstream decrease in grain-size within a fan is mainly due to a decrease in the competency of the stream. The deposits change gradually from coarse gravel in the proximal part of the fan to sand in the distal part, through clast-supported, fine-grained gravel, and sand-supported gravel in the intermediate areas (Gm $\rightarrow$ GP $\rightarrow$ Sp). A 'sieve texture' may be produced when the gravel lobes have a scanty supply of sand. The deposits are horizontally stratified or imbricate in the proximal parts and horizontally laminated in the distal sandy or muddy areas. When adequate proportions of mud are available in the distal areas, the clasts, being supported by a muddy matrix, may be oriented subhorizontally or even vertically (Rust and Koster 1984).

Braided River Facies: Two different facies, proximal and distal, are recognized within braided rivers and braid-plains. In the proximal areas gravels (facies Gm) accumulate by vertical accretion on longitudinal bars. These are either horizontally stratified or imbricate. Next in abundance are the cross-stratified gravel deposits (Gp) with minor amounts of sand (facies Sp and Sh). The distal braid-plains have autocyclic, fining-upward sequences. These sequences have cross-stratified gravels at the base, sandstone in the middle and mudstones at the top. Many such widespread braid-plains formed in the geological past in response to tectonic uplifts.

The deposits of the ephemeral Markanda River, India provide a well-studied example of a terminal fan. A network of radiating, abandoned distributaries mark the surface of this fan (Fig 7.25). The surfacial sediments of the fan consist of fining-upward channel and overbank deposits. The erosional surfaces of the channels are alternately overlain by coarser and finer sand facies. The former is horizontally bedded or cross-bedded. The latter is cross-laminated and flaser bedded. The channel sand occasionally passes onto horizontally laminated silt and massive mud facies upwards. From the proximal to the distal part of the fan there is a distinct decrease in the thickness of the channel sand. The overbank sediments consist of thin, climbing ripple-laminated sand, trough cross-laminated sand, interbedded mud, silt and sand facies passing into thick, massive mud facies with rootlets and desiccation cracks (Parkash, Awasthi and Gohain 1983).

Gravelly fan deposits, sandy, and muddy-interfluve deposits have been reported from the Ganga River of India. These depict a variety of geomorphic units, wide range of grain sizes, and channel patterns along 2000 km long course of the Ganga. (Singh and Bhardwaj 1991; Singh, 1996).

Sandy Alluvial Facies

Sandy Braided Stream: These sequences start with channel lag deposits at the bottom. The overlying sandy facies is cross-bedded in trough-lenticular and planar-tabular patterns. This in turn is overlain by finer grained cross-bedded and cross-laminated bar-top deposits, mostly consisting of siltstones and mudstones. Orientations of the successive sets of cross-beddings may differ widely, suggesting quick change in sediment trasportion direction. This facies model for sandy braided streams is based on observations on the Devonian Battery Point Sandstone of Quebec (Walker and Cant 1984). Coalescing bars and sand flats of braided streams often result in extensive sand sheets. These, unlike the sand deposits of meandering streams discussed below, are not confined by clays or shales.

Facies of Meandering Streams: Point bars are the main depositional areas within meandering stream channels. Fining-upward sequences are preserved in point bars by the process of lateral accretion. A typical sequence, exemplified by the Devonian Old

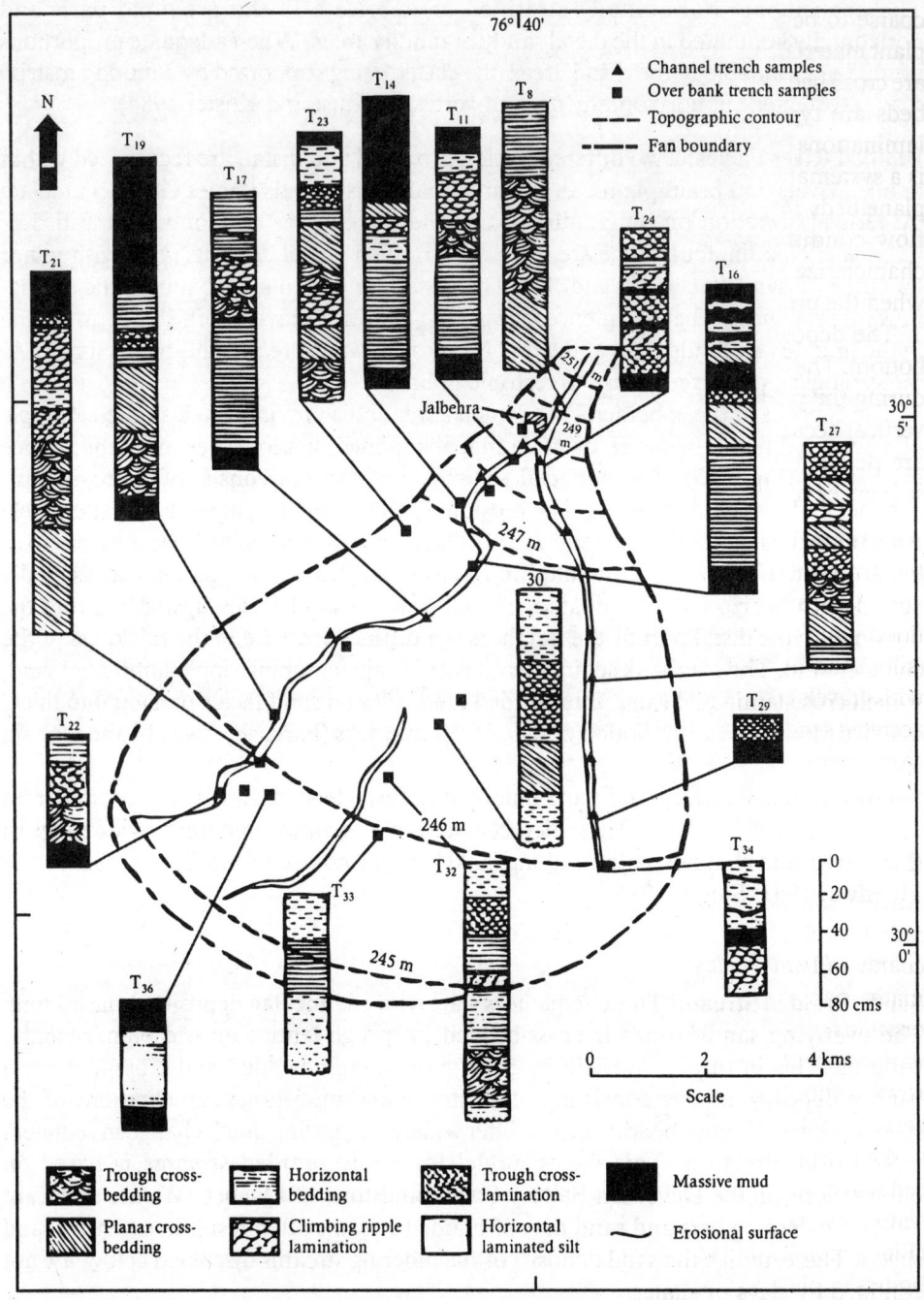

Fig. 7.25: Vertical sections of sedimentary structures at various locations of the Markanda terminal fan, India (after Parkash *et al.* 1983; with permission of the International Association of Sedimentologists).

Red Sandstone, starts with channel lag deposits at the base. This includes gravels too coarse to be moved by the flow as bed load except during high flood, water-logged plant material and chunks of mud eroded from the river bank. Overlying the lag deposits are cross-bedded sands constituting the bulk of the lateral accretion deposits. The cross-beds are typically of the trough-lenticular type. These are overlain by horizontal laminations (plane beds) which are the products of high-flow regime conditions. There is a systematic decrease in the scale of cross-bedding towards the top but occasionally plane beds may occur interbedded with the cross-bedded units due to fluctuations in flow conditions (Fig. 7.22 C). Sigmoid surfaces (large-scale epsilon cross-beds), characterized by a steep face, flattened towards the top and bottom, may be preserved when the process of lateral accretion is episodic instead of continuous (Allen 1970a).

The deposits within chute cut-offs start with a cross-bedded channel facies at the bottom. The latter are overlain by thick sequences of cross-laminated fine sand deposited during the phase of abandonment. The whole sequence is overlain by finer grained, vertical accretion deposits. In neck cut-offs thin sections of ripple-laminated fine sand are deposited during short periods of abandonment. These are overlain by thick sequences of vertical accretion deposits laid down during flood time (Fig. 7.24).

Anastomosing River Facies: Six different facies assemblages have been recognized in the gravelly, anastomosing rivers in western Canada (Smith and Smith 1980). These are : 1) peat bog facies, 2) backswamp facies (silt and clay with organic debris), 3) flood-pond facies (laminated silt and clay with sparse vegetable matter), 4) levee facies (silty sand and sandy silt with rootlets), 5) crevasse splay facies (thin layers of sand with gravels) and 6) channel facies (gravel and coarse sand). Thick deposits of vertically accreted sand bounded by backswamp facies are the essential characters of anastomosing river facies. Erosion-resistant fine sediments deposited within backswamps help in stabilizing the channel sands.

Recognition of Ancient Fluvial Deposits

Ancient alluvial fans may be identified in geological record by their geometry, which is radial in plan and wedge-shaped in cross-section. The deposits are characteristically coarse and unsorted and show a progressive decrease in grain-size downstream. Organic mater is absent due to the prevailing oxidizing conditions. Disarticulated and scattered organic remains may be found. A radial sediment transportation (palaeocurrent) pattern is another characteristic of alluvial fan deposits. In the absence of suitable cross-stratification, channel bar orientation, clast-size variation and pebble imbrication may serve as palaeocurrent indicators in braided stream deposits.

The deposits of meandering channels are recognized by typically fining upward sequence, starting from gravelly deposits at the bottom and ending in silty or muddy sediments on top. Coarse to medium grained sediments occupy the middle portion (Fig. 7.22). The meandering channel deposits, on the whole, form shoestring sand bodies in plan. The chute-and crevasse-splay deposits are lobate, while levees occur as disconnected lenses tapering towards the flood basin. Large-scale cross-beds of trough, lenticular type are common near the bottom of the sequence. These are overlain by smaller scale trough lenticles. Current lineation and ripple cross-bedding occur at top of the sequence. How closely the channel sands will occur within the envelop of overbank clay-silt depends on the rate of subsidence of the basin and the proportion of channel-fill and overbank deposits. A slow rate of subsidence together with a large proportion of channel-fill sand ensures close stacking of the sand lenses in a vertical section. When the situation is reversed, widely separated sand lenses will occur in an essentially clay-silt sequence (Fig. 7.26).

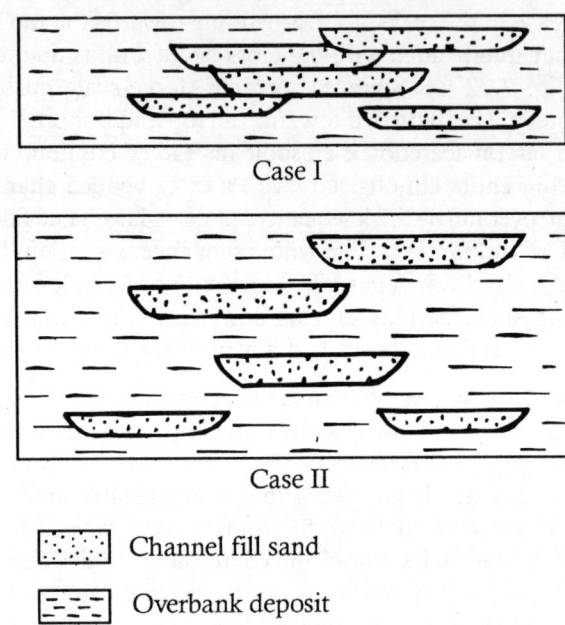

Case I

Case II

Channel fill sand

Overbank deposit

Fig. 7.26: Nature of stacking of channel sands as a function of subsidence and sediment supply. **Case I:** Stacking is close when the rate of subsidence is slow and the sediment supply is large (channel sand exceeds overbank deposits). **Case II:** Widely separated sand lenses occur when the rate of subsidence is fast and overbank deposits exceed the supply of channel sand. (Based on Nichols 1999).

Floodplain deposits are recognized by the presence of alternating layers of thin sandstone and mudstone together with *palaeosols*. The latter represent 'fossilised' soils. Like present day soils palaeosols display roots, burrows and mudcracks. In the Coal Measures of Europe and North America palaeosols contain layers of peat and other organic matters underlain by light coloured sandstone layers from which iron has been leached out. These palaeosols are known as '*seatearths*'.

Of all the ancient sedimentary facies the fluvial ones are possibly the most widely studied. These include the classical studies on the Moenkopi Formation (McKee 1954), Old Red Sandstone (Allen 1964, 1965B), Buntsandstein deposits (Mader 1985). Facies associations and palaeocurrent systems of the coal-bearing Indian Gondwana deposits have been the object of study of many Indian sedimentologists (Banerjee 1960, Casshyap, 1970, 1973; Sengupta 1966, 1970, to name only a few). In each of these studies two major facies associations were recognized: a coarse-grained sandy facies, and a finer grained facies. In the Permian Kamthi Formation of the Godavari Valley, India, the former consists of closely interwoven bands of quartz wackes of both high and low matrix variety. Grain-size frequency distribution of these rocks is comparable to the typical sediments of the fluvial bed loads. These occur as thin, elongate bodies, sinuous in plan and prismatic or lenticular in cross-section. The arenaceous facies represents the fluvial point bars produced by lateral accretion in meandering channels. The majority of cross-stratifications within the sandstones belonging to this facies are of trough, lenticular type. Consistency of palaeocurrent directions within these channel sands is often fairly large (66-75%, Sengupta and Rao 1966; Sengupta 1970). The sands have gradational contacts with thin lenticles of argillaceous bands, which occur within essentially arenaceous units.

The deposits constituting the finer facies of the Kamthi Formation are structureless, thinly laminated bands of clay, silt and silty sandstone having very large length/width ratio. The geometry and grain-size distribution of these sun-cracked, graded-bedded, fine-grained rocks are comparable to the suspension sediments of flood basins. These run parallel to the coarse-grained channel sandstone. The predominantly arenaceous units have intertonguing relationship with the argillaceous ones. The sequence of alternating arenaceous and argillaceous facies is repeated several times in time and space due to wandering of the river channel within its own flood plain. The pattern of sedimentation, which has evolved from this process of sedimentation, is schematically represented in Fig. 7.27 (Sengupta 2003). This diagram is designed to combine two diverging stratigraphic principles namely, the law of superposition depicting a change in the vertical direction, and the concept of lithofacies variation implying a lateral change. Such a situation is often encountered in ancient fluvial record.

A rhythmic pattern of sedimentation consisting of alternately dominating channel (lateral accretion deposits) and interchannel facies (vertical aggradation) has been also reported from the Barakar Formation of the South Karanpura Coalfield by Banerjee (1960). An idealized, standard sedimentary cycle within the Barakar (Lower Permian) Formation consists of the following in a vertically upward sequence - pebbly sandstone, coarse to medium sandstone, fine sandstone, interbeded assemblage of fine to medium sandstone, siltstone and shale, carbonaceous siltstone and shale, coal and shaly coal. Several such cycles are observed. Most of these cycles display a fining-upward (sometimes partially truncated) character. The lower units are characterized by large-scale cross-bedding, while the upper units contain only smaller scale cross-bedding and wavy or parallel lamination. Apparently the flow intensity progressively decreased with time and was considerably reduced during deposition of the carbonaceous unit. The presence of coal seams suggests stagnation of the depositional area during the late phase of sedimentation, allowing vegetal accumulation. This pattern of cyclic sedimentation has been interpreted as the result of channel wandering (apparently avulsion) and intermittent tectonism (Casshyap 1970). Where only immature, fine-grained sediments are available for transportation, as in the Gomti River of Uttar Pradesh, India, the sediments are transported and deposited mainly as suspension load. Under this situation the sedimentary successions in point bar and natural levee deposits exhibit similar bedding structures. Small ripple cross-bedding, climbing ripple lamination, horizontal bedding are the more common sedimentary

structures. Scour-and-fill, large-scale cross-bedding, antidune cross-bedding occur
only rarely under such a condition. Thick sedimentary sequences develop only
where several cycles of point bar and natural levee deposits follow one another
(Kumar and Singh, 1978).

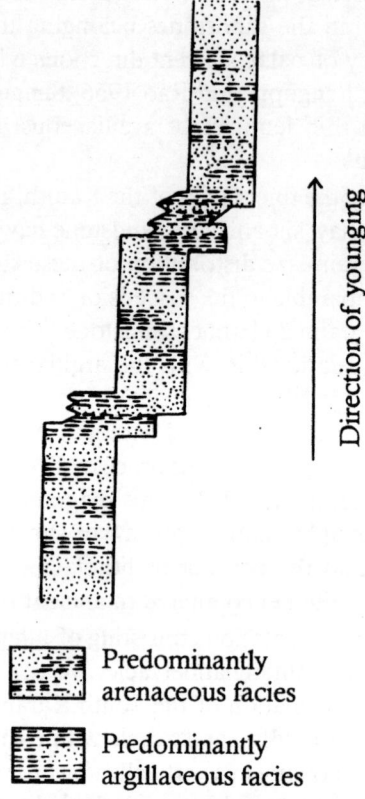

Predominantly
arenaceous facies

Predominantly
argillaceous facies

Fig. 7.27: Schematic representation of facies association in space and time in a typically
fluviatile sequence (from Sengupta 2003). With permission of Elsevier.

MIXED ENVIRONMENTS

Barrier-Island and Beach Facies

Bordering many a coastline of the world today are rows of long narrow islands. Between
these barrier-islands and the mainland lie lagoons rimmed by tidal flats. Connecting
the lagoons with the sea are tidal channels (inlets). Until about 1972 these barrier-
island complexes were believed to have developed wholly out of prograding barrier-
beach systems. Models for these systems, based mainly on studies in the Galveston
and Padre Islands, Texas, were also formulated (Dickinson *et al.* 1972).

More recent studies show that the barrier-island complex is a composite depositional
system involving several discrete environments—beach, tidal inlet, tidal delta and lagoon
(Fig. 7.28). A single generalized model cannot do justice to such a complex system.
Four separate facies models, applicable to the barrier-beaches, tidal inlets, tidal deltas
and lagoons, are discussed below, following Reinson (1984). Tidal Flat facies is discussed
separately.

Barrier-Beach Facies

The barrier-islands bordering coastlines have shorefaces (beaches) on the seaward side. Four subenvironments of sedimentation are possible within the barrier-beach system. These are : shoreface, foreshore, backshore-dune and washover flats (Figs. 7.28 & 7.29). Tidal inlets, tidal delta, and lagoons connecting barrier beaches and tidal flats are discussed under separate heads.

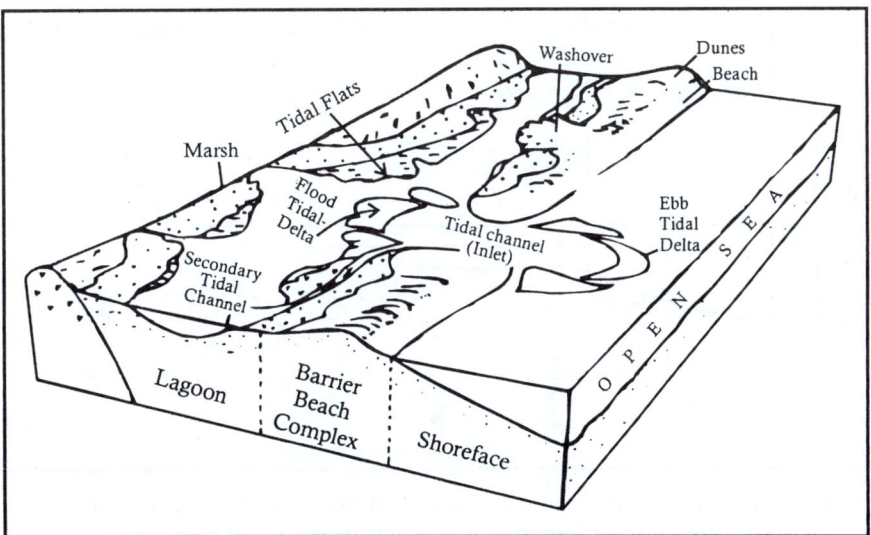

Fig. 7.28: Block diagram showing various subenvironments in a barrier-beach complex (after Reinson 1984). With permission of Geological Association of Canada.

Shoreface

Shoreface extends from the low-tide mark to a depth of about 10-20 m. Three zones of shoreface—lower, middle and upper, are recognized. The middle shoreface, which extends up to the breaker zone, is dominated by wave energy. The typical deposits in this zone are highly bioturbated, fine to medium-grained, clean sand, with silt and shell layers. Seaward-dipping, wedge-shaped, planar laminae and landward-dipping ripple laminae are the common sedimentary structures in this zone. Trough cross-laminae may also be present. Occasional storm surges may drastically lower the wave base in the middle shoreface, putting much of the sediment in suspension and subsequently depositing it seaward. This phenomenon is believed to be responsible for generation of the hummocky cross-stratifications found in geological record.

In the lower shoreface shelf and basinal processes are predominant although some wave action is also present. Plane beds of bioturbated, fine to very fine sand, with layers of silty mud are deposited in this zone. The degree of organic activity decreases towards the upper shoreface where the deposits are coarser. The complex hydraulic condition of the surf zone produces multi-directional trough cross-bedding. Ridges (bars) and runnels (troughs) develop on the upper shoreface when the sediments transported from the beach to the shoreface are returned and redeposited at the edge of the foreshore under quieter conditions.

Foreshore

Foreshore is the area exposed between tides. It has a steeper gradient than the surf zone. The deposits here are well sorted by swash-backwash action. Thinly laminated, wedge-shaped beach deposits dip gently towards the sea. In some beaches berms separate foreshores from backshores (Fig. 7.29).

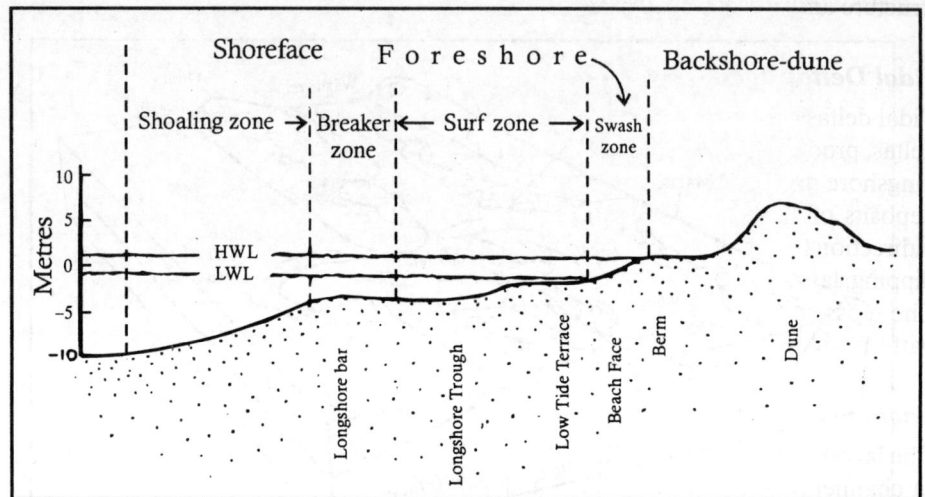

Fig. 7.29: Generalized cross-section of an ideal beach and shoreface (after Reinson 1984). With permission of Geological Association of Canada.

Backshore

Backshore occurs on the landward side of the foreshore. The sediments of the backshore include lag deposits and heavy minerals. These dip very gently towards the land. *Eolian dunes* with planar and trough cross-stratifications occur closer towards the land. Rootlets and burrows are common within these dunes.

Washover Flats

Thin lobate sheets of sand which flow over foredune barriers during high strom surges and extend into lagoons are called *washovers* (Fig. 7.28). These consist of medium to coarse sand and, rarely, gravels. Parallel lamination is the usual structure of the washover deposits but small-scale delta foresets may form when washovers plunge into lagoons.

Tidal Inlet Facies

Tidal inlets are narrow deep channels connecting the lagoons with the open sea (Fig. 7.28). Water flows vigorously through these channels during flood and ebb. Lateral migration of the tidal inlets causes the tidal deposits to accumulate as in a migrating stream channel. The sand body produced out of this process of migration is elongate parallel to the trend of the barrier islands. Its thickness is of the same order as the depth of the inlet.

The deposits within the inlets show fining upward sequences, starting with lag gravels at the bottom and ending in finer sediments at the top. The cross-beds within these deposits carry imprints of flood and ebb flows and are naturally bi-directional. Plane beds and ripple laminae occur within the upper deposits. Five major depositional units, starting from lag gravels at the bottom and ending in finer sand - and seaward-dipping laminae at the top, were recognized within the Fire Island Inlet, New York state (Kumar and Sanders 1974). These are the channel floor, deep- and shallow-channel, spit-platform and spit deposits. Each of these deposits is associated with a particular type of sedimentary structure and texture. This model should work broadly for all tidal inlet deposits.

Tidal Delta Facies

Tidal deltas generate at either end of tidal inlets by tidal flows (Fig 7.28). The ebb-tide deltas, produced at the seaward end on the tidal inlet, are modified by wave action and longshore drifts. The flood-tide deltas occur at the landward outlet of tidal inlets. The deposits of tidal deltas are essentially of sand size. The structures at the base are bidirectional megaripples. These are overlain, in the case of flood deltas, by landward dipping, large-scale ripples with planar-tabular and trough-lenticular internal structures. The cross-bedding set thickness decreases upwards. In ebb deltas, complex bedform patterns may be produced as a result of interaction with ocean waves.

Lagoonal Facies

The lagoonal deposits consist of alternating sandy and silty facies. The former is found in channel fills and washover sheets. The latter is produced within the lagoon under subaqueous conditions or derived from the adjacent tidal flats. The sandy deposits are rich in organic content. These are overlain by coal and peat produced within marshes and swamps at the margins of the lagoon.

Sedimentation in Mixed Environment: Indian Examples

Recent continental and marginal marine facies of the Godavari delta on the eastern coast of India provide an example of a mixed environment of sedimentation. The Gautami-Godavari River channel with its secondary distributaries forms a part of the subaerial topset beds of the Godavari delta. The two major distributaries of the Godavari River are interconnected by a reticulate system of tidal creeks. These creeks pass laterally into mangrove swamps. The backwater is a vast shallow lagoon, separated from the open sea on the south by a narrow barrier beach and on the east by a composite system of swamps, tidal creeks and beach. At least five barrier islands occur off the Vrudha-Gautami River confluence. These narrow, elongate islands run parallel to the mainland beach.

The sands of the offshore barrier islands of the Godavari delta are very well sorted and have an almost symmetrical grain-size distribution. Those of the mainland beach are moderately sorted and negatively skewed. The dune sands are well sorted and positively skewed. The silt and clay proportions are lesser in the beach and higher in the dunes (Naidu 1966). These observations provide clues to the identification of the subenvironments in a ancient barrier-beach complex.

The Proterozoic Vindhyan exposures of central India are arranged in several linear facies belts. Each of the facies in lithology, sedimentary structure and palaeocurrent pattern corresponds to a particular unit of sedimentation within a barrier-coastline model (see Banerjee 1974 for details). The Vindhyan sea was recognized as microtidal (tide range < 2m), occasionally rising to mesotidal level (2-4m). On the open shelf to the north-east a shore-normal, bipolar tidal pattern was identified, while within the embayment the pattern was much more complex. Strom surges, wind and wave-generated currents added to the complications (Banerjee 1980).

The parallel and cross-laminated, grey-black, pyrite-rich shales and siltstones of the Bijaigarh Formation (Kaimur Group) within the Vindhyans were interpreted as Proterozoic lagoonal deposits, while the micritic and dolomicritic lime-mud-bearing Bhander Limestone was thought to represent an ancient carbonate tidal flat (Banerjee 1974 ; Singh 1980, 1980a). Ripple-drift cross-laminations with rhythmically size-differentiated foreset laminae, believed to be characteristic of tidal environment, were also reported from the Bhander Formation (Chanda and Bhattacharyya 1974). Although these observations point to a barrier-coastline, tidal or lagoonal condition of deposition, alternative interpretations, recognizing at least a part of the Kaimur Group of rocks within the Vindhyan Supergroup as fluviatile deposits, are also known (Auden 1933 ; Bhattacharyya and Morad 1993).

Stratigraphic Sequences in Mixed Environments

Stratigraphic sequences developing out of regression (progradation), transgression and migration of barrier-island systems are broadly as follows (Fig. 7.30 A-C after Reinson 1984).

Regressive Model

A typical sequence developing out of regression of a barrier-island system has finer shoreface deposits at the bottom, grading gradually to coarser dune sands at the top through foreshore and backshore sediments (Fig 7.30 B). Abundant sediment supply is an essential condition for preservation of these deposits. This condition is satisfied in an inter-deltaic coastal setting. In fact, many a barrier-island sequence in stratigraphic record are associated with deltas.

Barrier Inlet Model

This is a fining-upward sequence with a thinning-upward trend in cross-bedding set thickness. It has an erosional base and is dominated by sand facies of tidal-channel and marginal spit-beach environments (Fig. 7.30C).

Transgressive Model

The facies sequence is variable, depending on the local condition. It is generally characterized by subtidal and inter-tidal back-barrier facies and does not show a definite fining or coarsening-upward facies. In the regressive model the lagoonal facies overlie the sandy facies but in the transgressive model they underlie or are incorporated into the lower part of the sequence (Fig. 7.30A).

Tidal Flat Facies

Tidal flats develop in coastal areas where wave action is not strong. Within the barrier-island system such an environment prevails near the lagoons (Fig. 7.28). The tidal flats of northern Europe, which are sheltered from the wave action of the North Sea by a row of barrier islands, may be cited as examples. Tidal flats may also occur on open sea coasts where large width and low gradient of the beach dampen wave action. The Mailiao Flat of Taiwan, the Inchon Bay of Korea and the Chandipur Flat on the eastern coast of India are examples of open-to-the sea tidal flats. Tidal flats can be divided into two groups, depending on whether the constituting materials are essentially siliciclastic or carbonates.

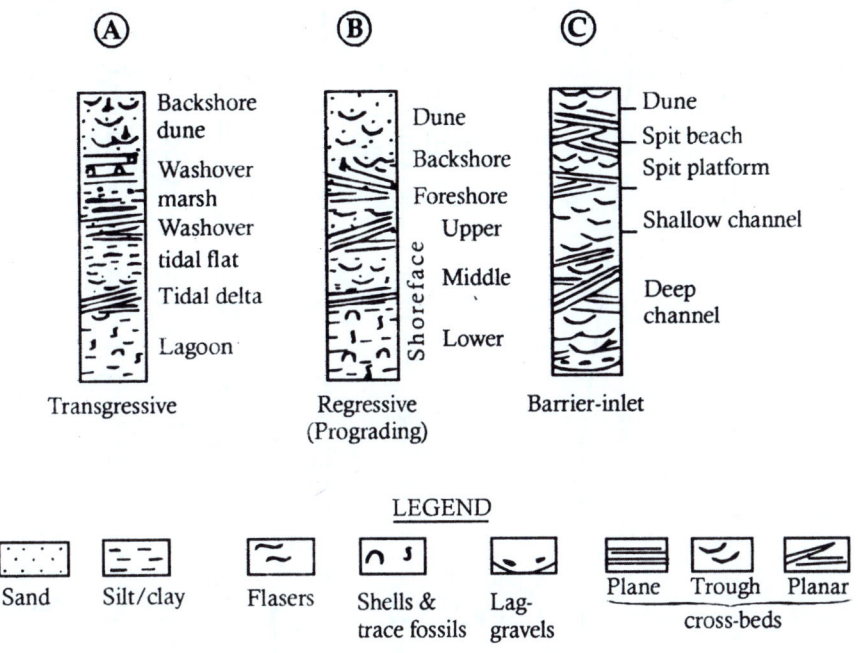

Ⓐ

| Backshore dune |
| Washover marsh |
| Washover tidal flat |
| Tidal delta |
| Lagoon |

Transgressive

Ⓑ

| Dune |
| Backshore |
| Foreshore |
| Upper |
| Middle | Shoreface |
| Lower |

Regressive
(Prograding)

Ⓒ

| Dune |
| Spit beach |
| Spit platform |
| Shallow channel |
| Deep channel |

Barrier-inlet

LEGEND

| Sand | Silt/clay | Flasers | Shells & trace fossils | Lag-gravels | Plane | Trough | Planar |

cross-beds

Fig. 7.30: Facies models of barrier island stratigraphic sequences (based on Reinson 1984). See text for details. With permission of Geological Association of Canada.

Siliciclastic Tidal Flats

Fine sand, silt and clay constitute the bulk of the siliciclastic tidal flats. The nature of sediment distribution in these flats largely depends on local geographic and climatic conditions. In the sheltered intertidal areas bordering the North Sea a mud flat composed of mud, fecal pellets and thin bands of sand occurs on the landward side of the flat. The proportion of sand increases towards the sea until the flat becomes wholly sandy near the sea coast. In-between the mud and the sand flats occur mixed flats composed of muddy sand. The biogenic content of the flat increases as one moves from land to the sea. A generalized cross-section across an open-to-the-sea tidal flat based on observations at St. Michel, France is shown in Fig. 7.31.

The surface structures in sheltered intertidal zones, like rain and hail imprints, desiccation cracks and bird tracks, indicate periodic subaerial exposure. Horizontal lamination in sandy flats and parallel lamination in silty flats are common. Combinations of straight-crested, bifurcating (in plan), tuning fork-shaped ripples,

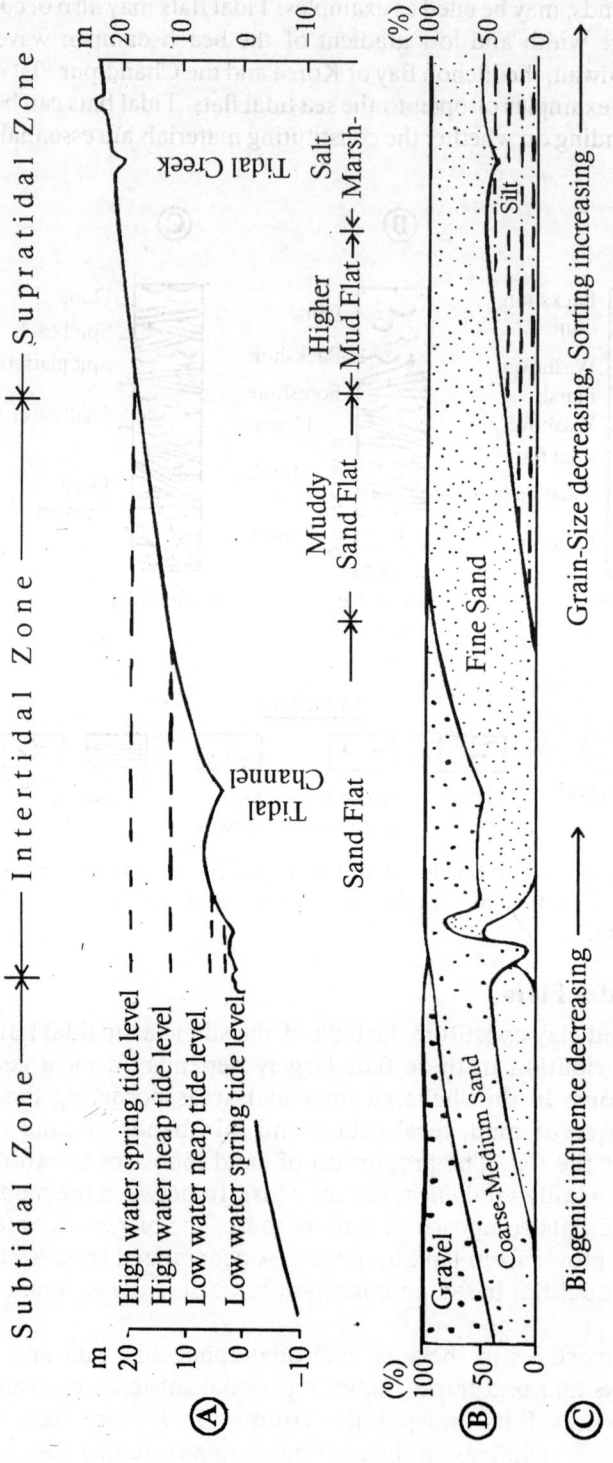

Fig. 7.31: Schematic section across an 'open-to-the sea' tidal flat (based on Larsonneur 1975).

interfering ripple systems produced by current and wave movements, and lenticular flaser beddings are some of the important characters.

The subtidal zones, which remain submerged during low-water, springtide level, are the sites of deposition of coarse sediments. The gullies traversing these flats contain coarse, channel-lag deposits. Megaripples and flaser beddings are found here. Escape traces of organisms are common but shells, in orientations which they occupied, when alive, occur rarely.

The supratidal areas located on the elevated landward side of the tidal flats normally remain exposed to the atmosphere. They are inundated only during exceptionally high spring tides. Salt marshes consisting of clayey sediments and occasionally peat are common in the supratidal areas. Coarser sediments are restricted to the channels traversing the marshes. The clayey sediments are often stabilized by growth of salt-resistant plants within the marshes.

Carbonate Tidal Flat

The carbonate tidal flats do not differ much from the siliciclastic flats except in composition. Based on studies in such classical areas as the Bahama Banks of the USA, Shark Bay of Western Australia and the Persian Gulf, James (1984) recognized two major carbonate flat facies: those associated with high-energy conditions and those related to low-energy environment. A summary of the characters of tidal flat carbonate facies based on James (1984) follows.

The intertidal carbonate deposits are characterized by the presence of algal mats, irregular or even laminations (*cryptalgal laminations*) with fenestral porosity and desiccation cracks. The algal mats are leathery sheets of blue-green algae growing on top of sediment surface. The growth of these mats is controlled mainly by rainfall, moisture, salinity and the presence of other organisms. Algal growth is luxuriant in areas of high rainfall and hypersalinity. In areas of normal salinity the algal growth is deterred by gastropods which feed on them. In areas of high rainfall the algal growth may extend even up to the supratidal areas.

Elongate, sheet-like cavities (designated as '*laminoid fenestrae*', '*loferites*' or '*bird's eyes*') left by growing algal colonies are preserved in the geologic record even after the algae have decayed. These allow identification of ancient algal colonies. The presence of algae can also be inferred from the occurrence of stromatolites—stratiform, crenulated, and arched organo-sedimentary structures produced by sediment-binding algae (see chapter 6). Desiccation cracks are created by subaerial exposure of the algal mats. The cracks between the desiccation polygons are filled by lime mud. Sediment characters of the different parts of an ideal carbonate tidal flat are as follows.

Lower Intertidal Zone

Burrowed, bioturbated sediments, desiccation polygons and luxuriant growth of algal mat characterize this zone, particularly when the condition is hypersaline. Oolites formed from aragonite needles and organic matter in the shallowest sea water, may be moved out of place by strong tidal currents and deposited in the lowest intertidal zone.

Middle to Upper Intertidal Zone

Sediments are grey-brown due to prevailing oxidizing condition. Algal mat growth is common. Fenestral pores and desiccation polygons, indicative of prolonged exposure, are present. Burrows of worms, crabs and crustaceans, and rootlets of salt-tolerant plants are common features.

Supratidal Zone

Surface crusts ('hard grounds') produced by lithification of sediments due to prolonged subaerial exposure are common. *'Teepees'* are produced by pushing up of the broken polygons on surface crusts due to the force of crystallization or penetration of plant roots. The clasts are often cemented by carbonates. Caliche, algal mats (in areas of high rainfall) and evaporites (in arid areas) are common features in some supratidal areas. The nature of carbonate and evaporite precipitation is largely dependent on the chlorinity of the ground water and the amount of freshwater flushing in the supratidal zone.

Stratigraphic Sequences in Carbonate Tidal Flats

High-Energy Carbonate Flats

The characters of high-energy carbonate flat deposits are essentially the same as those of the siliciclastic beaches—coarse-grained, poorly sorted carbonate sand with lime mud matrix in the lower foreshore, and well-sorted lime sands in the upper foreshore. The lower foreshore has small- to large-scale trough cross-bedding, oriented parallel to the shoreline. The upper foreshore has seaward-dipping, planar cross-beds. These sediments may also have *'keystone vugs'* or *'microcaverns'* (similar to fenestral structures) produced by gas escape. These are almost invariably cemented at a later stage. A typical vertical sequence formed in a high energy intertidal situation is shown in Fig. 7.32 C. The sequence, which starts with a lime-sand conglomerate at the bottom, is overlain successively by argillaceous limestone and bioturbated mudstone, to be capped, eventually by calcrete.

Low-Energy Carbonate Flats

Vertical sequences are developed either by progradation of the carbonate tidal flats or by shoaling of carbonate sands of barrier islands. The basal units of such sequences are composed of coarse grained deposits of the surf-zone. These are overlain by bioturbated wacke-stones and packstones with marine fauna and stromatolites of the subtidal-intertidal areas. The subtidal deposits are well sorted, oolitic, pelletoidal and skeletal lime sand occasionally with oncolites. Planar bedding and herring-bone cross-lamination, with set thickness decreasing upward and ending in small-scale ripples, on top, are the major sedimentary structures. The structures are often destroyed by worm burrows. In shoaling-upward hypersaline sequences, as in the Shark Bay of Western Australia, stromatolites may abound on top of a skeletal metazoan bioherm or biostrome (Fig. 7.32 A and B).

The deposits of the supratidal zone which overlie the subtidal-intertidal facies are evenly laminated. Flat-pebble breccia and desiccation polygons are common features in these deposits. Authigenic evaporates develop in arid regions flushed by hypersaline ground water. This is followed by precipitation of dolomite due to a rise of Mg^{++}/Ca^{++} ratio of ground water and subsequent dolomitization. Collapse breccias, produced by partial solution of evaporates, may cap the sequence (James 1984).

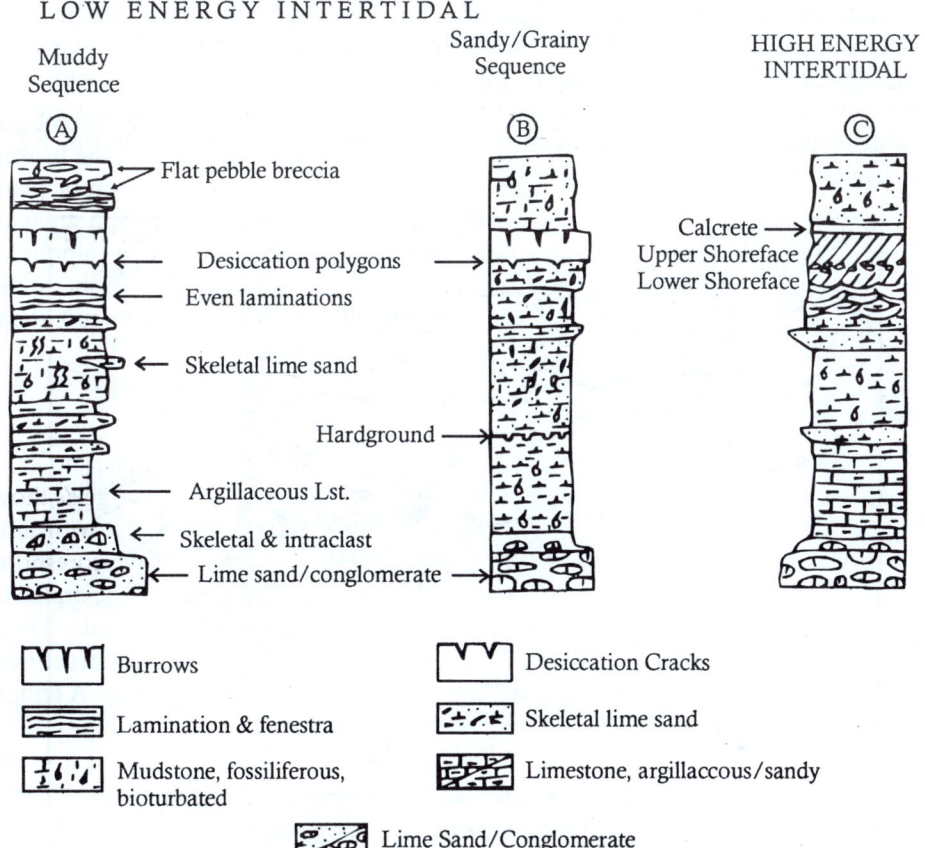

LOW ENERGY INTERTIDAL

Muddy Sequence (A)

Sandy/Grainy Sequence (B)

HIGH ENERGY INTERTIDAL (C)

Flat pebble breccia
Desiccation polygons
Even laminations
Skeletal lime sand
Hardground
Argillaceous Lst.
Skeletal & intraclast
Lime sand/conglomerate

Calcrete
Upper Shoreface
Lower Shoreface

Burrows

Lamination & fenestra

Mudstone, fossiliferous, bioturbated

Desiccation Cracks

Skeletal lime sand

Limestone, argillaceous/sandy

Lime Sand/Conglomerate

Fig. 7.32: Hypothetical sequences developed on (A) low-energy (muddy) subtidal unit, (B) high-energy (grainy) carbonate sand unit, and (C) high-energy intertidal carbonate unit (based on James 1984). With permission of Geological Association of Canada.

Tidal Flat Sedimentation: Indian Examples

The Chandipur tidal flat situated on the eastern coast of India provides an example of a beach-related siliciclastic tidal flat bordering the open sea (Fig. 7.33). This tidal flat enjoys tropical climate. The mean tidal range lies between 2 and 4 m; hence the setting is designated as macrotidal. The western part of the intertidal zone enjoys a non-barred setting, while several sand bars occur on the eastern part.

Nearer the land the non-barred planar flat is composed of laminated sand frequented by burrows of ocypodid crabs. As one moves towards the sea the laminated sand grades to silty sand which is occasionally bioturbated by polychaetes. Nearer to the sea occurs a more silty facies. The silty sand has ripple-drift lamination, scour and fill, flaser bedding and climbing ripple lamination. The silty facies has ripple-drift lamination, scour-and-fill structures and hummocky cross-stratifications (HCS). The HCS develop only during monsoon weather frequented by cyclonic storms. In the barred flat laminated sand, megaripple lamination and flaser beddings predominate over other structures (Mukherjee, Das and Chakrabarti 1987).

The Rann of Kutch, located on the western coast of India, enjoys intertidal-supratidal environments. The extensive Rann ($\sim$ 30,000 km^2 in area) is flooded for about three months every year by sea-water 2m deep as a result of monsoon storm surges. Fine silt, clay, gypsum and a small amount of carbonate constitute the deposits in this Rann (Glennie and Evans 1976).

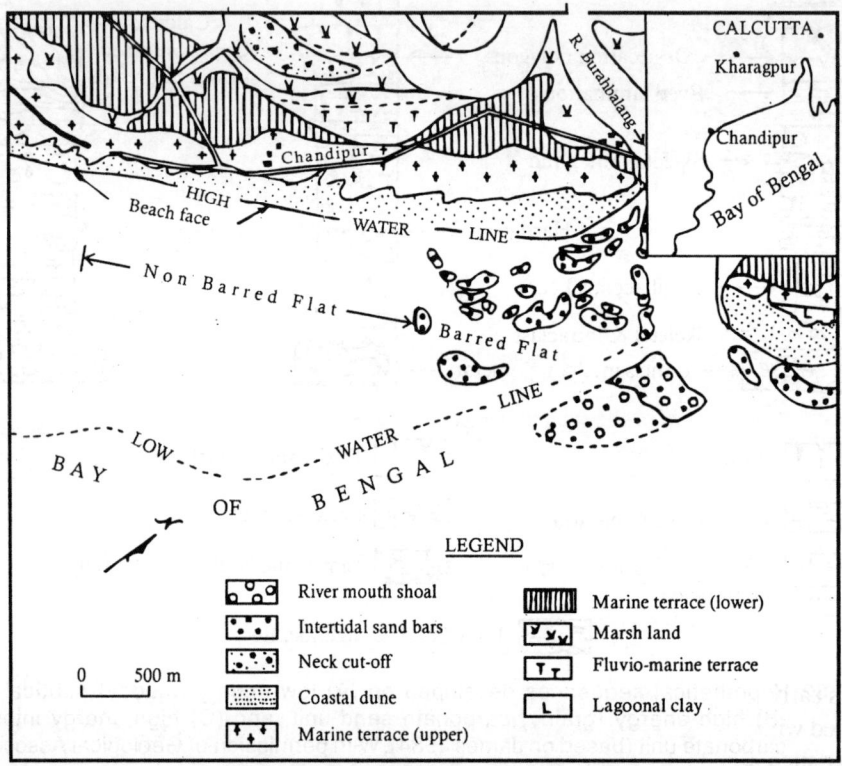

Fig. 7.33: Geomorphic units within the tidal flat around Chandipur, India (after Mukherjee *et al.* 1987; Chakrabarti 1991, unpublished DST report)

Deltaic Environment

A delta is defined as 'a deposit, partly subaerial, built by a river into or against a permanent body of water' (Barrell 1912). The morphology of a delta depends on the ratio of sediment supply and the energy of water transporting and depositing it. When the sediment supply is large, deltas appear as protrusions of shorelines into the sea or large lakes. Arcuate, cuspate, lobate, and bird-foot deltas develop progressively as the sediment supply is increased with respect to the wave energy. When the proportion of sediment supply is very low, the erosive power of wave energy predominates and indentations are produced on the coast in the form of estuaries (Fig. 7.34).

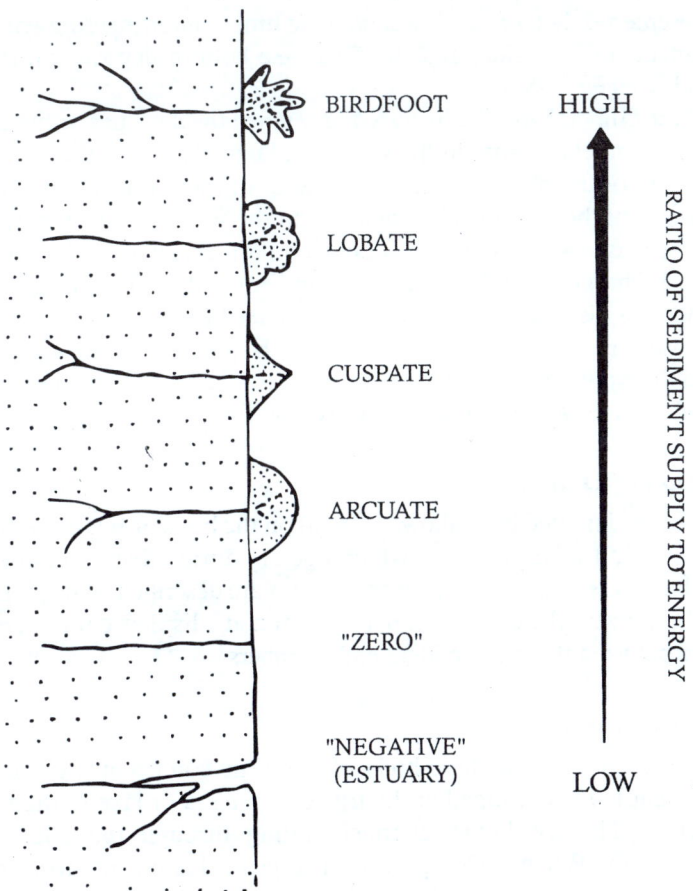

Fig. 7.34: Diagrams illustrating changes in shapes of subaerial deltas depending on wave energy and sediment input (after Curray 1975, *In* A.G. Fisher and S. Judson 1975, eds.).

As early as 1885, G. K. Gilbert recognized that topset, bottomset and foreset beds are formed when streams debouch into still bodies of water in lakes. The slope of the foreset, under such a condition, is guided by the angle of repose (~ 30°), whereas the top- and bottomsets remain nearly horizontal. Since 1885, deltas showing well-demarcated top- , bottom- , and foreset units have come to be known as *Gilbert-type deltas*.

In deltas of the classical Gilbert-type it is the foreset that characterizes the delta, the top- and the bottomsets being extensions of the fluvial and lake-bottom sediments respectively. The foreset in a Gilbert-type delta consists of coarse sediments deposited suddenly as the flow of the stream is deterred on reaching the quieter body of the lake. The foreset remains undisturbed only when the water level in the lake is stable for a fairly long period of time. When the water level fluctuates irregularly, complex patterns may develop due to interfingering of the top-, bottom- and foresets. When the level of water changes in a series of steps, the delta may take a step-like form. In cases of extremely irregular fluctuation of water level, the foresets, which characterize the Gilbert-type delta, may be completely obliterated. The whole delta deposit in such a case is represented only by a thick wedge of sediment interfingering with the coarser fluvial sediments on one side, and the finer lake bottom deposits on the other side (Dunbar and Rodgers 1957).

Deltas of the Gilbert-type are produced only when the densities of the effluent water in the stream channel and that in the sea or the lake into which this water is released are of the same order, allowing uniform mixing in three dimensions (Busch 1974). When the water in the stream is less dense than the water in the embayment, a sheet flow results because the mixing now is only 'two-dimensional'. When the magnitude of sediment discharge is small, an arcuate sand bar is deposited at the mouth of the stream. When the discharge is high, a cuspate or bird-foot type of delta may result. If the effluent water in the stream, because of its high suspended load, has a much higher density than the water in the ocean, a channel is eroded on the continental shelf in the form of a submarine canyon (Bates 1953, see also Fig. 7.34).

Classification of Deltas

Deltas have been classified by Galloway (1975) on the basis of relative influence of the fluvial, oceanic and tidal processes which operate during delta building (Fig. 7.35). The facies characters of these three types of deltas are described below, following Miall (1984a). Many natural deltas, it must be realized, display characters which are intermediate between these three ideal end-members.

River-Dominated Deltas

As a river debouches into a static body of water its competence decreases and the coarser sediment load is dropped at the fringe of the sea or lake in the form of spits, bars and islands. The distributary channels cutting through the alluvial plain are bordered by levees. When the proportion of suspended sediment carried by the stream is high, the river mouth is extended towards the sea by sedimentation. The deposits are oriented at a high angle to the coastline to form bar-finger or shoestring sand bodies, as in the bird-foot delta of the Mississippi River (Fig. 7.36A). The space between the distributary channels of such a delta is filled by marshes. At a high stage of flooding the distributary channels may be breached by crevasses. Crevasse splay deposits are formed when coarse sediments, pushed out of the crevasse, splay out on the adjoining marshes. Unstable conditions may arise when the rate of sediment deposition is exceptionally high in a delta. Frequent failures on the seaward side of the delta in such a case might result in faults contemporaneous with sedimentation. Such faults are called 'growth faults'.

A lobate, fan-shaped delta is produced when a heavy load of sediment is discharged by a stream on a plain of low gradient. In the Mississippi delta for example, a major delta lobe was created every time the Mississippi River discharged its sediment load towards the sea. The lobes thus produced are preserved even after the distributary channels responsible for their sedimentation abandoned their earlier courses to take a shorter one to the sea. Seven such deltaic lobes have been mapped in the Louisiana coastal area (see Fig. 7.37).

While deltaic sedimentation was going on along the Louisiana coast, the sea encroached on the land on several occasions to flood the delta surface. Complex patterns of fluvial, fluvio-marine, paludal and marine sedimentation developed within the

Mississippi delta out of the interaction between the delta and the sea. The fluvial sediments were deposited in stream channels, natural levees, point bars and abandoned stream courses. Brackish fluvio-marine deposits were laid down at the mouths of distributary channels. Paludal sedimentation took place in marshes, swamps, tidal channels and lakes. The marine sediments were confined to beach, foreshore and deeper water reefs (Kolb and van Lopik 1966).

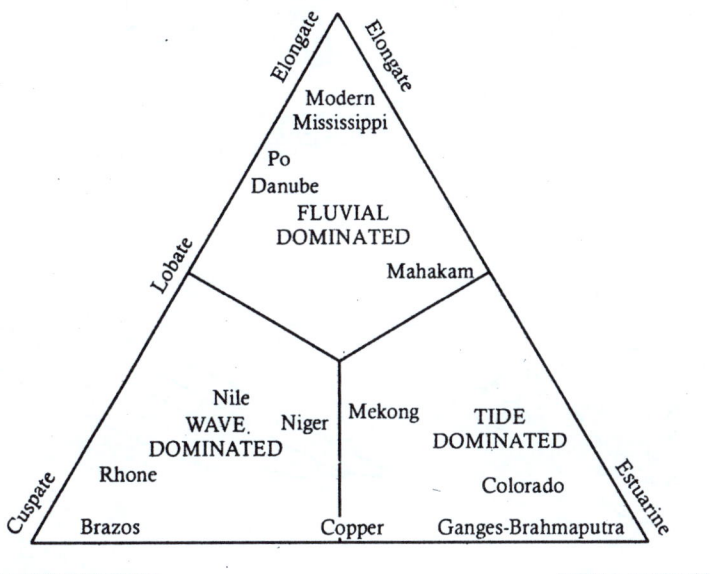

Fig. 7.35: Variation of shapes of deltas as a function of sediment input, wave energy flux, and tidal energy flux (after Galloway, 1975).

Wave-Dominated Deltas

In areas of strong wave action the delta sediments are continuously reworked and the sand bodies are reoriented into a series of beach ridges paralleling the coastline. These are called 'strike-oriented' ridges. In areas dominated by longshore drifts the entire delta assumes an asymmetric ('skewed') shape. The delta of the Nile is a typical example of wave-dominated delta (Fig. 7.36B).

The Rhone delta, perhaps the best-studied wave-dominated delta in the world, has a thick sequence of prodelta shelf mud at the bottom. This is overlain by delta front sand, often called 'coastal barrier sand' (Galloway and Hobday 1983). Stacking and coalescence of individual sand bodies here resulted to delta front sand sheets of enormous thickness. The thickness has further increased by growth faulting. The delta front sand dips very gently towards the sea and shows a larger degree of marine influence than the sands of river-dominated deltas. The wave-dominated delta front sequence coarsens upward like the sequence produced by progradation of a marine shoreface. In areas influenced by strong winds the coarsening-upward facies is covered by finer grained, well-sorted, coastal dune sands. The uppermost part of the Rhone delta sequence is criss-crossed by fluvial distributaries.

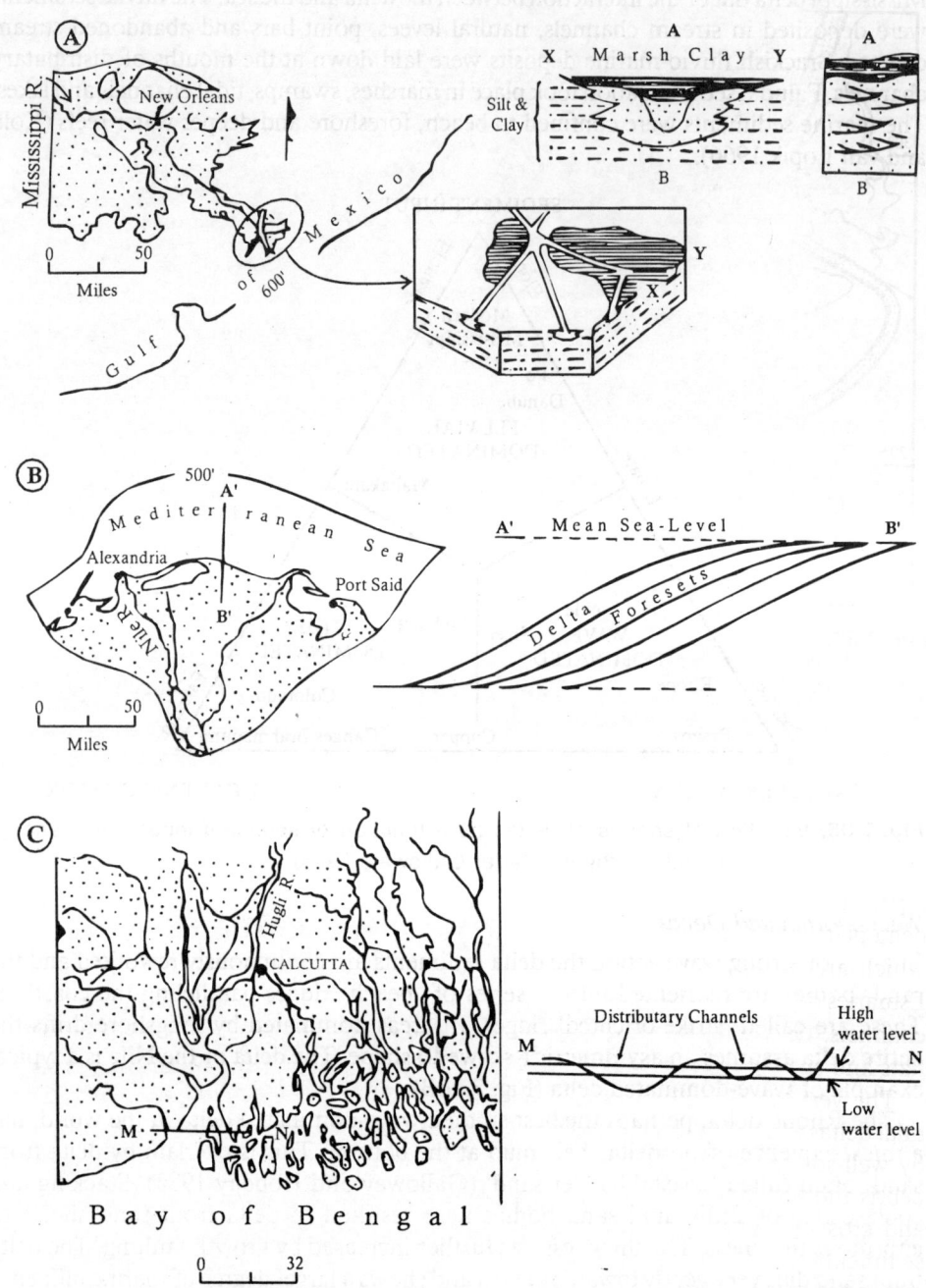

Fig. 7.36: Deltas in plan and profile. (A) the Bird-foot delta, Mississippi, dominated by fluvial action, (B) the Nile delta, dominated by wave energy, and (C) the Bengal delta-estuary system, dominated by tidal energy. (Based partly on Smith Jr., 1966).

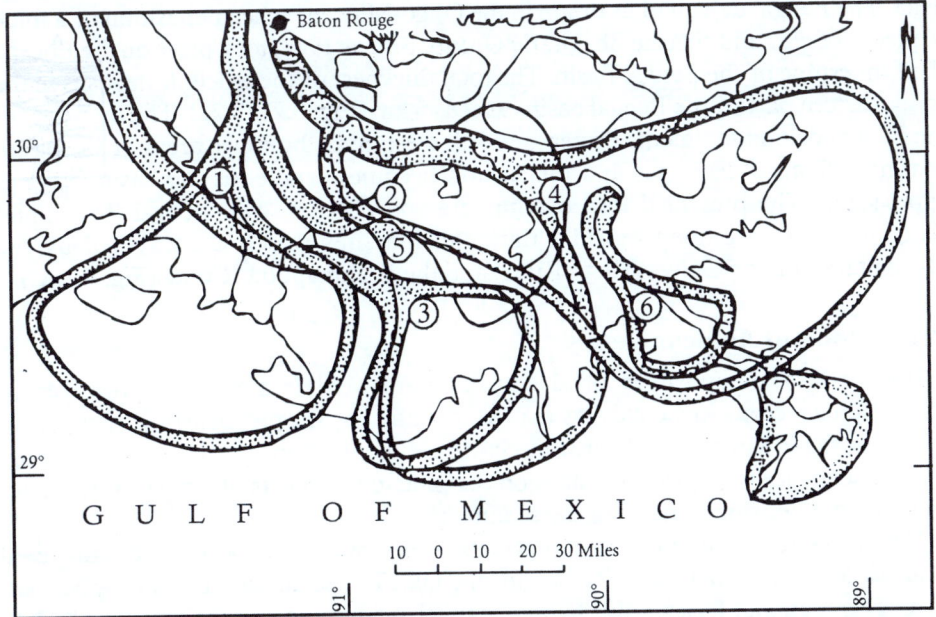

Fig. 7.37: Deltaic lobes of the Mississippi River (after Kolb and van Lopik 1966). The flow lasted for 500 to 1000 years in each case. Approximate time of flow before 1950 : (1) around 5000 years, (2) around 4000 years, (3) around 3000 years, (4) around 2000 years, (5) and (6) between 2000 and 1500 years, (7) about 500 years to present day.

Tide-Dominated Deltas

Tidal currents, like waves, rework the deltaic sediments and rearrange them into a series of elongate bars extending ('dip-wise') from the channel mouth to the subaqueous delta platform. This platform is produced by deposition of mud, silt and muddy sand, which are carried downstream by suspension currents form the prodelta deposits. Estuaries develop when the distributaries find outlets into the sea through funnel-shaped channels. On land these tidal estuaries may take sinuous forms, cutting across tidal flats and supratidal marshes (Fig. 7.36C).

The vertical sequence in the Colorado estuarine channel, one of the well-studied tidal deltas of the world, starts with a basal scour filled with lag deposits. This is overlain by well-sorted channel-fill sand containing mud, silt drapes, clasts, macerated plant debris and shells. The sedimentary structures in this sequence include ripple laminations and cross-beddings with bi-directional dips. Laterally, the coarse-grained estuarine sediments grade into finer delta plain deposits. The sediments in the tidal current sand ridges coarsen upward. The bottom of the sequence is occupied by mud of prodelta shelf facies. These are overlain by sand with interbedded mud lenses. Ripples, cross-bedding and parallel laminations are the common sedimentary structures. The sequence is capped by finer grained tidal flat deposits.

The history of sedimentation in the Bengal Delta, the largest tide-dominated delta in the world, with an area exceeding 59,000 sq. km. is comparable to that of the Colorado delta. The Bengal delta had its origin in the Oligo-Miocene when deposition of a thick sequence of clay and silt (the Bhagirathi Group; Biswas 1963) took place on the Eocene shelf limestone of the Bengal Basin. The total thickness of these deltaic deposits in the south-eastern part of the Bengal basin exceeds 5 km (Fig, 8.21, 8.22). The thickness of the Bengal deltaic sediments decreases north-west, near the fringe areas of the basin. Marine influence decreases and the section becomes sandier as the basin fringe is approached. The present delta, covering large portions of West Bengal (India) and Bangladesh, is occupied by tidal flats and marshes which are travesed by the distributaries of the Ganga (Hugli River) and the Brahmaputra Rivers. (Fig. 7.36C).

Recognition of Ancient Deltas

Being the largest repository of clastic sediments, deltas are well represented in stratigraphic record. Ancient deltas may be identified by the presence of a thick wedge or lobe of shallow-marine to lacustrine clastic sediments showing typically coarsening-upward sequences. The wedge grades on one side to sediments of deeper water facies and on the other side to non-marine sediments.

The character common to delta deposits is an upward-coarsening progradational sequence leading to a marine, delta margin sand unit. This sequence is overlain by deposits of an aggradational facies - distributary channel sand and crevasse-splay sand bodies. The typical vertical succession in such a delta, as recorded by Galloway (1975), is :

Top   4. Open marine shelf mud or carbonates
         3. Thick sand body separated from the coarsening-upward sand body by thin layers of mud.
         2. Typically coarsening-upward sand body which thins and pinches out laterally
Bottom   1. Prodelta mud blanket

In some cases the sandy zones (2 and 3 above) are replaced by a single, thick sand body which is narrow and elongate in plan and shows a fining-upward profile. In all cases the delta platform and delta plain deposits are criss-crossed by distributary channel sands. These lens-shaped sand bodies show fining-upward sequences.

8

TECTONICS AND SEDIMENTATION

SEDIMENTARY BASINS

Sediments exposed to atmosphere are likely to be eroded soon after their deposition by various natural agencies. For preservation in the geological record it is essential that the sediments be pushed below the base level of erosion. The preservation potential of a deposit is determined by the rate of subsidence of the area of deposition vis-à-vis the rate of sedimentation. The areas of the earth's crust which sink rapidly to allow preservation of a thick sediment load are called *basins*. Basins occupy large areas measuring thousands of square kilometres. Thinning of sediment layers towards the basin fringe is a sure indication of subsidence concomitant with sedimentation. In modern tectonic parlance a basin is a 'prism of rock forming a thick sedimentary succession' (Dickinson 1974). A model for basin evolution must explain, not only the mechanism of initiation of a depression in the earth's crust but also the process of uplifting of the adjacent areas, in order to provide a source for the sediment fill.

Basin subsidence was long imagined to be the effect of gravitational loading during sedimentation. Precise computations have shown, however, that the crust of the earth would not buckle to the extent needed for generation of large sedimentary basins due to the weight of the sediment load alone, but this could happen in the case of smaller basins. For more than a century the origin of large sedimentary basins was explained in terms of the classical geosynclinal model. Since the advent of the plate tectonics concept in the mid-sixties of the 20th century, there has been an equally enthusiastic attempt to explain the origin and development of large sedimentary basins in terms of plate tectonics. These ideas are discussed below in the chronological order in which they evolved.

GEOSYNCLINAL CONCEPT

The geosynclinal concept had its origin from the discovery by James Hall in the mid-19th century that the thickest sediment accumulation in the Appalachians is along the line of maximum depression (Fig. 8.1 A). It was recognized that large-scale sedimentation, which eventually led to the development of mountain belts, occurred along linear troughs along the continental borders. These troughs, separated from the ocean basins by elevated areas ('*borderlands*'), were designated '*geosynclines*' by Dana in the late 19th century. A geosyncline need not always be restricted to the continental-oceanic boundary. The Alpine-Himalayan belt, an area of thick sedimentation and intense orogenic activity, is located between two continental blocks. Hence modern workers define a geosyncline as a basin of sedimentation located marginal to, or between, two cratons (Aubouin 1965). *Cratons* are the tectonically stable areas constituting the nuclei of the continental crust, while geosynclines are the locales of intense tectonic and igneous activity, both syn- and post- orogenic. The essential implication in the concept of geosyncline, therefore, is an idea of mobility. Mobility of the earth's crust in certain long, linear belts was imagined to be responsible for the accumulation of an unusually thick sequence of sediments.

A typical geosyncline of the above-mentioned model consists of an amagmatic zone of relatively shallow-water sedimentation bordering the craton. This zone, which receives sediments mainly from the cratonic source, has been termed the *miogeosyncline*. Away from the craton lies the deeper part of the geosyncline, the *eugeosyncline*. Separating the miogeosyncline from the eugeosyncline is a ridge, originally termed the *geanticline* but renamed the miogeosynclinal ridge by Aubouin in 1965 (Fig. 8.1 B). The eugeosyncline receives sediments from the geanticline on one side and from the tectonically active *borderland* located on the other side. In marked contrast to the miogeosynclinal deposits, the eugeosynclinal sediments contain abundant tuffaceous material, the products of volcanic activity of the island arcs bordering the eugeosynclines. The eugeosynclinal sediments are also permeated by basic flows erupting on the ocean floor in the form of pillow lavas. The eugeosynclinal sediments of extraordinary thickness, traditionally related to orogenic events, are broadly classified into two groups, pre and post-diastrophic. Since the nineteenth century the former has been designated as *flysch* and the latter as *molasse*.

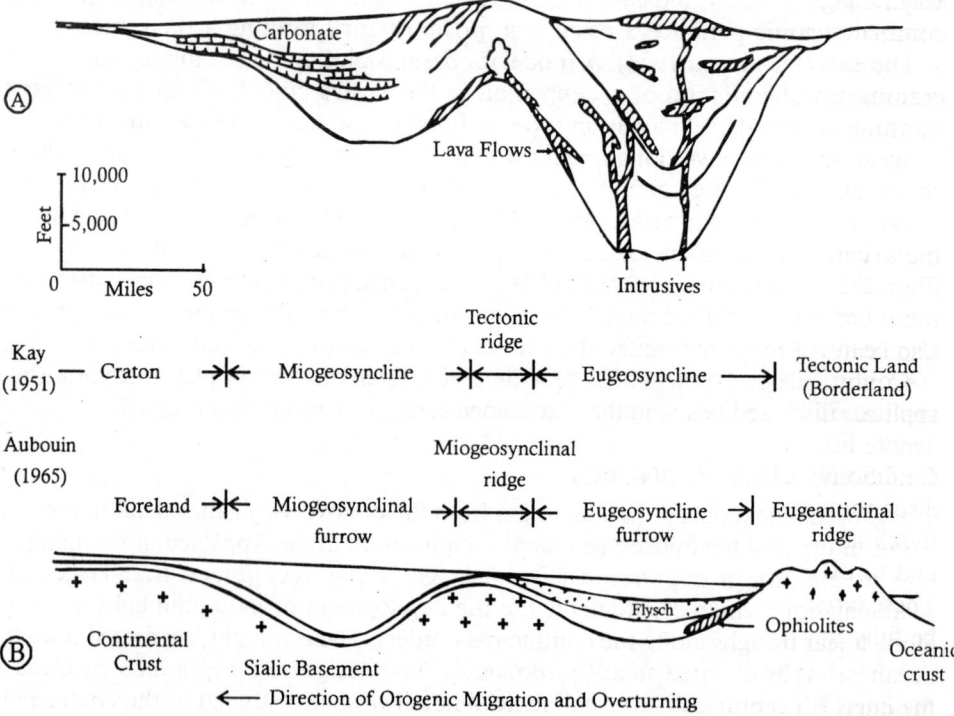

Fig. 8.1: Geosynclinal terminology (A) after Kay 1951, (B) after Aubouin 1965. (A) represents restored cross-section across the Appalachians from Eastern New York to Maine. The deposits are mostly Cambro-Ordovician in age.

Flysch has traditionally been described as a thick succession of argillaceous sediments of deep-water origin interrupted by dark gritty sandstones. Pillow lavas (altered to '*greenstones*') are common within this sequence. Many flysch deposits start with massive graded beds at the bottom and end in pelitic layers on top through a sequence of parallel lamination and rippled beds (the Bouma sequence). The presence of such sequences in beds of deep-water origin, as evidenced by deep-water biota, led to the conclusion that the dark sandstones sandwiched between the argillaceous sequences were transported and deposited by turbidity current activity.

The shale-siltstone-fine-grained sandstone-graywacke sequence of the Simla 'Series' (Precambrian) of the Himachal Pradesh and Garhwal Himalayas were interpreted as flysch (Sen, Bhattacharya and Ray 1969; Valdiya 1970). The Precambrian metasediments of Ghatsila in eastern India consisting of a rhythmically bedded sequence of quartzites and mica schists, originally deposited as a thick sequence of sandstone and shale, bear all the imprints of a typical flysch. Penecontemporaneous deformation structures such as load casts, pull-aparts, convolute laminations and slump balls have been reported from these metasediments (Naha 1961; Gaal 1964).

Molasse is the product of post-orogenic sedimentation. The deposits, generally several thousand metres thick, consist of many sequences of fining-upward, fluvial and deltaic cycles. The conglomerates contain well-rounded clasts. The fluvial cycles include immature, lithic and feldspathic sandstones, mudstones, coal, caliche, freshwater carbonates and evaporites. In areas of marine transgression the environment may range from paralic to lagoonal. Persistent, well-sorted conglomeratic sandstones dominate the areas of marine transgression.

The extensive Cenozoic molasse of the northern Alps consists of clastic wedges, grading upward from autochthonous flysch, through thin-bedded mudstone, to transitional paralic sequences. The succeeding rocks merge to marine deposits laterally because sedimentation kept pace with subsidence to produce near-sea-level surfaces where exchange of marine and non-marine conditions were possible (van Houten 1974). The Miocene-Pleistocene molasse of the Himalayas (the Siwalik Group of India) consists of more than 6000 m of fluviatile conglomerates, sandstones and shales (Figs. 8.2 and 8.3). The palaeodrainage of the Siwaliks was mostly from north to south but northerly drainage systems, controlled by intrabasinal tectonic highs, have also been identified (Tandon 1976, Tandon 1990; Tandon, Kumar and Singh 1985).

Confusion has sometimes been created in sedimentological literature by varied application of the terms flysch and molasse. The terms have been variously used to denote lithofacies, depositional environments, and/or tectonic (syn-or post-tectonic) conditions. The confusion created has prompted some sedimentologists to recommend discontinuation of both terms. It should be remembered, however, that these terms, flysch as well as molasse, have thoroughly pervaded the sedimentological literature and to discard them at this stage would only add to the confusion already created. These terms may thus be retained but the specific sense in which they are used should be fully understood.

Modern Geosynclines

Following the classical studies on geosynclines in the late nineteenth and early twentieth century, many a sedimentary basin was identified as a geosyncline. In fact, Marshall Kay, who proposed a comprehensive scheme of classification of geosynclines

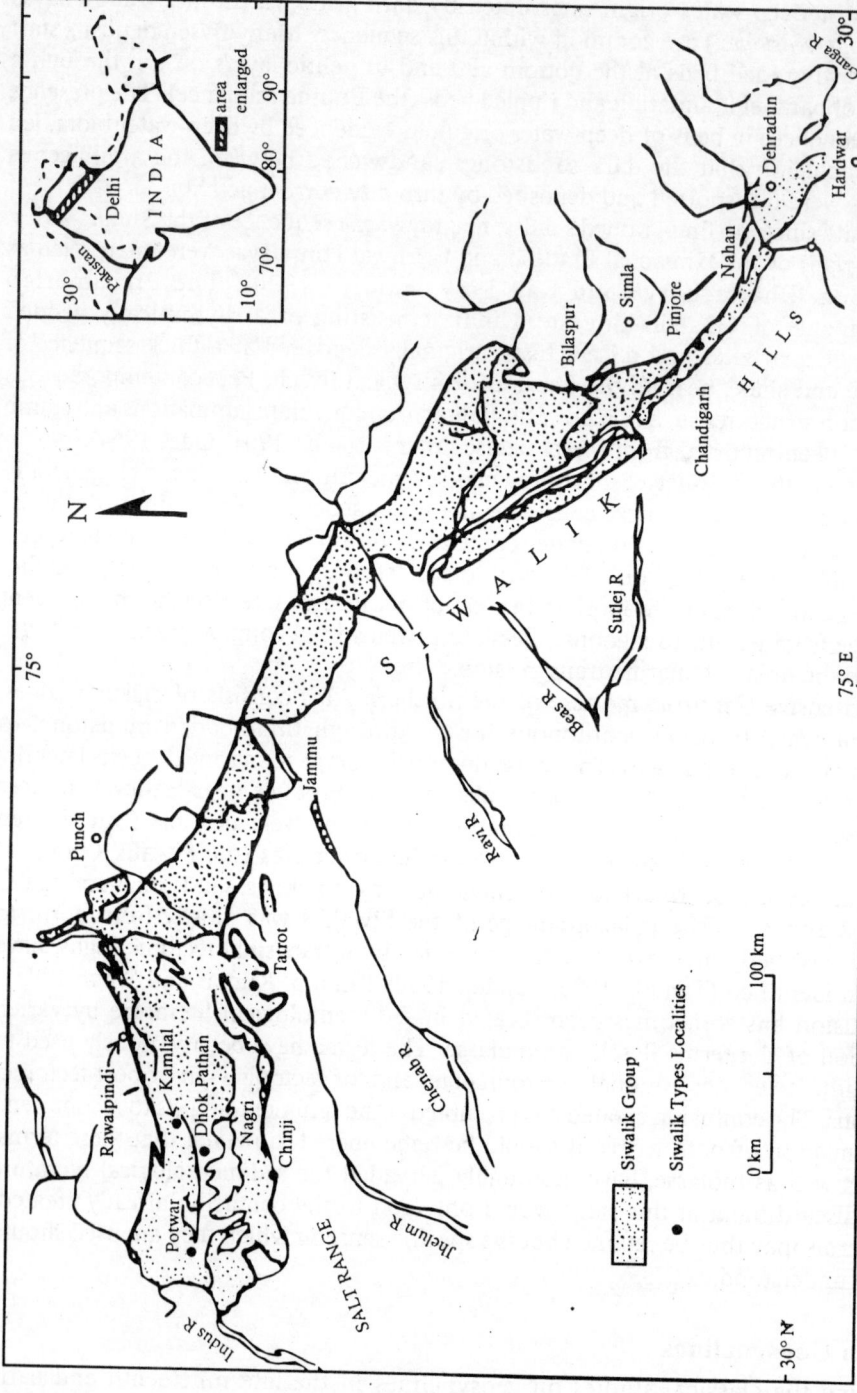

Fig. 8.2: The Siwalik Basin between the rivers Indus and Ganga (after Tandon 1990). Reprinted with permission of the author.

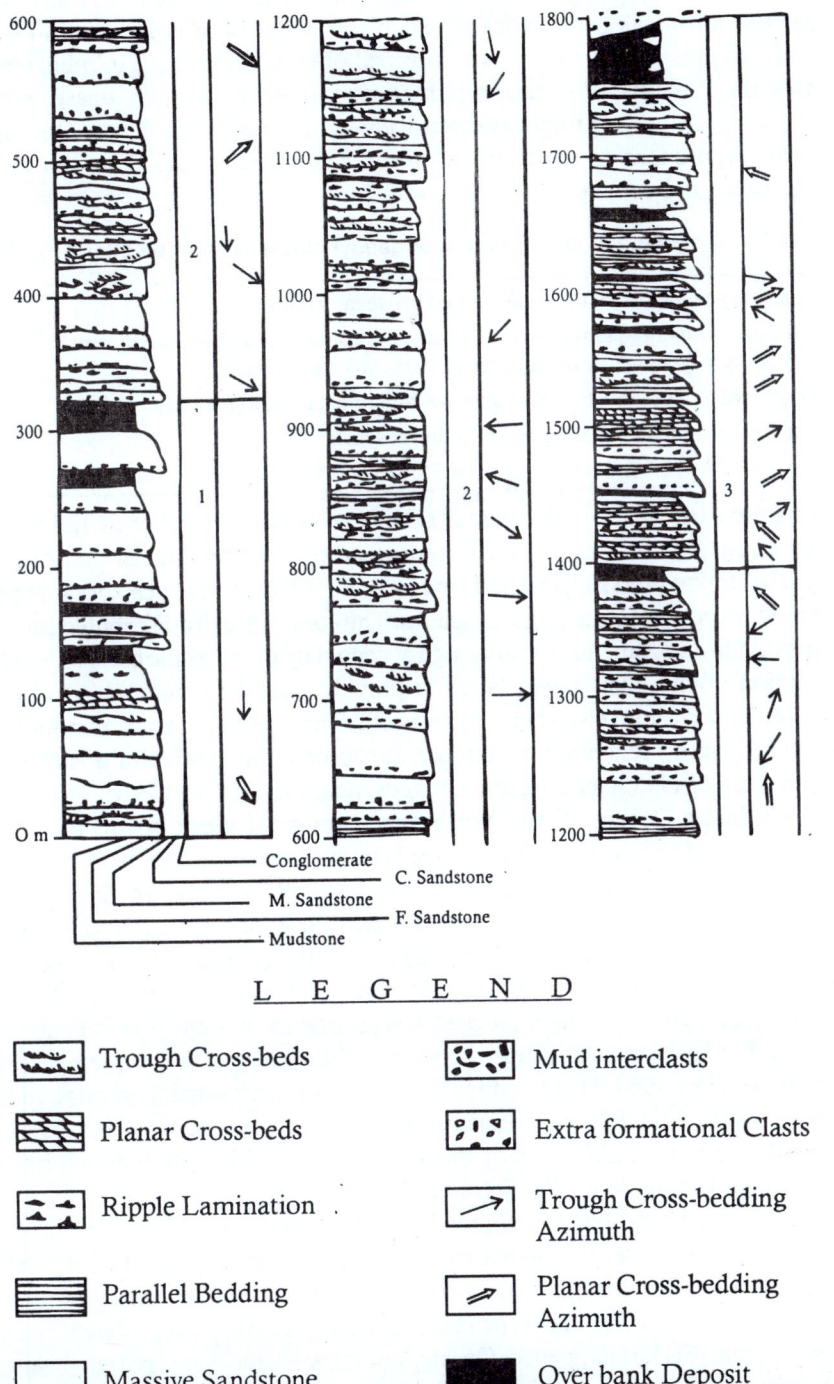

Fig. 8.3: The Middle Siwalik lithological columns, Dehra Dun region (after Kumar and Nanda 1989). With permission of the Geological Society of India.

in 1951, preferred to designate every sedimentary basin located within, outside or adjacent to cratons as a geosyncline. Such a scheme of classification (Table 8.1) apparently accepts a geosyncline as a universal phenomenon, both in time and space. Following the Principle of Uniformitarianism, one would, in such a case, expect the geosynclines to be present today as extensively as in the geological past. Our difficulty in recognizing modern geosynclines, however, has raised doubts about the very basis of the geosynclinal concept.

Table 8.1: A scheme of classification of geosynclines (simplified after Kay 1951)

1. Orthogeosynclines (linear troughs located between cratons)

    (a) Eugeosynclines (actively subsiding areas, volcanics present with sediments)
    (b) Miogeosynclines (less active tectonically, volcanics absent)

2. Parageosynclines (ovate troughs located within or adjacent to cratons)

    (a) Exogeosynclines (extensions of orthogeosynclines)
    (b) Autogeosynclines (isolated basins within cratonic uplifts)

During the early years of this century the deep oceanic trenches of Indonesia and the bordering volcanic arcs were intensively studied by Dutch geologists and geophysicists following a suggestion from Haug in 1900 that these deep oceanic trenches are the modern counterparts of geosynclines. The concentrically distributed Indonesian Arcs (the Banda and Sunda Arcs) although accepted by many as the basis for explaining the evolution of the East Indies, failed to provide a model comparable to the large-scale geosynclines found elsewhere in the world. The sediment layers identified within the Java and Japan Trenches, for example, are comparable neither in thickness nor in character to those characterizing ancient geosynclines.

In the Atlantic Ocean, off the North American coast, thick prisms of submarine sediments (named '*continental terrace*') have been interpreted as the modern analogues of *orthogeosynclines* (Drake, Ewing and Sutton 1959). Similar thick sequences of suboceanic sediments have been termed '*continental rise*' elsewhere (Dietz 1963). Some of these submarine deposits are enormously thick—the estimated thickness of the Bay of Bengal Deep-Sea Fan, or example, exceeds 12 km (Curray and Moore 1971). Many authors would like to associate these deep-sea sediments with the phenomenon of sea-floor spreading and would imagine them to be the modern counterparts of classical geosynclines (Dietz 1963; Dietz and Holden 1974). In contradiction to this hypothesis, it has been pointed out that the sediments of the continental rise were derived from the continental side only, while those of the ancient geosynclines were drawn from both the foreland and borderland (Pettijohn 1975).

The idea of whole ocean basins being considered modern representatives of geosynclines being dragged towards mobile belts has been questioned by Pettijohn (1975, p. 577) on the ground that in that case the pelagic carbonates would have ended up in deep oceanic eugeosynclinal sequences. There is no evidence, according to Pettijohn, that this has happened. Of late, several workers have reported the presence of CO_2-rich inclusions in granulites (see for example, Hansen, Newton and Janardhan 1984; Newton 1988). It has been suggested that one of the possible sources of CO_2 is the deeply buried carbonate sediments. This leads one to wonder whether the source behind the carbonate fluid was the pelagic carbonate which was actually carried to the mantle by the process of subduction.

PLATE TECTONICS CONCEPT

A model of global tectonics, called the *plate tectonics* model, emerged during the two decades of intensive exploration in the sea following the Second World War. During this period palaeomagnetic studies provided evidence of secular changes in palaeolatitudes, thereby confirming large-scale movements of the continents. Detection of newer oceanic crusts with '*magnetic stripes*' being added to the ocean-floor on either side of the midoceanic ridges since the Mesozoic, confirmed the idea of sea-floor spreading. The major dynamic features of the earth such as orogenesis and seismicity could be connected to the phenomenon of gliding of the rigid lithospheric plates over the asthenosphere. It was also recognized that the lithosphere is consumed at the *Benioff subduction zones*, on the landward side of the deep-sea trenches, so that the addition of newer crust at the midoceanic ridges does not change the net global surface area. Following these findings six major and a few minor lithospheric plates were identified and the plate boundaries were divided into three groups, depending on the nature of their movements (McKenzie and Parker 1967; La Pichon *et al.* 1973). These are: (1) Diverging plate boundaries at the midoceanic ridges—accreting margins where new oceanic crusts are added to the ocean-floor. (2) Converging plate boundaries (called *subduction* or Benioff zones) at the submarine trenches defined by the plane where the deep foci earthquakes are located. The lithosphere disappears below the mantle at these boundaries. (3) Transform fault or shear boundaries where the plates glide past one another without addition or destruction of the crust.

The continental margins, on the basis of their coincidence (or not) with the plate boundaries, are divided into two groups, 'active' and 'passive'. The western margin of the continent of South America, which is actively involved in plate movement, is an example of an 'active' margin, while the eastern margins of North and South America, not coinciding with any plate boundary, are 'passive' margins (Fig. 8.4).

The setting of a sedimentary basin can now be described in terms of three factors: 1) the substratum of the basin, 2) proximity of the basin to the plate margin and 3) the type of plate margin involved in basin development (Dickinson 1974). In this model the continental and the oceanic crusts are the two end-members of a series of transitional types of substrata. In terms of proximity to plate margin, Dickinson (1974) divided sedimentary basins into two groups : 1) those located within the plates ('*intraplate basins*') and 2) those lying at the zones of plate interaction. Three different types of plate junctures described earlier, namely, divergent, convergent and transform, were also recognized. For each of the four broad classes of plate settings (*intraplate, divergent, convergent and transform*), Dickinson recognized four subclasses based on the nature of substratum.

Dickinson's model provided a basis for classification of sedimentary basins from the newer perspective of plate tectonics, but the real situation is even more complex than that stated above. Because of the movements involved over long spans of time, anomalous crusts, inherited from the previous plate junctures, may be incorporated into the substratum of intraplate basins. Furthermore, substrata of both continental and oceanic types may be present in the basins located on the zones of divergent,

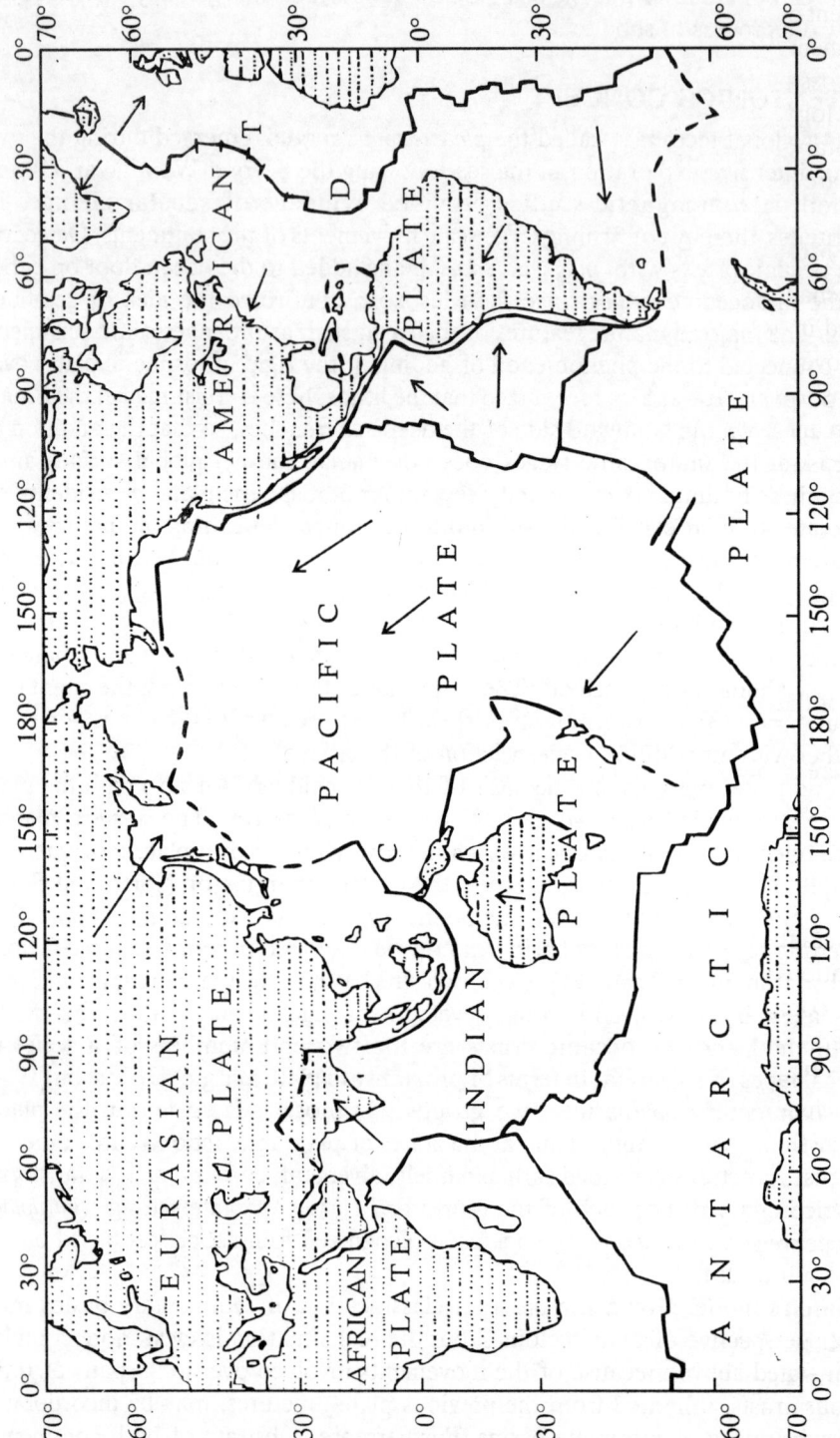

Fig. 8.4: Map showing the major plate boundaries of the world. D - diverging boundary, C - converging boundary, T - transcurrent faults.

convergent or transform plate boundaries. For these reasons a direct correlation between the sedimentary basins classified according to the geosynclinal and plate tectonic models is not possible. As Reading (1982) very aptly commented: 'The old geosynclinal terminology can now usually be abandoned provided it is realized that many sedimentary basins fall into more than one category and that interpretations of the ancient are never more than working hypotheses'.

PLATE MOVEMENTS AND BASIN FORMATION

The mechanism of origin and evolution of large sedimentary basins can now be discussed in terms of *diverging* (extensional), *converging* (compressional) and *strike-slip* plate movements. Some of the current models of basin development under these situations are briefly presented below. The readers may refer to Reading (1986) and Shannon and Naylor (1989) for more detailed discussions. The models discussed here are based largely on theoretical studies and partly on actual geological-geophysical observations on basin architecture, lithic fill and lithospheric conditions. They should be taken only as tentative suggestions. Not one is fully proven.

Basins in Extensional Setting

The simplest type of extensional model envisages upwelling of the asthenosphere during passage of the continents over hot spots leading to doming of the lithosphere. Subsequent cooling of the crust and erosion of the dome leads to generation of sedimentary basins (Fig. 8.5). In another model thinning of the continental crust by rapid stretching of the lithosphere is contemplated to initiate subsidence and thermal anomaly. Slow cooling, following withdrawal of stretching, leads to thermal contraction and subsidence (McKenzie 1978; Jarvis and McKenzie 1980). Seismic evidence of thinning of the crustal layer below the North Sea provides support for this model. The phases of crustal subsidence and marine onlap worked out in the North Sea on the basis of this model match actual observations. At a later stage, isostatic adjustments may introduce complications to the basins developed in this way.

More complex, fault-controlled models for basin development have also been postulated. These models are based on detection in the Rocky Mountains and the Colorado Plateau of major zones of shear traversing the whole lithosphere (Fig. 8.6). These include the uniform simple shear, non-uniform pure shear and mixed shear models. A shear traversing the lithosphere offsets the regions of maximum thinning of the crust and the mantle. The basins developed on such a shear are naturally asymmetric in shape. Moreover, complications may arise due to differential response of the upper brittle crust and the lower ductile mantle to shear movements (Wernicke 1981).

The initial site of rifting is controlled to some extent by the composition of the crust and the mantle. The vertical strength profile of the lithosphere is influenced by these compositions. For a dry, granulitic lithosphere, the strength profile increases downwards, through the crust, only to decrease sharply at the ductile mantle. This makes it difficult for the upper brittle section to stretch. The presence of water pushes up the ductile level, thereby making stretching of the lithosphere easy.

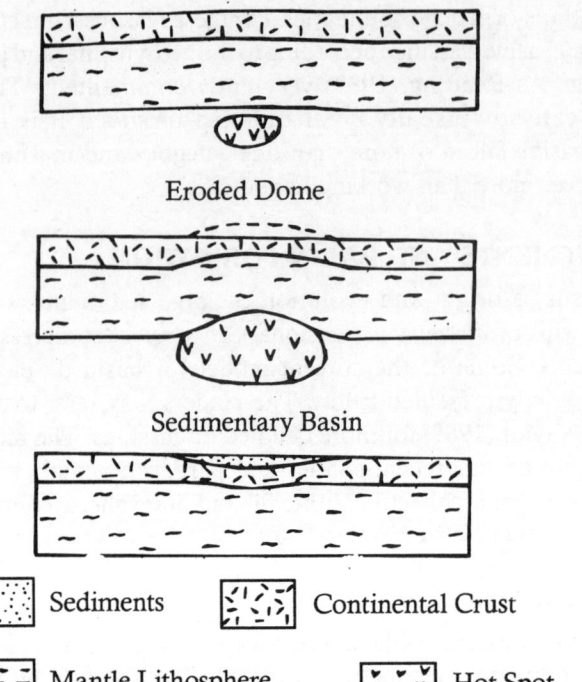

Eroded Dome

Sedimentary Basin

Sediments   Continental Crust

Mantle Lithosphere   Hot Spot

Fig. 8.5: Mechanism of development of sedimentary basins in response to thermal doming (after Shannon and Naylor 1989). With permission of Graham & Trotman.

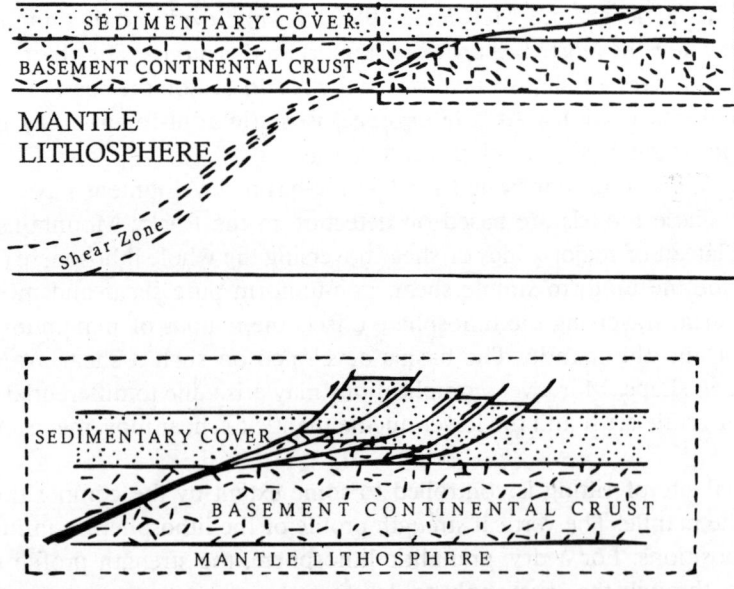

Fig. 8.6: Diagrams illustrating the mechanism of development of 'thin-skinned thrusting' in areas of compression (after Shannon and Naylor 1989). With permission of Graham & Trotman.

The processes responsible for formation of grabens within the continental crust and also for continental rupture creating new sea-floor are similar in many ways. Both are associated with tensile faulting, crustal thinning, warming of the mantle below the rift zone and updoming of the outer flank of rifts. Thermally controlled subsidence occurs as the anomalously warm mantle below the rift cools. Lateral heat flow may cause transient uplift of the graben flanks (Jarvis 1984). The uplifted basin flanks, in turn, provide sediments to the basin.

In another model of basin development by extensional process, three-armed, radial rift systems ('*triple junctions*') are contemplated to be produced by convective currents originating in the mantle. Large ocean basins are believed to be produced by spreading of the two active rift arms, while re-entrance of the third arm into continental platforms initiates rift basins. These abandoned rift basins have been termed '*aulacogens*' (meaning 'born as furrows') by Russian geologists. The tectonic movements in aulacogens are vertical and the sediment transport direction is mainly longitudinal, along the basin axis. Alkalic (basalt-rhyolite) rather than tholeiitic volcanic activity is another characteristic of the aulacogens (Hoffman *et al.* 1974). The Benue Trough is generally cited as a classical example of a rift arm abandoned during separation of Africa and South America during the Cretaceous (Fig. 8.7).

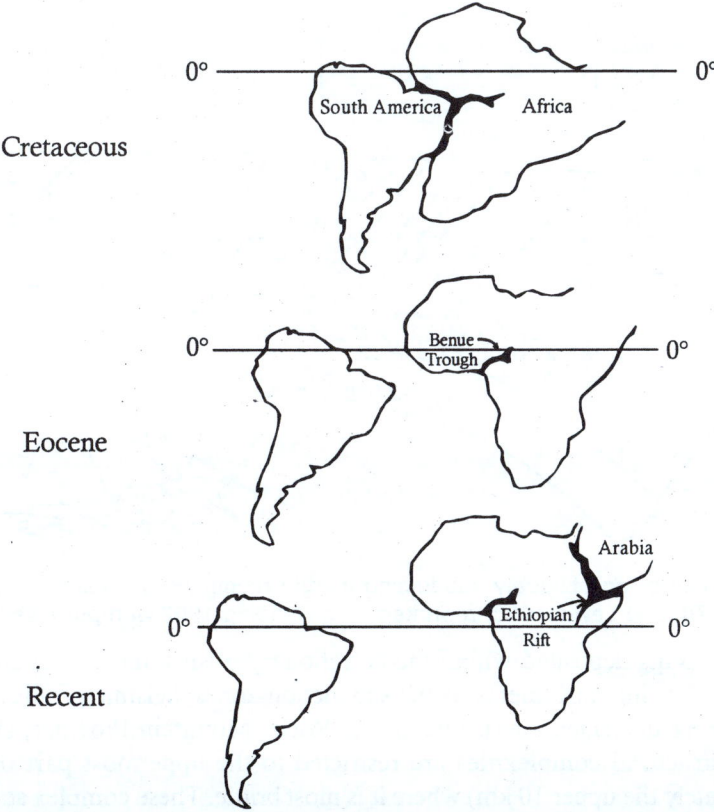

Fig. 8.7: Maps showing stages of origin of Benue trough as a rift arm abandoned during separation of Africa from South America (after Hoffman *et al.* 1974). With permission of SEPM (Society for Sedimentary Geology).

Basins in Compressional Setting

Seismically active subduction zones mark the converging plate margins. On the landward side of these subduction zones, which dip towards the continents, lie volcanic island arcs. On the seaward side occur deep-sea trenches. Orogenic belts are produced out of collision of the lithospheric plates. Intrusion of the subducted material into the lower crust causes doming of the igneous and metamorphic terranes at the zone of convergence, into what are called '*arc massifs*'. These elevated areas at the basin margins serve as the source of sediments while the sediments themselves are deposited in one of the following five areas: within the ocean, on the down-going oceanic plate, within the deep-sea trench, in small basins on the slope of the inner wall of the trench, seaward of the volcanic arc (*outer arc* or *forearc basin*), and behind the volcanic arc (*back-arc basins*).

Thick sedimentary piles are stacked in the form of accretionary prisms in the forearcs. As the subduction zone migrates towards the sea the forearc sediments are successively thrust upwards (Fig. 8.8). Small sedimentary basins may develop over these accretionary prisms. Those lying between the volcanic arc and the trench slope break are generally very wide (50-100 km). These are the forearc or outer-arc basins. Behind the volcanic arc occur the back-arc basins.

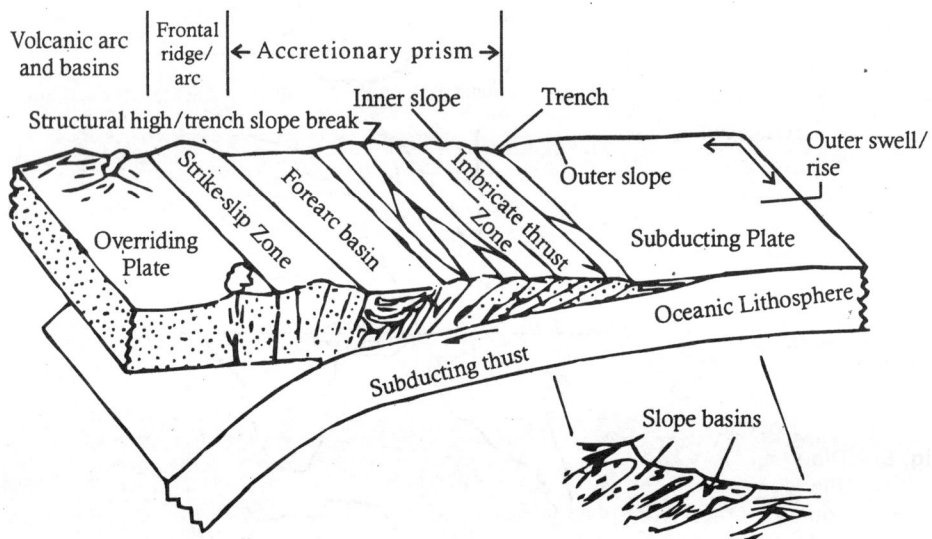

Fig. 8.8: Model of an obliquely subducting margin (complied by Reading after Walcott 1978, and Lewis 1980, reprinted from Reading 1982 with permission).

The sediments deposited within the accretionary prisms undergo intense tectonic deformation during movements of the subduction zones. Seismic profiles run across many of these deformed areas, such as the Rocky Mountain Province, showed that most the structural complexities are restricted to the uppermost part of the crust (approximately the upper 10 km) where it is most brittle. These complex structures are replaced underneath by only a few simpler, poorly defined thrusts going down almost to the base of the lithosphere. This phenomenon has been termed 'thin-skinned compressional tectonics'.

Basins in Strike-slip and Transform Fault Zones

Strike-slip faults may occur either at the major plate boundaries or at the border of microplates. *Transform faults* occur as fracture zones running across the midoceanic ridges. The overall movement in a strike-slip fault is essentially horizontal and parallel to the fault trace but the dip-slip component may also be important locally. This oblique movement causes the strike-slip motion to be either divergent (*transtensile*) or convergent (*transpressive*). The former is likely to produce normal faulting, leading to basin development and volcanism. The latter produces reverse faults, thrusts, uplift and folding. With the passage of time, a divergent strike-slip fault may be converted to a convergent one.

Curving, splitting and offsetting of strike-slip fault trends may lead to local zones of extension and compression. The latter produces sedimentary basins while the former leads to uplifted areas which eventually serve as the source of sediments (Reading 1982; also see Fig. 8.9).

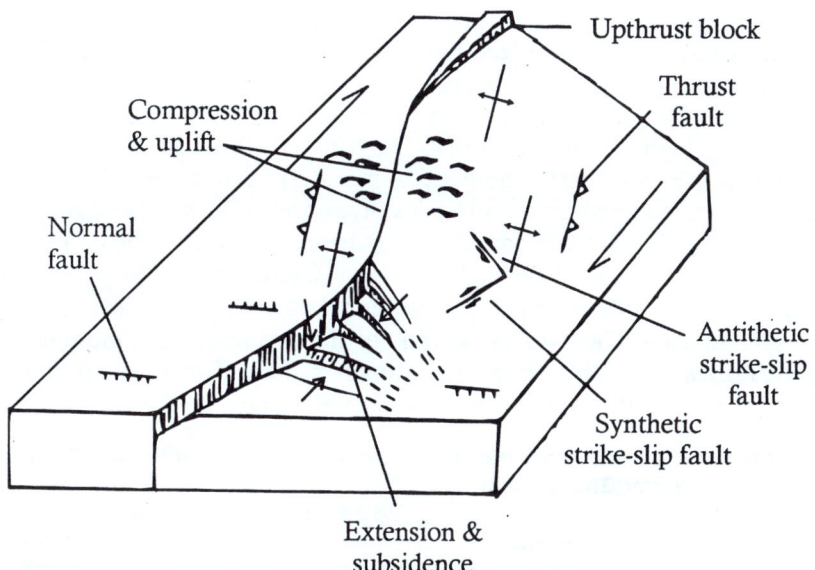

Fig. 8.9: Diagram illustrating the mechanism of basin development from dextral strike-slip fault. Both extensional basins and compressional uplifted areas serving as source of sediments may be produced as a result of such faulting. Direction of sediment transport is indicated by thick arrows (after Wilcox *et al.* 1973, modified by Mitchell and Reading 1978, reprinted from Reading 1982 with permission).

TECTONIC CONTROL OF SANDSTONE COMPOSITION

The technique of interpretation of tectonics from sandstone mineralogy and chemistry is based on the belief that the detrital framework mineralogy of a sandstone is dependent largely on the tectonic setting of its provenance. This technique of interpretation should be used with caution, however, because the mineralogy of a sandstone may be considerably modified by sorting processes during transportation. Moreover, diagnostic processes may also obliterate the original detrital mineralogy and chemistry of a sandstone.

Quartzose sands from continental cratons are widespread within interior basins, platform successions and opening ocean basins. Arkosic sands from uplifted basement blocks are present in rift troughs and wrench basins related to transform faults. Volcaniclastic lithic sands and volcanic-plutonic sands derived from magmatic arcs are found in trenches, forearc basins and marginal seas. Recycled orogenic sands, rich in quartz or chert and other lithic fragments, occur in closing ocean basins, diverse successor basins and foreland basins. These sands are derived from subduction complexes, collision orogens and foreland uplifts. Using triangular QFL and QmFL plots of quartz (mono- and polycrystalline varieties), feldspar (both plagioclase and K-feldspar) and lithic fragments (volcanic/metavolcanic and sedimentary/metasedimentary), Dickinson and Suczek (1979) could discriminate these tectonic settings (Fig. 8.10).

Bhatia (1983) could discriminate the Palaeozoic sandstones from various tectonic settings of eastern Australia with the help of their major element geochemistry. Plots of Fe_2O_3 + MgO versus TiO_2, Al_2O_3/SiO_2, K_2O/Na_2O, and $Al_2O_2/(CaO + Na_2O)$ were utilized for this purpose. The oceanic island arc sandstones, derived from calc alkaline andesites, are characterized by higher proportions of TiO_2, Al_2O_3, Na_2O, and Fe_2O_3, and a relatively lower proportion of SiO_2 and K_2O than all other sandstones. The continental island arc sandstones, derived from felsic, volcanic rocks, are characterized by higher proportions of SiO_2, K_2O and the K_2O/Na_2O ratio (≈ 0.6) and lower Fe_2O_3 + MgO than the oceanic island arc sandstones. The sandstones from the active continental margins (Andean-type and strike-slip basins) are derived mainly from the uplifted basement. These sandstones, reflecting the composition of the upper continental crust, have high SiO_2 and K_2O content with $K_2O/Na_2O \approx 1$. The passive margin sandstones are generally enriched in SiO_2 and depleted in Al_2O_3, TiO_2, Na_2O and CaO. The K_2O/Na_2O ratio of these sandstones is more than 1, but these sandstones may show large variations in composition, sometimes even overlapping the compositions of active continental margin sandstones (Table 8.2).

Table 8.2: Abundances and proportions of various oxides in sands of various tectonic settings (after Bhatia 1983)

	Oceanic island arc (OIA)	Continental island arc (CIA)	Active continental margin (ACM)	Passive margin (PM)
$Fe_2O_3^{(t)}$ + MgO	8-14	5-8	2-5	(2.89)*
	(11.73)	(6.79)	(4.63)	
TiO_2	0.8-1.4	0.5-0.7	0.25-0.45	Depleted
	(1.06)	(0.64)	(0.46)	
Al_2O_3/SiO_2	0.24-0.33	0.15-0.22	(0.18)	(0.10)
	(0.29)	(0.20)		
K_2O/Na_2O	0.2-0.4	0.4-0.8	(0.99)	(1.6)
	(0.39)	(0.61)		
$Al_2O_3/(CaO + Na_2O)$	(1.72)	(2.42)	(2.56)	(4.15)
SiO_2	(58.83)	(70.69)	(73.86)	(81.95)
				Enriched

* figures in parentheses indicate the average values.

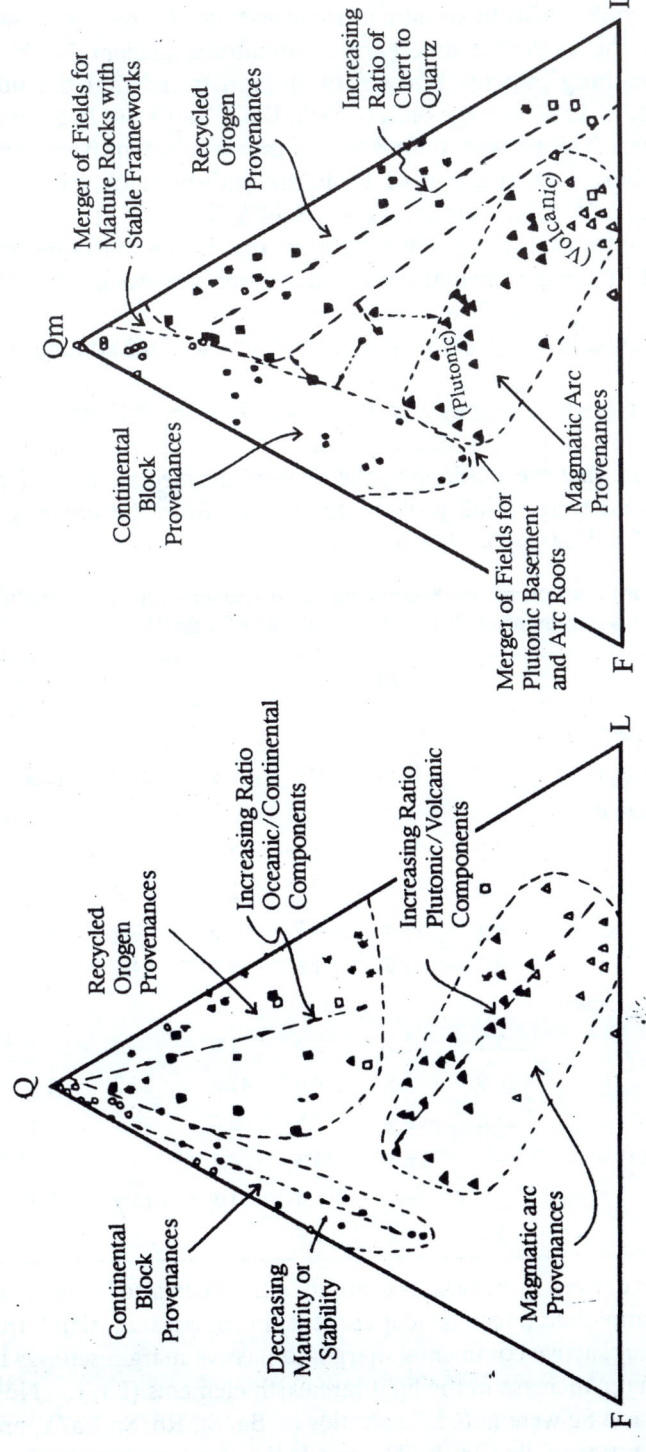

Fig. 8.10: QFL and QmFL plots showing derivation of sandstone suits from different tectonic provenances (after Dickinson and Suczek 1979) Q = total quartz grains, Qm = monocrystalline quartz grains, F = total feldspar grains, L = total unstable lithic fragments, all polycrystalline. With permission of American Association of petroleum Geologists.

In another study the framework mineralogy and volatile-free chemistry of graywackes were related by Crook (1974) to specific tectonic settings. Using data from deep-sea sands, taken to be the modern counterparts of arenites in ancient flysch, and sands from rivers debouching into the sea, Crook demonstrated that the quartz-poor graywackes (quartz < 15%, average SiO_2 = 58%, K_2O/Na_2O << 1) are indicative of magmatic island arcs. The average composition of these rocks approximate the average composition of these arcs and that of tholeiitic andesites. Quartz intermediate graywackes (quartz = 15-65%, average SiO_2 = 68-74%, K_2O/Na_2O < 1) are indicative of Andean-type continental margins and have approximately the same composition as the upper level of the continental crust. Quartz-rich graywackes (quartz > 65%, average SiO_2 = 89%, K_2O/Na_2O > 1) are characteristic of Atlantic-type continental margins and are similar in composition to the sand fraction of the continental platform cover.

Quartz content (19% average) and K_2O/Na_2O ratio (<1) of representative greywacke samples from the Chitradurga Schist Belt (> 2600 m.y.) of India (Tables 8.3 and 8.4) indicate that these belong to the quartz-intermediate type of graywacke of Crook (1974) and were, in all probability, derived from the upper level of the continental crust (Srinivasan et al. 1989).

Table 8.3: Modal composition of the Keshavapura Graywackes from Chitradurga Schist Belt (KD1 samples after Srinivasan et al. 1989, Table 1)

	3	3A	4	5	6	7	M (Avg.)
Quartz 'Q' as clusters and polycrystalline augens	18.2	12.9	20.7	25.9	16.7	19.3	18.9
Polycrystalline detrital grains	5.9	10.9	15.7	5.5	7.3	9.9	9.2
Monocrystalline grains	5.2	1.6	–	2.1	1.4	2.0	4.5
Plagioclase (P)	6.1	14.3	4.7	4.3	10.9	8.2	8.0
K-Feldspar (F)	4.7	1.9	1.6	1.2	1.0	1.9	2.0
Rock fragments, volcanics + glass (R)	2.1	2.7	3.4	1.0	–	–	1.5
Biotite	30.9	6.5	9.0	47.8	36.9	44.1	29.2
Chlorite	25.0	8.9	5.5	5.0	9.9	6.3	6.1
Sericite/Muscovite	–	36.6	34.8	–	–	–	11.9
Calcite	–	–	1.6	3.3	13.6	6.0	4.1
Opaque minerals	1.5	2.9	1.9	3.0	1.1	1.0	1.9

The graywackes derived from various tectonic settings can also be distinguished on the basis of their immobile trace element content. From oceanic island arc through continental island arc, active continental margin to passive margin settings in eastern Australia, a systematic increase in the light rare earth elements (La, Ce, Nd), Th, Nb and decrease in V and Sc were noted. The ratios of Ba/Sr, Rb/Sr, La/Y, and Ni/Co increase while the ratios of Ba/Rb, K/Th, and K/U decrease in the same direction

(Bhatia and Crook 1986). Using a similar approach Srinivasan *et al.* (1989) suggested an island arc setting and a mixed felsic and mafic source (as indicated by the high proportion of Zr as well as Cr and Ni; see Table 8.5) for the Keshavapura Graywackes of the Chitradurga Schist Belt of Karnataka, India. Discrimination between oceanic and continental island arcs was, however, not possible.

Table 8.4: Chemical composition of the Keshavapura Graywackes from Chitradurga Schist Belt (KD1 samples after Srinivasan *et al.* 1989, Table IIA)

Wt%	3	3A·	4	5	6	7	8
SiO_2	61.05	58.05	61.16	67.66	66.70	64.20	58.80
TiO_2	0.64	0.68	0.64	0.46	0.46	0.57	0.80
Al_2O_3	14.10	14.33	14.13	13.93	14.05	14.62	14.63
FeO	6.44	6.48	5.20	3.36	2.92	4.20	7.12
Fe_2O_3	3.01	4.02	3.93	3.94	3.32	3.83	4.04
CaO	4.70	4.81	4.71	3.72	4.24	2.57	1.94
MgO	6.05	6.25	4.55	1.72	1.42	2.42	4.38
Na_2O	3.05	3.30	3.79	2.70	3.66	3.43	3.21
K_2O	1.50	1.15	1.09	1.75	2.00	2.41	3.04
MnO	0.08	0.10	0.06	0.03	0.03	0.04	0.08
P_2O_5	0.13	0.13	0.12	0.08	0.08	0.09	0.10
LOI	0.52	1.19	0.66	0.53	0.71	0.71	0.70

BASIN CLASSIFICATION AND DESCRIPTION

The concept of plate tectonics prompted classification of sedimentary basins in terms plate movements (see, for example, Miall 1984; de V. Klein 1987). The comprehensive scheme of classification of Shannon and Naylor (1989) reproduced in Table 8.6 relates sedimentary basins to crust type, nature of plate movement, dominant structural movement and location of the basin with respect to the plates. The characters of some of the major basin types are discussed below with reference to this scheme of classification, drawing examples from India (Fig. 8.11) and elsewhere. Many of these groupings, based on scant information, are necessarily tentative. Many of the older sedimentary basins are polygenetic because they passed through several stages of development at different times of their existence.

Downwarp Basins

These large, elongate basins develop by downwarping (sagging) of the cratonic shelves bordering oceans. In these basins sheets of stable shelf quartz arenites and carbonates overlie deeper water argillaceous sediments. The West Canada Basin of North America, the Saudi Arabian Platform of the Middle East, and the Vindhyan Basin of central India are some of the typical examples of downwarp basins.

The Canadian Shield slopes smoothly towards the west until it reaches the eastern boundary of the Rocky Mountain thrust belt. This downwarping, which took place in the Late Proterozoic, accommodated sediments ranging in age from the Cambrian to the Lower Jurassic. The deposits, which wedge out on the Canadian Shield, attain a maximum thickness of about 6 km on the western side bordering the Rockies. Marine and nearshore sediments, ranging in age from the Cambrian to the Lower Jurassic, were initially deposited on this downwarped cratonic shelf. These were followed by the deposition of thick marine and non-marine sediments during the Middle Jurassic to the Lower Tertiary on a foreland basin which developed east of the orogenic belt. The whole sequence, from the Cambrian to the Lower Tertiary, is punctuated by four major unconformities caused by uplifts related to westward movement of the North American plate. The sedimentary facies and structural elements of the West Canada Basin were controlled by these tectonic pulses (Shannon and Naylor 1989).

Table 8.5: Trace and rare earth element composition of Keshavapura Graywackes from Chitradurga Schist Belt (KD1 samples after Srinivasan et al. 1989, table IIB)

(ppm)	3	3A	4	.5	6	7	8
Zr	141	153	172	221	228	217	188
Y	32	24	27	44	42	45	44
Sr	395	382	467	687	466	378	288
Rb	53	40	41	65	70	76	82
Ba	406	293	327	389	461	448	599
V	140	147	121	72	76	106	155
Cr	423	414	280	114	122	161	193
Ni	69	80	50	34	24	24	56
Cu	47	68	48	55	67	51	51
Zn	105	106	102	91	84	103	108
La	–	–	23.40	–	52.14	–	33.65
Ce	–	–	51.18	–	81.77	–	67.34
Pr	–	–	5.8	–	8.17	–	6.95
Nd	–	–	21.62	–	30.83	–	26.42
Sm	–	–	4.16	–	4.79	–	4.84
Eu	–	–	0.76	–	0.78	–	0.76
Gd	–	–	4.72	–	5.11	–	4.56
Tb	–	–	0.66	–	0.65	–	0.64
Dy	–	–	2.94	–	3.12	–	3.31
Er	–	–	1.70	–	1.54	–	1.27
Tm	–	–	< 0.03	–	0.30	–	0.24
Yb	–	–	1.12	–	1.40	–	1.03
Lu	–	–	0.22	–	0.30	–	0.24
K/Rb	232	235	217	220	234	259	304
Rb/Sr	0.13	0.10	0.08	0.09	0.15	0.20	0.28
Ba/Sr	7.66	7.32	7.97	5.98	6.58	5.89	7.30
Zr/Y	4.4	6.37	6.37	5.02	5.42	4.82	4.27
V/Ni	2.02	1.83	2.42	2.10	3.16	4.40	2.76
Cr/Ni	6.13	5.17	5.60	3.30	5.08	6.70	3.44

Table 8.6: Basin classification based upon genesis and tectonic location (after Shannon and Naylor 1989), with examples of basins inserted.

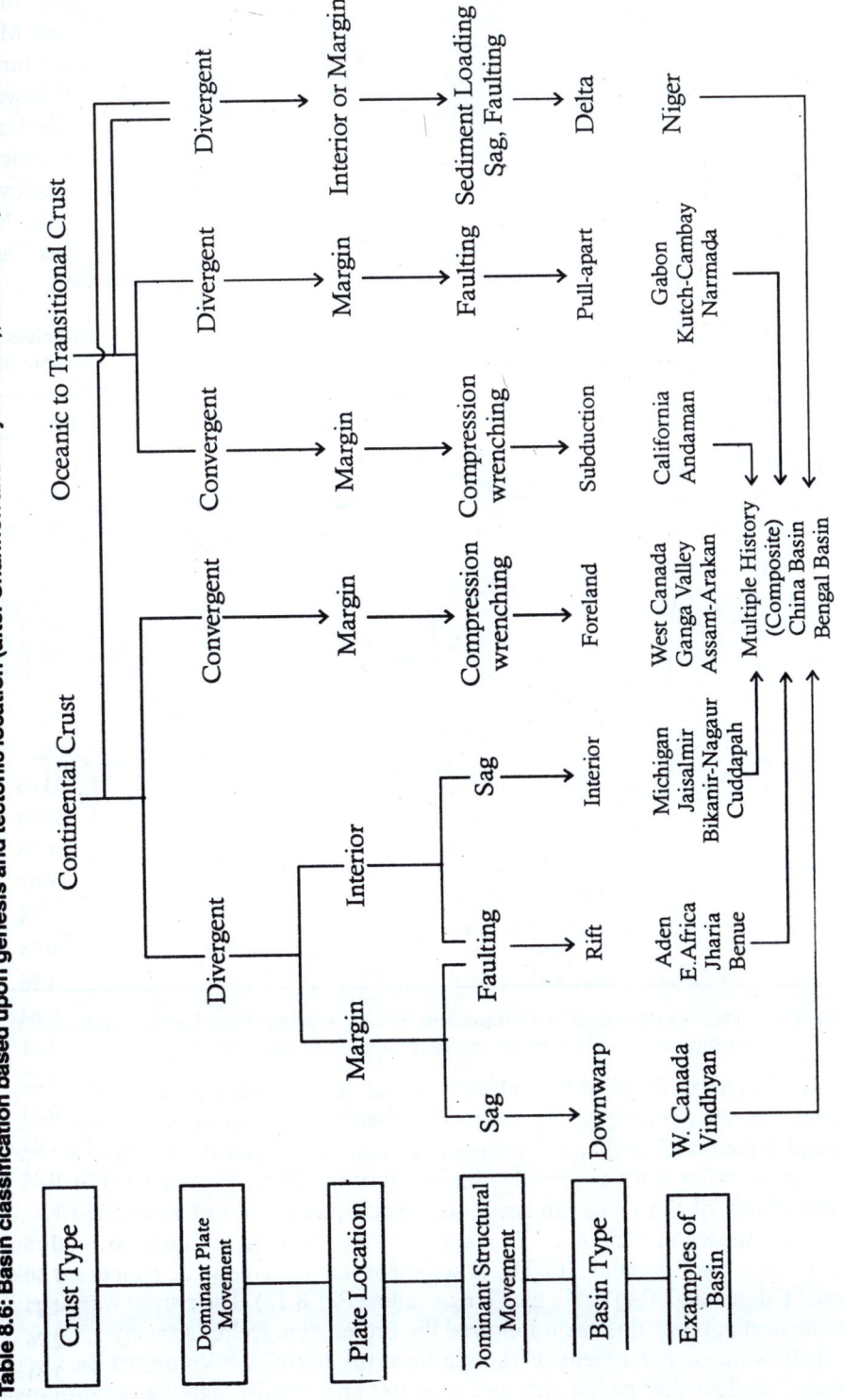

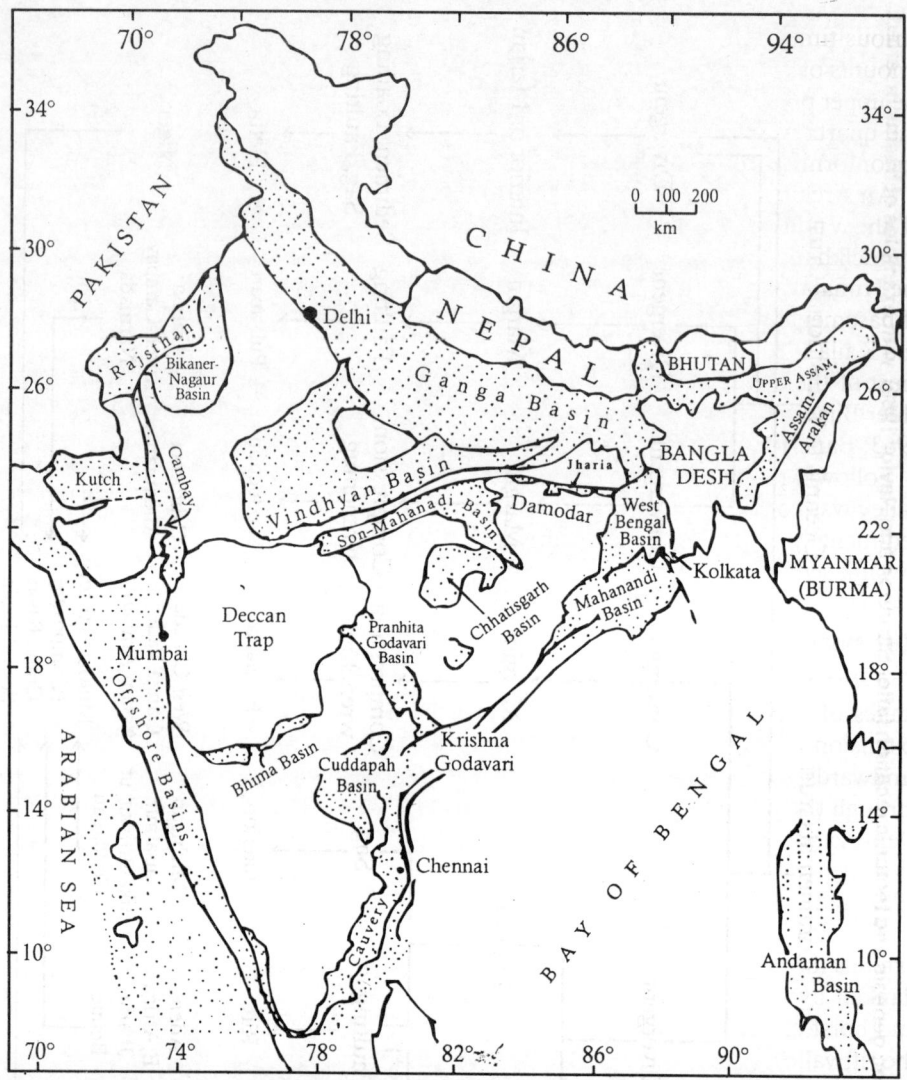

Fig. 8.11: Sedimentary basins (stippled) of India (redrawn from Berger *et al.* 1983). With permission of Schlumberger Technical Services, Inc.

The Vindhyan Basin of central India, which is morphologically similar to the West Canada Basin, apparently started as a downwarp on the northern fringe of the Indian Shield. Like the Canadian basin the floor of the Vindhyan Basin slopes smoothly to the north until it abuts against the thrust belt of the Outer Himalayas (Fig. 8.12). The exposed part of the Vindhyan basin, which occupies an area of more than 120,000 sq km is composed of 5250 m of Precambrian (ranging to early Cambrian) sediments in the thickest part. Another 1000 m of sediments are believed to be concealed below an area of about 40,000 sq km in the Ganga Valley (Fig. 8.12). The crustal warping leading to the formation of this basin followed the Proterozoic Delhi Orogeny.

Following encroachment of the sea from the north, the Vindhyan Basin enjoyed beach, shallow marine, barrier bar, tidal flat and locally, lagoonal environments at various times. An essentially quartz arenite-carbonate type of sediment (with minor amounts of shale, chert and pyroclastics) is found in the lower part of the sequence. In the upper part a much larger proportion of shale occurs, together with some limestone and quartz wackes. The lower sequence, which is separated from the upper one by an unconformity, is structurally more deformed than the upper one.

An Archaean granitic massif (the Bundelkhand granite) separates the exposed part of the Vndhayn Basin into the eastern and the western sub-basins (Fig. 8.12). The Bundelkhand massif extends to the north, below the Gangetic alluvium, almost up to the Himalayan foothills. Geophysical surveys have indicated the presence of a number of basement faults subdividing the Ganga Basin into several horsts and grabens (Narain and Kaila 1982). The major palaeocurrent pattern was northerly or northwesterly during most of the time. The sediments in the eastern part of the basin were supplied by the older Bijawar rocks bordering the southern fringe of the basin (Banerjee and Sengupta 1963; Banerjee 1974; Srivastava *et al.* 1983).

Following collision of the Indian and the Eurasian plates, the present day Ganga valley was converted into a foreland depression and the northern part of the Vindhyan sediments was covered by predominantly non-marine, Upper Miocene to Pleistocene Siwalik sediments derived from the newly developed Himalayan orogenic belt.

Rift Basins

The Red Sea, and the Gulf of Aden are examples of rift basins formed during early stages of sea-floor spreading. The cause of rifting in these areas was lithospheric extension accompanying thermal doming. The rifted blocks were faulted and titled landwards as the Red Sea Basin began to form during the Tertiary. Basalt erupting through the fissures as a result of mantle upwelling floored the Red Sea and Danakil Depressions. About 3000 m of volcanics and volcaniclastic rocks have accumulated in these depressions. These pass laterally into marine pyroclastics and evaporites (Reading 1982; see Fig. 8.13).

The East African rift system, the most important of all the continental rift systems, extends for a length of about 3000 m and is about 40-50 km wide (Fig. 8.14). The rift, flanked by steep escarpments, is often broken and off-set by cross-fautls, volcanoes and horst blocks rising well above the plateau. The drainage flows mostly away from the rift valley, keeping the valley starved of sediments. Only small quantities of clastic sediments, derived from the fault scarps and uplifted blocks, are transported by streams along the rift. Alluvial fans and lakes constitute the main areas of sedimentation. A number of lakes (e.g., Lake Albert, Lake Tanganyika) occur along the rift. These are mostly freshwater lakes but alkaline-saline lakes, whose chemistry is controlled by the nearby carbonatite volcanics, also occur along the rift (Reading 1986).

The Permo-Triassic Gondwana basins occupying three well-defined belts in the eastern part of the Indian Shield may be cited as Indian examples of intracontinental rift basins. The most well-studied of these Gondwana basins, the Jharia Basin in eastern India, was initiated by a general downwarping of the basin floor. At a later phase, the basin was intersected by a large number of parallel faults reaching the basement as a result of crustal tension. The smaller subbasins later merged into a single major depression, which subsided as a narrow intrabasinal graben bounded by contemporaneous normal faults. Sinking of the inselbergs along with rising of a basement ridge, reactivation of old faults and development of new post-depositional

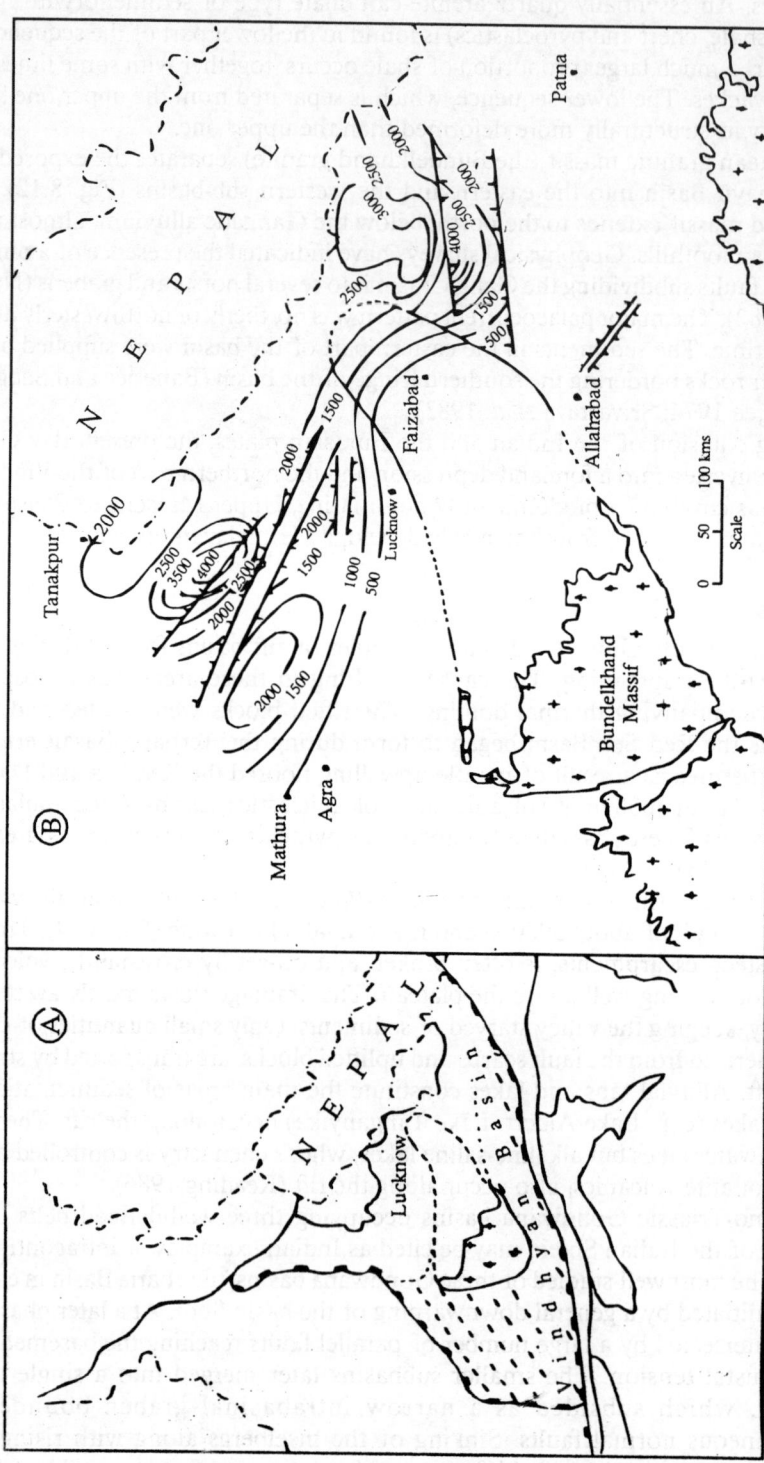

Fig. 8.12: (A) Exposed areas of the Vindhyan Basin (B) Isopach map of the Vindhyan deposits in the subsurface of the Ganga plains based on seismic-cum-geological information. Contours indicate thickness of the Vindhyan sediments in metres (generalized and redrawn after Narain and Kaila 1982). With permission of H. Narain.

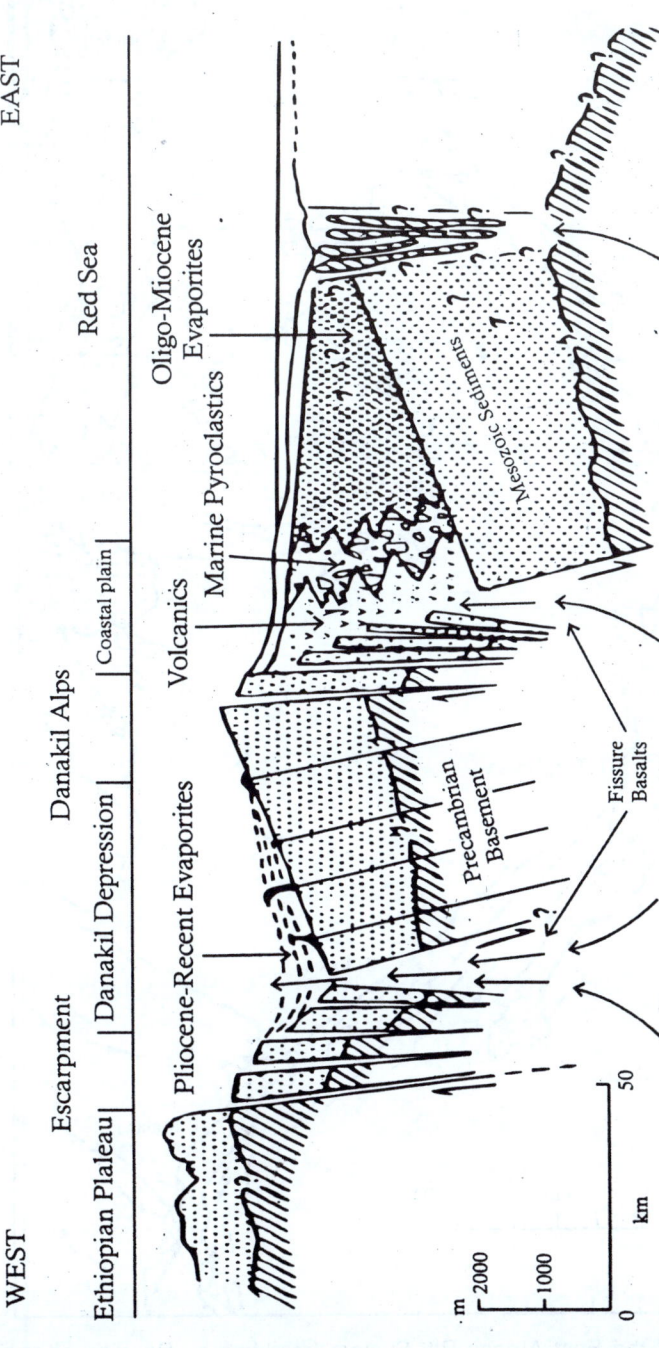

Fig. 8.13: Section across the Red Sea and Danakil depression (modified by Reading after Hutchinson and Engels 1970, reprinted from Reading 1982 with permission).

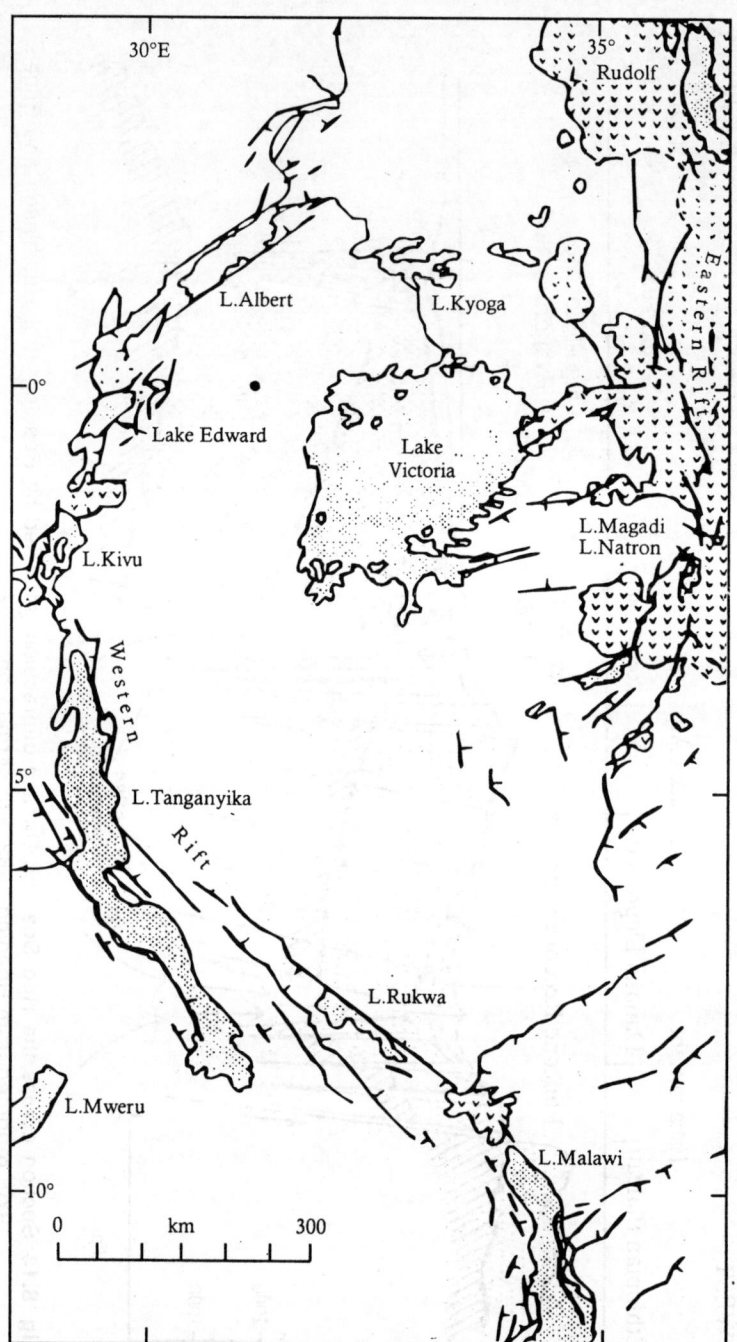

Fig. 8.14: Map of the East African Rift System (modified by Reading after King 1970, reprinted from Reading 1982 with permission).

faults caused the basement to rise at either ends of the basin, thereby providing a source of sediments. Faulting, caused by crustal tension, continued until the deposition of the youngest sediments within the basin (Fig. 8.15).

Sedimentation within this fault-bounded graben took place by fluvial and fluvio-lacustrine processes. At a later stage the fluviatile clastics were intruded by dykes and sills of dolerites and lamprophyres. Emplacement of the lamprophyres, which are restricted to the basinal areas only, post-dates deposition of the major part of the Gondwana sediments. The tensional cracks, active since the early phase of basin formation, reached the mantle at an advanced stage of basin development, causing release of pressure, lowering of temperature and consequent intrusion of lamprophyre dykes in zones of maximum subsidence (Ghosh and Mukhopadhyay 1985).

Failed Rifts and Aulacogens

The 1000 km long and 100 km wide Benue Trough in the Gulf of Guinea in West Africa is generally cited as the best example of an aulacogen formed during drifting of Africa and South America. This trough is filled by more than 5 km of fluvial, deltaic and marine Cretaceous sediments. These are overlain by another sequence of fluvial, deltaic and submarine fan deposits extending into the Atlantic Ocean (Fig. 8.16). The Cretaceous aulacogen phase was separated from the Post-Cretaceous sediments by a phase of folding, followed by erosion (Reading 1986).

Interior Basins

Interior basins occupy large circular or elliptical areas within cratons. They are generally symmetrical in profile. Their lithic fill includes clastics and carbonates.

The Michigan Basin, located in the east central part of the United States, is an example of an interior basin. This basin was initiated by thermal events. Deep crustal metamorphism leading to phase transformation (gabbro to eclogite) might have led to subsidence. This was followed by thermal sag throughout the Lower Palaeozoic in a setting of low geothermal gradient. The sagging was effected by periodic movements along basement faults throughout the Palaeozoic. The basin contains clastic and non-clastic sediments ranging in age from Precambrian to Jurassic (Shannon and Naylor 1989).

The Jaisalmer and Bikaner-Nagaur Basins of Rajasthan, India are interior basins. Bikaner-Nagpur, which was initially a stable shelf up to the Early Palaeozoic, later developed into a restricted ('silled') basin to accommodate about 2 km of Cambrian-Triassic sediments (Berger *et al.* 1983; see Fig. 8.11).

The Cuddapah Basin occupying a crescentic area of about 42,000 sq km in the south-eastern part of the Indian Shield is another example of an interior basin. A maximum thickness of about 10 km of clastic and non-clastic Proterozoic sediments is estimated to have been deposited in this basin. The total load of accumulated sediments and emplaced igneous material in this basin cannot account for a subsidence of 10 km. The main driving force for generation of this basin seems to be thermal heating and cooling, followed by isostatic loading. Three major unconformities within the sedimentary sequence of this basin indicate crustal uparching as a result of isostatic adjustment to rising temperature in the lithosphere. Rising temperature not only caused thermal expansion but also reduced the density and thickness of the lithosphere. A maximum crustal thickness of about 38 km below the basin, with a thinning of about 4 km towards the eastern fringe, is suggested by gravity data.

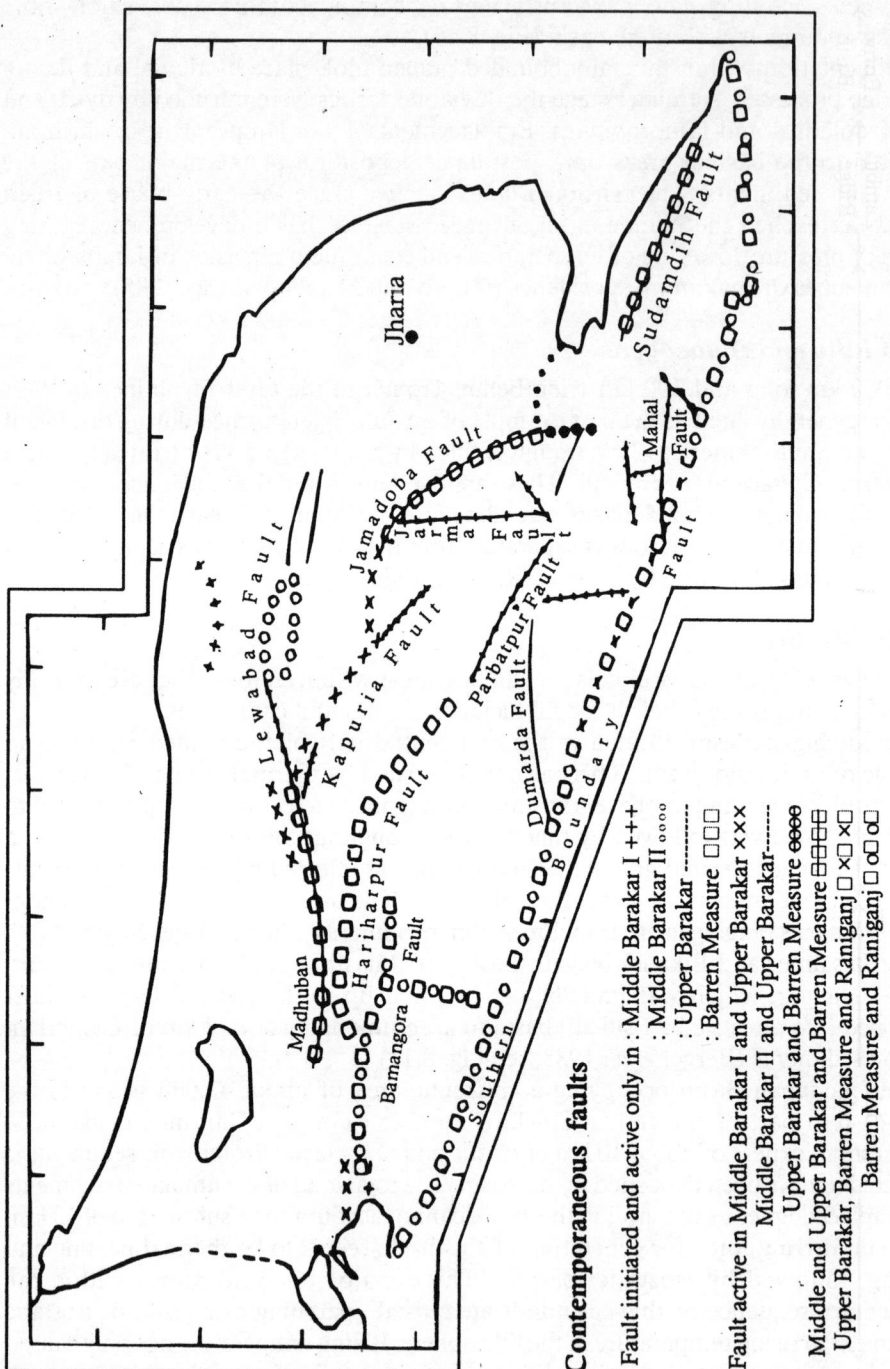

Contemporaneous faults

Fault initiated and active only in : Middle Barakar I +++
                                      : Middle Barakar II ooooo
                                        : Upper Barakar ▭▭▭
                                        : Barren Measure ▢▢▢

Fault active in Middle Barakar I and Upper Barakar ×××
        Middle Barakar II and Upper Barakar ──────
        Upper Barakar and Barren Measure ϴϴϴϴ
        Middle Barakar and Barren Measure ▯▯▯▯

Middle and Upper Barakar and Barren Measure ▯×▯×▯
Upper Barakar, Barren Measure and Raniganj ▯ ×▯ ×▯
        Barren Measure and Raniganj ▯ o▯ o▯

Fig. 8.15: Contemporaneous faults of the Jharia Basin, India (after Ghosh and Mukhopadhyay 1985). With permission of Geological, Mining & Metallurgical Society of India.

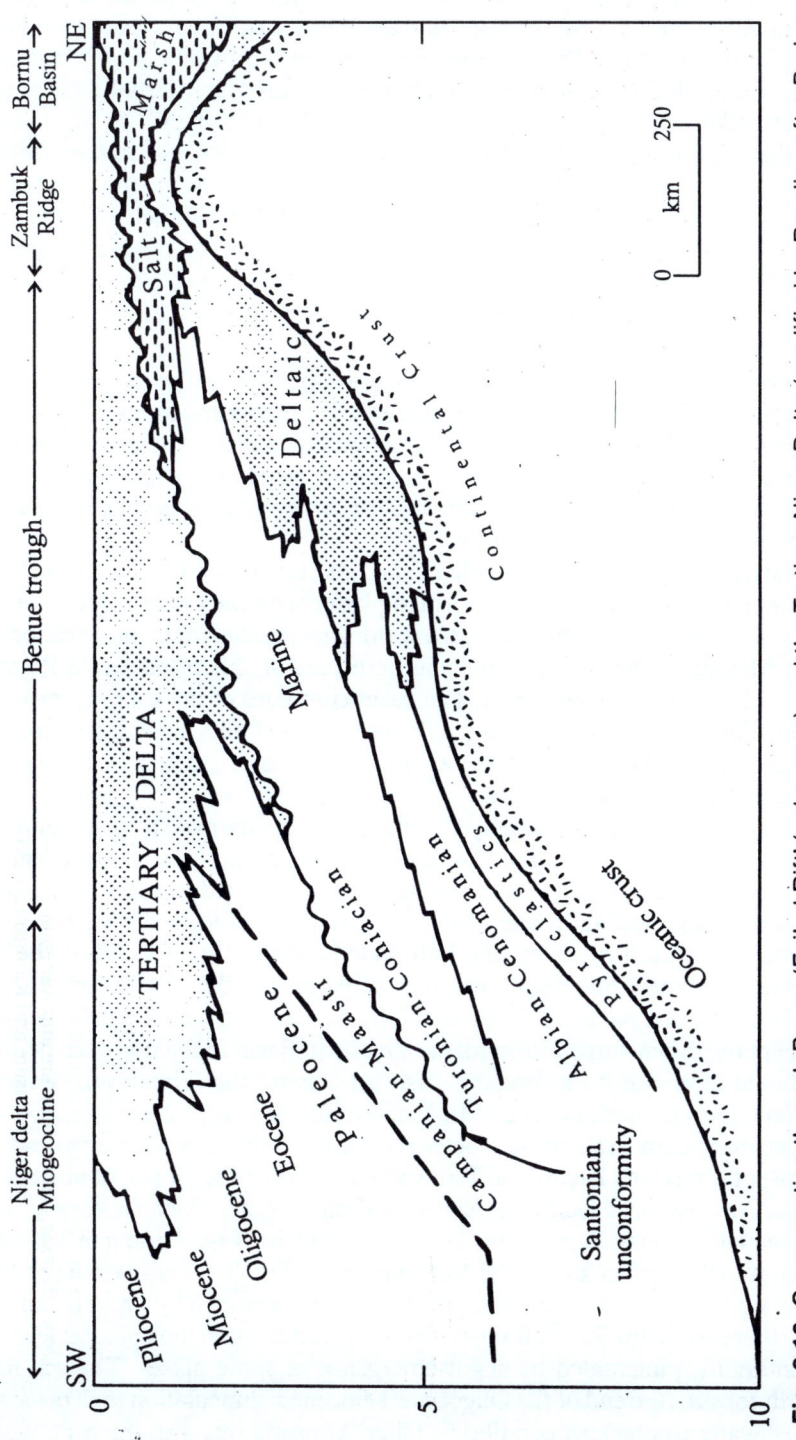

Fig. 8.16: Cross-section through Benue 'Failed Rift' (aulacogen) and the Tertiary Niger Delta (modified by Reading after Burke et al. 1972; Petters 1978, reprinted from Reading 1982 with permission).

Step faults, tilting and local variations in the depth of the basin-floor revealed by Deep Seismic Sounding (DSS) profiles run across the Cuddapah Basin were possibly caused by flexural loading. These deep-seated faults also provided passage for upward migration of mantle fluids (kimberlite-carbonatites) and other partial melt products, namely, calc alkaline, alkaline and felsic magmas. That the role of tensional faulting was diminishing during the later phase of depositional history is shown by dwindling of igneous activity after lower Cuddapah sedimentation. Compressional rather than tensional forces were active during the later (and post-depositional) phases of basin evolution. This is indicated by the presence of thrust faults in the eastern margin of the basin (Bhattacharji and Singh 1984).

Foreland Basins

Foreland basins generate at the toe of the overriding plates in the areas of continent-ocean or continent-continent collision. Subsidence in the foreland basin is caused largely by subduction. The lithosphere in front of the overriding plate is depressed also by the weight of the thrust slices. As the basin subsided, a relatively low but wide bulge developed on its cratonic side. The older surface, where not covered by deep water, is eroded in the form of an unconformity.

The western part of the West Canada Basin consisting of a folded and thrust belt on the western margin and a broad undeformed platform on the eastern side is described as a foreland basin (Cant 1989). The basin is elongate parallel to the orogenic belt. The preserved Mesozoic rocks occupy the southern part of this basin while Palaeozoic carbonates occupy the northern part. The basin continues to the south to merge with the Western Interior Basin of the United States. Reverse faults involving the basement divided the foreland basin in the latter area into a series of smaller basins during the Late Cretaceous Laramide Orogeny.

The West Canada Basin initially developed as a downwarp to accommodate sediments ranging in age from the Cambrian to the Lower Jurassic. Apparently as a result of the Laramide orogeny, the same basin was converted to a foreland basin and received sediments ranging in age from the Middle Jurassic to the Lower Tertiary. Farther south, in South America, the Laramide basins in the foreland of the Andes resulted from shallow dipping subduction, causing drag at the base of the lithosphere (Cant 1989). A similar picture of basin development can be traced in the Ganga Valley, India, where the downwarped Proterozoic Vindhyan Basin in the northern part of the Indian Shield gave way to a foreland basin, following the Himalayan orogeny, to accommodate Siwalik sediments of Miocene-Pleistocene age.

The Assam-Arakan Basin located between two thrust belts in north-eastern India is an example of a foreland basin. The Tertiary sediments of the Upper Assam belong to two distinct depositional facies - a relatively shallow-water shelf facies and a deep-water geosynclinal facies. Between the Himalayan and the Naga-Disang thrust complex occurs an autochthonous zone ('the foreland spur'; Fig. 8.17) containing sediments ranging in age from Eocene to Pleistocene. These are intersected by a number of gravity faults. Farther east, in the Patkai Range, Tertiary sedimentation took place under deeper water conditions, punctuated by slight emergence at some places. The sea receded southwards towards the end of the Oligocene. Miocene sedimentation took place mostly under freshwater to brackish conditions. Oligo-Miocene was also the time when the deposits of the Assam-Arakan Basin were overthrust towards the north-west, over the north-eastern extension of the Indian Shield. The outermost of these thrusts, the Naga Thrust Belt, consists of a succession of six thrust sheets. In the thrust strip nearest to the foreland spur there are anticlinal structures, such as the one on which the Digboi Oilfield is located (Evans 1932; Berger *et al.* 1983).

Simplified Section Across Upper Assam along X—Y

(B)

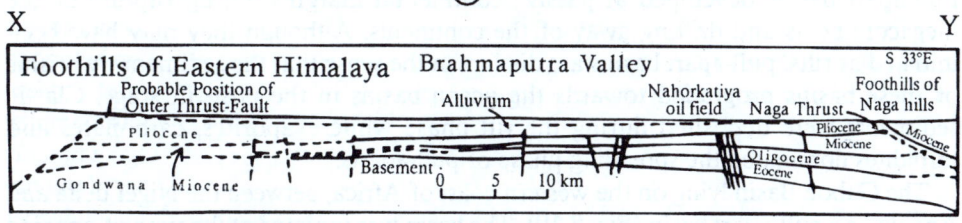

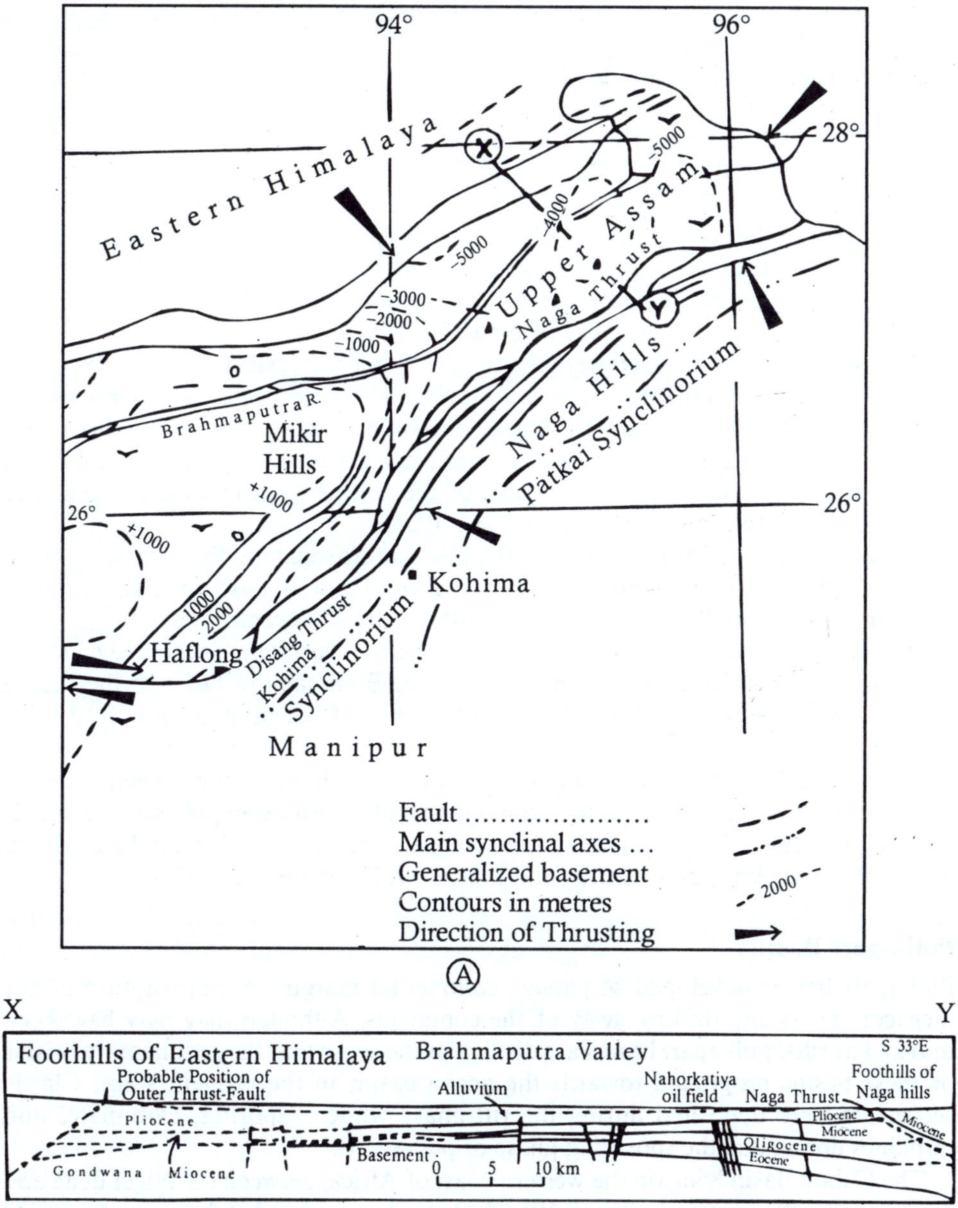

Fig. 8.17: (A) Foreland spur of north-eastern India, (B) NW-SE cross-section across the foreland spur shown in (A). (Redrawn from compilations by Berger *et al.* 1983). With permission of Schlumberger Technical Serivices, Inc.

Subduction Basins

The California Basin located at the junction of the North American continental plate and the Pacific oceanic plate provides a well-documented example of a subduction basin which experienced compressive as well as wrench movements. Mesozoic sedimentation in California took place in forearc basins related to· subduction movements, to be followed by sinistral, strike-slip basin development during the Neogene. The oceanic-continental plate collision and changing direction of plate movement brought complications to the area.

In complexity of character the California Basin is comparable to the deep-sea basin which runs for a length of about 3000 km from the offshore of Sumatra to the mainland of Burma via the western Andaman Sea (Fig. 8.18). The forearc basin off Sumatra contains about 4 km of sediments including deep-water turbidities, montmorillonitic clays and volcanic ash. As the Sumatran continental block subsided with time and the subduction zone migrated southwards, a gradual widening of the forearc basin took place. The basin shape is controlled by a number of strike-slip faults. In the Andaman Sea a central volcanic arc separates two troughs:— a forearc basin on the western side and a system of rift valleys on the eastern side. The latter, associated with several strike-slip faults, is the spreading centre of several pull-apart basins (Rodolfo 1969). The total sediment thickness in the eastern Andaman basin is small (1.5 km), indicating that the Irrawaddy delta, the main source of the sediments in this basin, came into existence only in recent years (Reading 1982, 1986).

In the mainland of Myanmar (Burma) a volcanic arc separates the forearc from the back-arc. The estimated sediment thickness is about 12 km in the former basin and 17 km in the latter. Both basins are possibly underlain by an oceanic crust.

The largest sediment cover on a presumably oceanic crust today is the Bengal Deep-Sea Fan extending from the mouth of the Ganga-Brahmaputra Delta to the Indian Ocean, covering an area of 3,000,000 km^2 of the Bay of Bengal floor (Fig. 7.10). The average thickness of the·fan sediments here is 7.5 km but the maximum estimated thickness is of the order of 12 to 15 km. Sedimentation in this fan has been going on since the beginning of the collision of the Indian and the Eurasian plates at a rate of 2-10 cm/1000 years. The phases of fan sedimentation are correlatable with the phases of upheaval of the Himalayas (Curray and Moore 1971; Moore *et al.* 1974).

Pull-apart Basins

Pull-apart basins developed at passive continental margins during rupture of the megacontinents and drifting away of the continents. Although they may have been initiated as·rifts, pull-apart basins are distinct in the sense that the sedimentary prisms of these basins prograded towards the ocean basins in the post-rift stage. Clastic sediments were deposited during the rift phase while evaporites, carbonates and turbidites dominated the spreading phase of pull-aparts.

The Gabon Basin lying on the western coast of Africa, between the Niger delta and Angola, is a pull-apart basin (Fig. 8.19). The basin was initiated at the time of opening of the South Atlantic, following migration of South America from Africa. The onshore part of the basin has a maximum width of about 200 km. A Precambrian horst block separates the basin into two parts—a deeper basin opening to the Atlantic in the west, and a shallower, interior eastern basin. The Atlantic basin, an extensive monocline, contains more than 8000 m of sediments ranging in age from the Jurassic to the Miocene. A thick evaporite formation of Aptian age has resulted in structural discordance between the sequences above and below. In the eastern basin rocks ranging in age from the Jurassic to the Middle Cretaceous, up to a maximum thickness of 5000 m, occupy a faulted syncline.

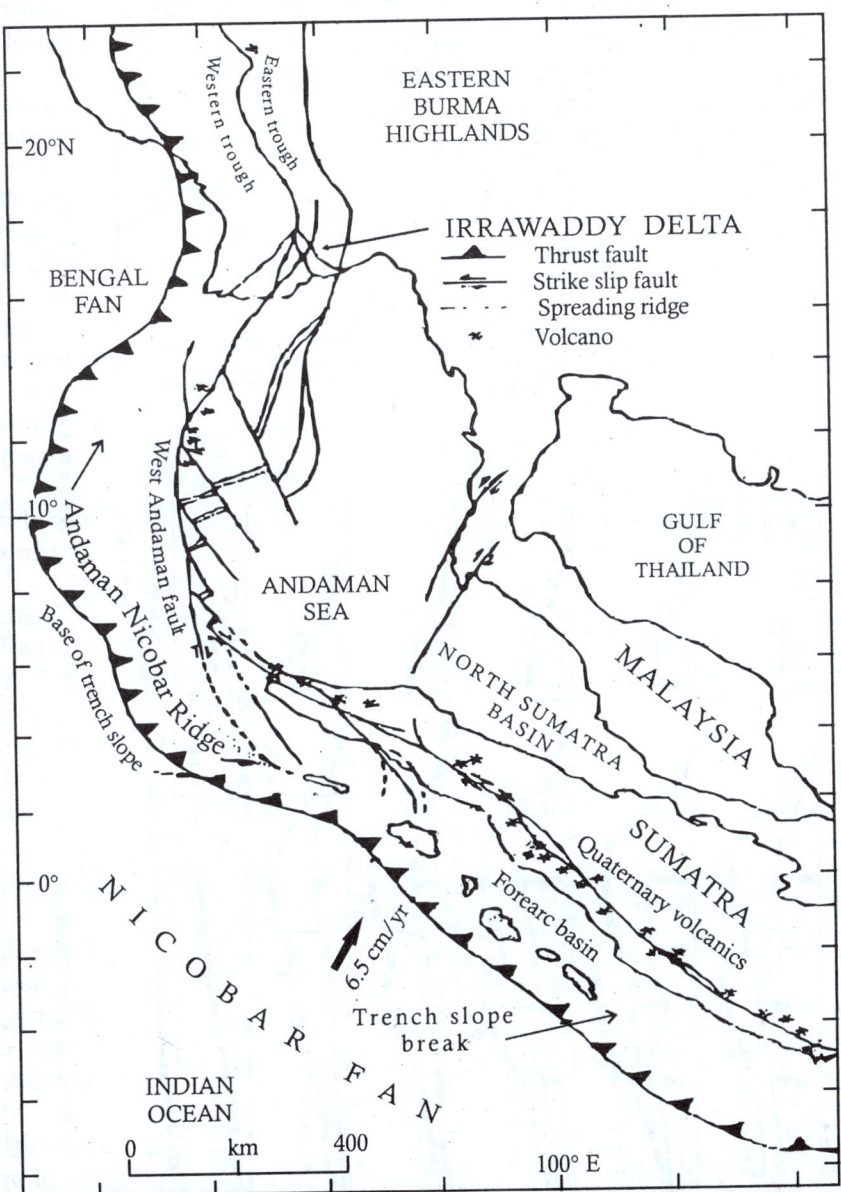

Fig. 8.18: Arc-trench system of Burma-Andaman-Indonesia showing strike-slip faults, volcanic arc, forearc and backarc areas (complied by Reading 1982 after Curray *et al.* 1979; Kraig *et al.* 1980). Reprinted with permission.

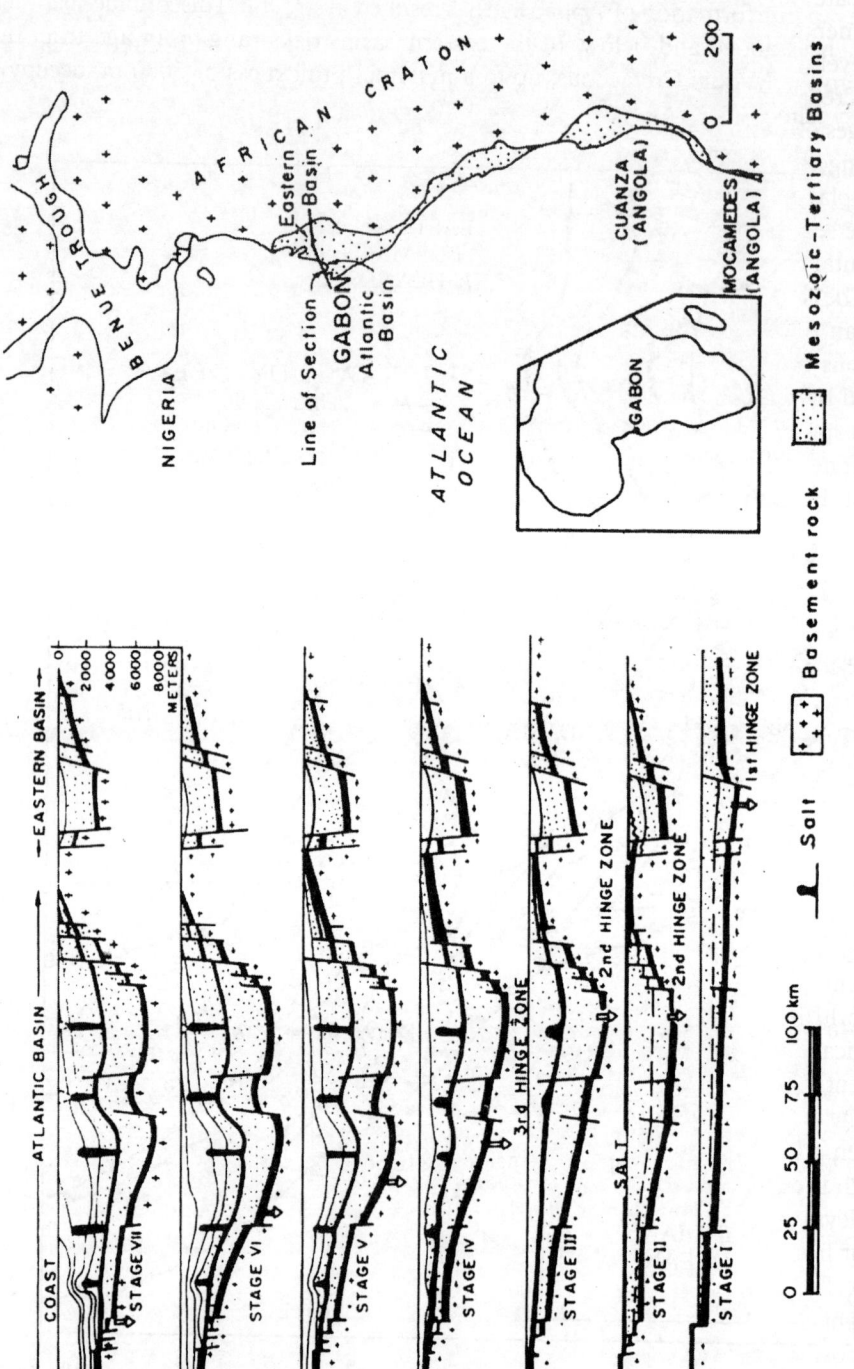

Fig. 8.19: Gabon 'Pull-Apart Basin' of West Africa, and cross-sections showing stages of development of Gabon Basin (complied from Shannon and Naylor, 1989; Brink 1974). With permission of Graham & Trotman and American Association of Petroleum Geologists.

The Late Jurassic-Early Cretaceous phase of rifting and basin development initiated by continental drifting, was followed in the Gabon Basin by a long post-rift phase of basin development. The rate of subsidence, which was high during the Jurassic to the Lower Cretaceous, gradually declined during the Upper Cretaceous and the Tertiary. The stages of basin development can be traced in the sections shown in Fig. 8.19. The three hinge lines, each representing a cluster of faults seen in the sections, had major roles to play in the development of the basin (Shannon and Naylor 1989).

Three sedimentary basins, at Kutch, Cambay and Narmada, occur in the western continental margin of India. The faults bordering these grabens open seawards (Fig. 8.20). The sediments within these basins, ranging in age from the Upper Bathonian/Callovian to the Neocomian (? Santonian), are essentially clastic, interspersed by calcareous or marly bands at some places in Kutch. A major marine transgression, followed by regression and deltaic environments with periodic marine influence, are the interpreted depositional environments in Kutch (Biswas 1981; Biswas and Deshpande 1983). These rift basins are supposed to have opened up successively as a result of the counterclockwise drifting of the Indian craton after its detachment from Gondwanaland in the Late Triassic or the Early Jurassic. Rifting is inferred to have developed from north to south in the following order : Kutch (Early Jurassic), Cambay (Early Cretaceous) and Narmada (Late Cretaceous), (Biswas 1982). These rift-bounded basins opening towards the sea satisfy the characters of the pull-apart basins of Shannon and Naylor (1989).

Delta Type Basin

In the Shannon and Naylor (1989) scheme (shown in Table 8.6), deltas have been classified as sedimentary basins. Deltas, as discussed in Chapter 7, represent a particular type of deposit produced by interaction of marine and non-marine agencies. They may develop whenever a sediment-water mixture is funnelled through a channel into a relatively quieter body of water provided that the rate of sediment input is higher than the rate of its dispersal. Deltas may develop in continental, transitional or oceanic crust, and on any type of plate margin, convergent or divergent. Although all deltas of the world have many broad similarities, their characters differ depending on the geological and tectonic setting and local climatic conditions, which control the rate of sediment supply.

In the Niger Delta, cited by Shannon and Naylor (1989) as a typical delta type of sedimentary basin, a mean annual supply of 2×10^7 m^3 of sediment is maintained through the Niger River and its tributary, the Benue River. These rives flow through rift valleys believed to be the 'failed arm' (aulacogen) which extended into the continental interior from the 'tripple junction' formed during separation of South America from Africa. The rifting, initiated during the Late Jurassic, led to deposition of a wedge of sediment along the Benue Trough until the Early-Middle Cretaceous. This was followed by deposition of a coastal prism of sediments during the Tertiary. The depositional centre of this wedge gradually shifted towards the sea as the pre-Tertiary depression was filled. The Oligo-Miocene delta complex, for example, is located wholly offshore on the rapidly subsiding oceanic crust (see Fig. 8.16).

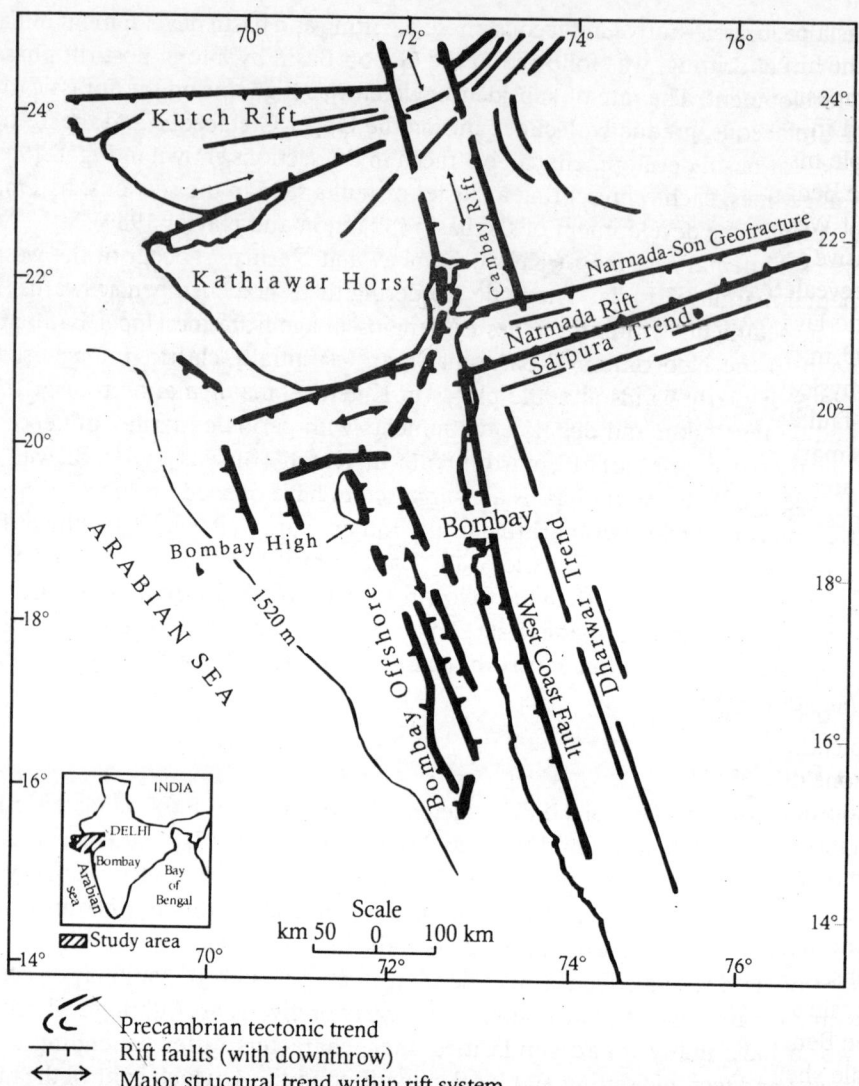

Fig. 8.20: Kutch, Narmada and Cambay rift basins on the western coast of India (modified after Biswas 1982). With permission of American Association of Petroleum Geologists.

Composite Basins with Multiple History

The practice of compartmentalizing sedimentary basins to specific associations of crust type and plate movement, followed in Table 8.6, is artificial to some extent. Many sedimentary basins have passed through a variety of lithospheric movements during their span of existence and hence are polygenetic in character. The Benue Basin, for example, developed into a delta, starting from an aulacogen. The West Canada Basin and the Vindhyan Basin, both of which were initiated as downwarps, experienced foreland depositional conditions following orogenic movements in the bordering areas. Some of the foreland basins again, like the one in north-eastern India, can be looked upon as a part of the adjacent subduction complex. Shannon and Naylor (1989) have cited the history of development of some of the complex basins in China as examples of composite basins. We discuss below the history of development of the Bengal Basin, located in eastern India and adjacent Bangladesh, as an example of a basin with a multiple history.

The Bengal Basin occupies an area of more than 90,000 km^2, covering most of the present West Bengal, India and Bangladesh. The basin is open to the sea in the south. Extensive geophysical surveys and deep drilling in the alluvium-covered plains of Bengal have revealed a thick section of Cretaceous and Tertiary sediments on a basement of basaltic lava flows. The lava flows are of the same age as the Rajmahal Volcanics (~ 100 m.y.) exposed on the north-western and north-eastern fringes of the basin. Geophysical surveys also suggested the presence of Gondwana sediments in several block-faulted troughs below the basalt flows. A series of fault-bounded buried basement ridges mark the western fringe of the basin. East of this marginal zone of fault lies the shelf area of the basin with a low, homoclinal dip towards the south-east. The total thickness of Mesozoic and Tertiary sediments on this shelf exceeds 9 km in the eastern part of West Bengal and is of the order of 16 km in the deepest part of the basin located in south-eastern Bangladesh (Fig. 8.21). Except for a few normal faults this sequence is structurally undisturbed.

The most prominent marker horizon within the whole sedimentary sequence of this basin is a bed of Nummulitic (Eocene) limestone. This bed dips uniformly at the rate of 1½° over the whole of the Bengal shelf but in the subsurface of Kolkata exhibits a conspicuous basinward flexure traversing the entire basin in a NE-SW direction. Beyond this zone of flexure, which has been interpreted as the 'hinge zone' of the basin or as the Eocene shelf break, lies the deeper part of the basin with a greater rate of subsidence and a more argillaceous lithofacies. In the younger Tertiary sediments a similar change in lithofacies, from the arenaceous sediments of the stable shelf, to a dominantly argillaceous, deltaic sediment downdip, is noticed (Fig. 8.22). Except for phases of marine transgression occurring during the Upper Cretaceous (locally), Eocene (extensively) and Miocene (in the eastern parts only), the Tertiary sedimentation in this basin took place under freshwater, estuarine or deltaic conditions. The broad division of the Bengal basin into three major zones, (*i*) a western scarp-fault zone, (*ii*) a stable, middle shelf zone, and (*iii*) a deeper basinal area (Sengupta 1966a) is confirmed by more recent deep seismic sounding (DSS). This study also shows the presence of nearly normal thickness of continental crust in the western part of the basin. This changes to a significantly thinned continental crust in the eastern part (Kaila *et al.* 1992, Reddy *et al.* 1995).

The evolution of the Bengal Basin took place in several stages. Block faulting within the Bengal plains and the adjacent areas was followed by deposition of freshwater Gondwana sediments of Permian to Early Jurassic age. During the Early Cretaceous a major rifting of the Bengal plains took place consequent to the break-up of Gondwanaland and the drifting away of Antarctica from India. The 'failed arm' of the aulacogen presumably rifted the entire Bengal plains while the other two arms mark the present boundaries of the eastern Indian coastline. The deep-seated faults located by seismic surveys between the Rajmahal and the Garo Hills (Rao and Sengupta 1964) are perhaps the only detectable remnants of this 'failed rift' in the subsurface.

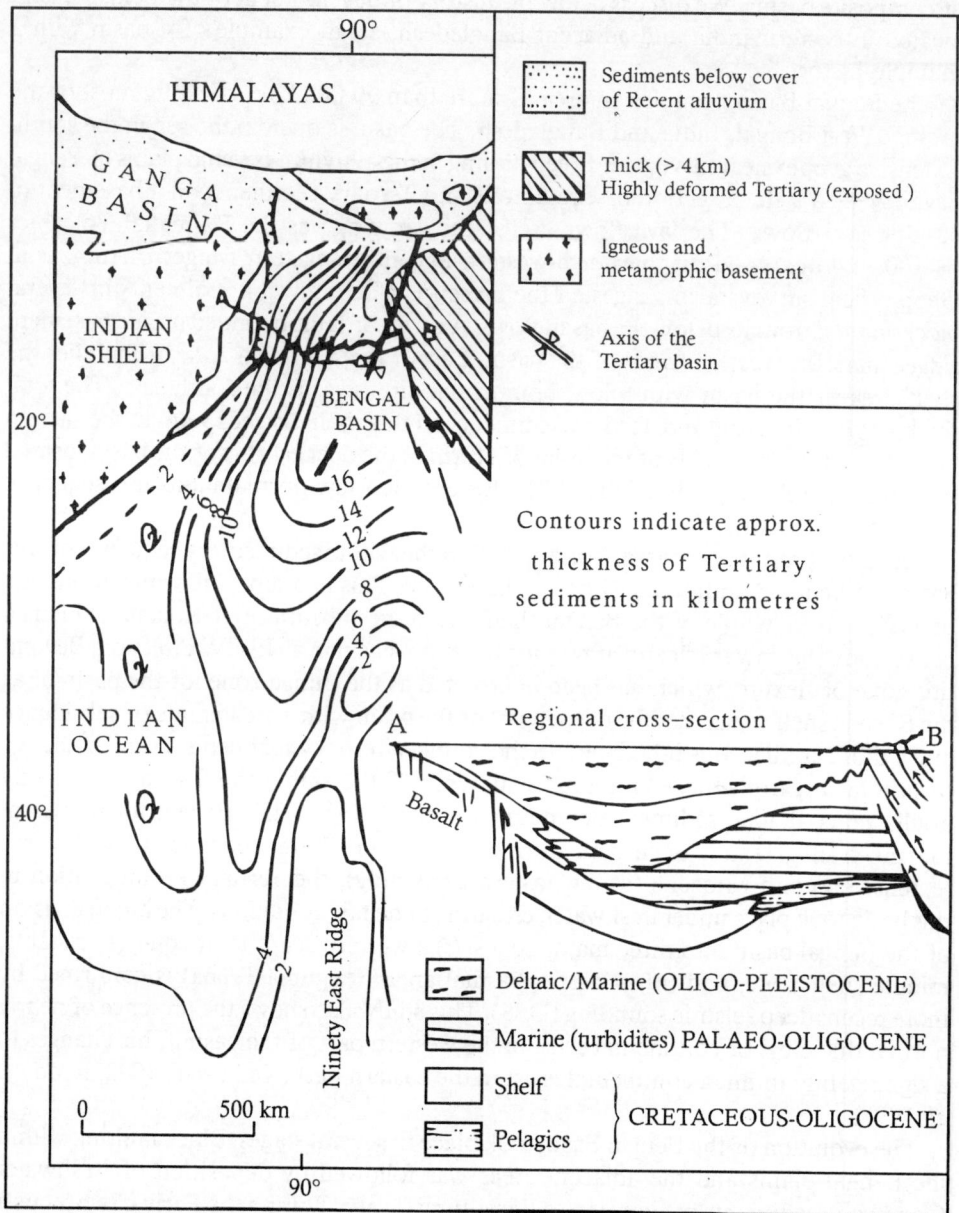

Fig. 8.21: Tertiary basins of Bengal and adjacent areas (compiled and redrawn after Paul and Lian 1975).

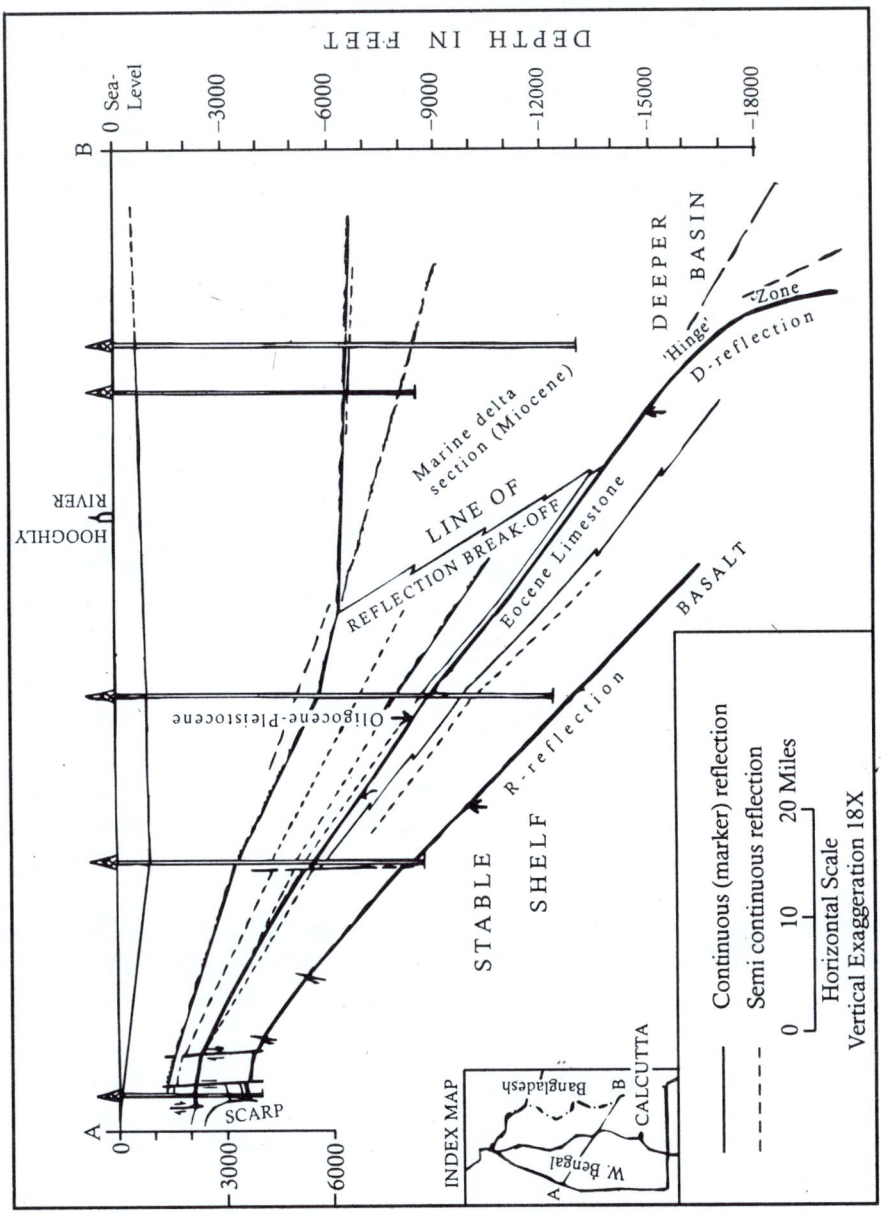

Fig. 8.22: Cross-section across western part of the Bengal Basin showing continuity of the seismic reflection horizons in the subsurface (redrawn from Sengupta, 1966a). With permission of American Association of Petroleum Geologists.

Rifting was associated with extensive eruption of flood basalt throughout the Bengal plains with the Kerguelen Hotspot possibly serving as the heat source (Mahoney *et al.* 1983; Baksi *et al.* 1987). The eruption of basalt was followed by freshwater, brackish or estuarine sedimentation over the Bengal plains. This situation continued until the Lower Eocene when extensive marine transgression caused deposition of Nummulitic limestone over the whole Bengal plains extending north-east up to Assam. In the regressive phase which followed, most of the basin experienced deltaic sedimentation from the Miocene through the Pleistocene, when an enormous thickness of essentially argillaceous sediments was deposited, particularly in the south-eastern parts of the basin (Sengupta 1966a; Roybarman 1983). This phase of deltaic sedimentation is apparently continuing in the south and south-eastern parts of the present West Bengal and Bangladesh. Offshore explorations suggest continuation of the Bengal basin sediments southwards, into the present Bay of Bengal. The upper three sediment layers of the deep-sea fan in the Bay of Bengal are presumably the offshore continuation of the Miocene, Pilocene and Quaternary sediment layers of the Bengal basin inland (Curray and Moore 1971).

9

STRATIGRAPHY AND SEDIMENTATION

CLASSICAL CONCEPTS IN STRATIGRAPHY

Estimation of Geological Time

The ultimate aim of a sedimentological analysis is to unravel the history of depositional events leading to the development of particular stratigraphic sequences. Studies on sedimentology and stratigraphy are, therefore, inseparable. Evolution of the planet earth, both in the organic and inorganic sphere, through geologic time, has been the main concern of stratigraphy. Estimation of geological time, both in relative and absolute terms, has naturally been a matter of prime' concern for stratigraphers as well as sedimentologists.

Steno's Law of Superposition, and Cuvier's studies on faunal assemblage made estimation of the relative ages of rock layers possible even in the early years of the nineteenth century. A layer of sediment occupying a higher position than the others, as also a layer containing an assemblage of organic remains with advanced anatomical features, is clearly younger than the rest. A novel method of age determination in absolute terms, introduced by Charles Lyell, was based on the principle that the younger rock layers should contain larger proportions of living organic species compared to extinct ones (see Chapter 1).

In an early method of age determination in terms of years, Lord Kelvin, a physicist, assumed that the earth had cooled at a uniform rate from a state of white heat to a condition wherein condensation of water vapour in the form of oceanic water was possible. The age of the earth obtained by this method in the late nineteenth century was a gross underestimate because the process of radioactivity, which replenishes much of the heat lost through radiation, was still unknown. This phenomenon, discovered jointly by H. Becquerel and Marie and Pierre Curie in 1896, and explained by Rutherford and Soddy in 1902 in terms of transmutation of atoms with a constant 'half-life', provided, for the first time, a dependable method for age determination of rocks in absolute terms. The *radiometric method of dating* is very much refined today and a host of radioisotopes, such as $U^{238} \rightarrow Pb^{207}$, $K^{40} \rightarrow Ar^{40}$, $Rb^{87} \rightarrow Sr^{87}$, are employed in radiometry. A similar method, using the ratio of radiogenic (C^{14}) and non-radiogenic carbon (C^{12} or C^{13}), can date woody matters, which are less than 50,000 years old.

A more recent method of dating sediments uses the principle of *magnetostratigraphy*. Magnetically susceptible particles in sediments occurring in the deep ocean-floor reorient themselves, in a statistical sense, to the direction of the earth's magnetic field existing at the time of their deposition. By matching the magnetic polarity of an ancient sediment with the standard, radiometrically dated, magnetic stratigraphic column, some idea of the time of depositon of the sediment can be obtained. Problems are often created by bioturbation causing reorientation of sediment grains by deep-water burrowing organisms. On land the technique of *magnetic stratigraphy* can be used for dating ferruginous detrital (non-authigenic) particles laid down under oxidizing conditions provided they have not been disturbed subsequent to deposition.

Dual Hierarchy in Stratigraphy

By the end of the nineteenth century it was recognized that the strata constituting the earth's crust could be grouped either as mappable rock units or on the basis of the time span they cover. This concept of '*dual hierarchy*' in stratigraphy has continued to influence all our stratigraphic thinking since then. The problem of correlation between visible rock units and inferred time units was a difficult one. In the absence of a direct method for dating, the pioneers were obliged to use fossils as indicators of relative ages of rocks. This practice was later challenged on the ground that growth in the organic world is controlled not only by the trend of long-ranging organic evolution, but also by short-ranging local environmental conditions. Moreover, a parallel relationship between the rock and biostratigraphic units, an essential prerequisite for dating of rock layers by their fossil content, may not exist.

In an elegant solution to this problem, spans of time are recognized not by their fossil content, but by the deposits accumulated during specific time intervals (e.g. 'systems' deposited during 'periods' of time). The Schenck and Muller (1941) scheme of classification of stratigraphic units, reproduced in Table 9.1, is based on this principle. Although the geological time units and time-stratigraphic units are directly correlatable in this scheme, the rock unit terms have been deliberately written across to make direct correlation between rock and time-stratigraphic units impossible. This trend of thinking eventually led to the classification of stragraphic units into two distinct groups : 1) observable units consisting of litho-and biostratigraphic units and 2) inferential units consisting of chronostratigraphic units and geologic time units. This subdivision was the basis for creation of the *Code of Stratigraphic Nomenclature*. The categories and ranks of stratigraphic units defined in a modern version of this code, as proposed by the North American Commission for Stratigraphic Nomenclature in 1983, is reproduced in Table 9.2.

Table 9.1: Schenck-Muller (1941) Scheme of Stratigraphic Classification

Geologic Time Units	Time Stratigraphic Units	Rock Stratigraphic Units
Era	–	
Period	System	
Epoch	Series	
Age	Stage	
–	Zone	

Rock Stratigraphic Units (written across): Group, Formation, Member, Bed

Table 9.2: Categories and ranks of units defined in the North American Stratigraphic Code (1983)*

A. MATERIAL UNITS

Lithostratigraphic	Lithodemic	Magnetopolarity	Biostratigraphic	Pedostratigraphic	Allostratigraphic
Supergroup Group	Supersuits Suite	Polarity Superzone			Allogroup
Formation	*Lithodeme* (Complex)	*Polarity zone*	*Biozone* (Interval, Assemblage or Abundance)	*Geosol*	*Alloformation*
Member (or Lens or Tongue) Bed(s) or flow(s)		Polarity Subzone	Subbiozone		Allomember

B. TEMPORAL AND RELATED CHRONOSTRATIGRAPHIC UNITS

Chronostratigraphic	Geochronologic Geochronometric	Polarity Chronostratigraphic	Polarity Chronologic	Diachronic
Eonothem	Eon	Polarity Superchronozone	Polarity Superchron	
Erathem (Supersystem)	Era (Superperiod)			
System (Subsystem)	*Period* (Superperiod)	*Polarity Chronozone*	*Polarity Chron*	*Episode*
Series	Epoch			Phase
Stage (Substage)	Age (Subage)	Polarity Subchronozone	Po'arity Subchron	Span
Chronozone	Chron			Cline

(Diachron)

* Fundamental units are italicized.

The 'material units' and the 'temporal and chronostratigraphic units' of Table 9.2 correspond respectively to the observable and inferential units of the conventional scheme. Of the material units again, the lithostratigraphic units are used for categorizing and ranking of deposits of sediments in the field. The fundamental lithostratigraphic unit, called a formation, is defined as 'a body of rock identified by lithic characteristics and stratigraphic position; it is prevailingly but not necessarily tabular and is mappable at the earth's surface or traceable in the subsurface.'

Subsurface explorations in large sedimentary basins conducted during the last few decades by deep drilling and geophysical techniques have provided a new perspective to stratigraphic concepts. The conventional as well as the newly developing ideas are briefly summarized below. Readers may refer to Dunbar and Rodgers (1957) for a masterly treatment of the traditional stratigraphic concepts. The newer concepts are summarized by Walker (1984a) and Reading (1986)

RELATIONSHIPS BETWEEN SEDIMENTARY UNITS

Vertical and Lateral Relationship

Vertical Relationship

Sediments accumulate within basins by aggradation, progradation and lateral accretion. *Aggradation* is the process of vertical filling, as in the flood basin of a river. *Progradation* is the filling from the margin, as in the case of an advancing delta front. *Lateral accretion* is caused by the process of accumulation against the margin of a channel, as in a river point bar (Figs. 9.1, 9.16). The vertical relationship between succeeding sedimentary units produced in this way may be either conformable or non-conformable. A conformable relationship may be either sharp or gradational. Two types of gradations are possible: a continuous gradation or an intercalated contact. A body of sand grading vertically up into a shale section through mixtures of sand and shale in various proportions, is an example of a continuous gradation. A sand body passing vertically up into a body of shale through an intermediate layer of closely spaced lenticles of sand and shale is an example of intercalation.

An unconformable relationship between two sedimentary bodies implies a break in the process of sedimentation on a large scale. This is generally caused by tectonic uplift of the area of sedimentation leading to subaerial exposure and erosion of the previously deposited units. When the area subsides again, the newly deposited layers are said to be unconformably overlying the preceding ones. Various types of unconformable contacts are possible. When stratified rock layers overlie igneous or metamorphic bodies, the relationship between the two is said to be non-conformable. The *unconformity* in this case is readily recognizable. While mapping a non-conformity one should be certain that there is no intrusive relationship between the two sets of rocks. The possibility of contact metamorphism is also to be eliminated.

When the older beds within a sequence of sediments are tilted or deformed, before being buried by the younger layers, the resultant break is called an angular unconformity. Such unconformities are easily detectable during field mapping. When the older and younger layers lie parallel to each other with only an erosional break between the two, the relationship is designated a *disconformity*. This is often marked by a layer of conglomerate consisting of erosional products of the stratigraphically older unit. The term *paraconformity* was introduced by Dunbar and Rodgers (1957) to describe the situation wherein a time gap occurs without a detectable physical break between the upper and the lower units. Such time gaps between the two depositional units can be identified by palaeontological or radiometric method (Fig. 9.2).

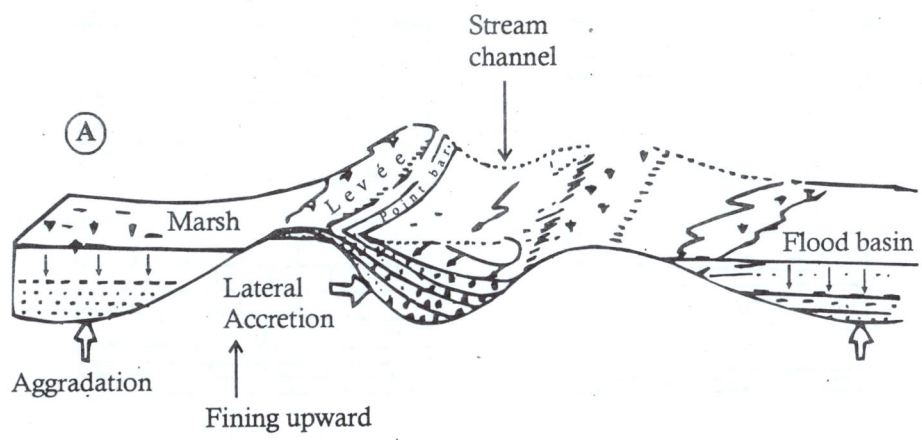

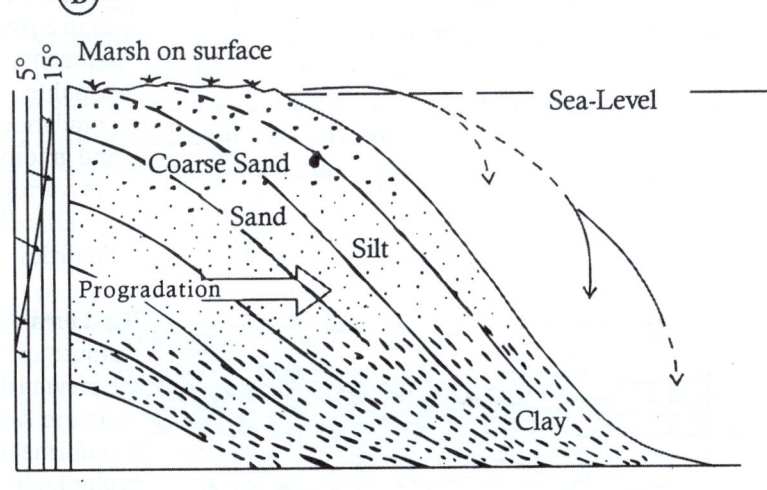

Fig. 9.1: Various mechanisms of sediment accumulation : (A) fining-upward sequences produced in a fluvial regime by lateral accretion within point bars and aggradation within flood basins, (B) coarsening-upward sequences produced by progradation of a delta (based on illustrations by Galloway and Hobday 1983 ; Scruton 1960 *in* Pettijohn 1975). Note 'blue pattern' in dipmeter log at the left.

Recognition of the different types of unconformities can help in deciphering the depositional history of a sedimentary basin. A non-conformable relationship is expected at the basin fringe where layers of sediments overlie the igneous or metamorphic basement. Periodic withdrawal of the sea from the deeper parts of a basin or a temporary phase of subaerial exposure, may produce a disconformable relationship between the underlying and the overlying layers. Paraconformable relationships are to be expected at the centre of the basin where the sedimentary pile is very thick.

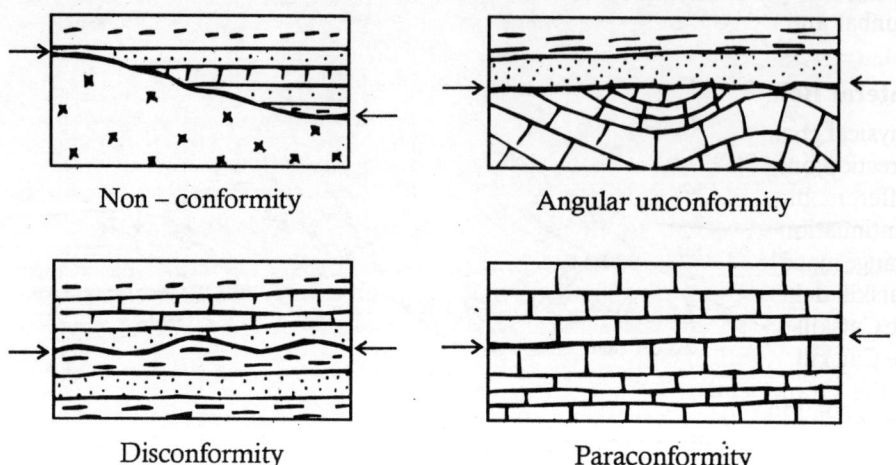

Non – conformity   Angular unconformity

Disconformity   Paraconformity

Fig.9.2: Four types of unconformities (based on Dunbar and Rodgers 1957).

In the case of a gradual marine transgression over an unconformity, the younger layers overlap the older ones and are onlapped by the still younger sediments. The younger layers in this case are progressively displaced away from the centre of the basin. In a typical case of marine transgression on land, this results in a fining-upward sequence. In the event of a gradual marine regression, the sea is progressively displaced towards the basin centre and the younger layers are shifted away from the land area. The relationship between the successive sedimentary layers in this case is offlapping. Pinchouts are common in onlaps but truncations occur frequently in offlaps (Fig. 9.3).

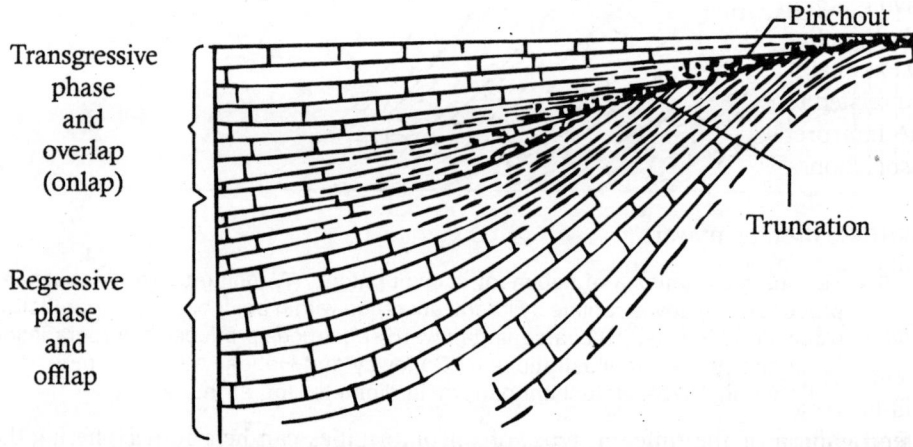

Fig. 9.3: Onlap and offlap relationships produced by marine transgression and regression. During a transgressive phase each unit onlaps the preceding one (based on Krumbein and Sloss 1963).

While an unconformity represents a major break in sedimentation, on a regional scale over a protracted period of time, many small breaks of short duration (*diastems*) may occur within a basin. Erosion and dispersal of loose sediments at sea bottom by sheets of moving water, a process termed '*sublevation*' (Latin: sublevare-to lift up) by Dunbar and Rodgers (1957) is an example of such a break.

Lateral Relationship

Physical changes between consecutive sedimeutary units do not take place in vertical direction only. Lateral changes in depositional conditions ('facies changes') may cause different but contemporaneous rock types and faunas to be deposited in strike continuation of each other. One of the classical areas in which demonstration of such a change made the concept of facies variation acceptable in North America is the Devonian Catskill delta of New York State. Careful mapping in this area in 1933 indicated that the Catskill delta facies and the Chemung Group of rocks are of the same age, although the Catskill, resting on top of the Chemung, appears to be younger (Fig. 9.4). Similarly, in the Grand Canyon of Colorado, isochronous deposition of several Cambrian formations was demonstrated by McKee (1945, 1949). Several such classical examples of facies variation are to be found in Dunbar and Rodgers (1957). In the Permian Barakar Formation of the Southern Karanpura coalfield of India, Banerjee (1960) demonstrated how a rhythmic pattern dominated alternately by coarse-grained, sandy channel facies and fine-grained (coal-shale) interchannel facies could develop by shifting of ancient river channels. Gradational and interfingering contacts between homotaxial channel and interchannel Permo-Triassic fluvial deposits were also established in the Godavari valley of southern India by Sengupta (1970, 2003, see Fig. 7.27).

Lithosome

The rock units involved in the process of lateral (facies) change, like the situations described above, show up as three dimensional masses of essentially uniform lithology having a gradational or intertonguing relationship with adjacent masses of different lithology. Such a body of rock is designated as *lithosome*. Shapes of lithosomes, defined by external morphology and three-dimensional geometry (length, breadth and width), are related to specific tectonic and environmental settings and may provide clues to the interpretation of ancient environmental conditions. For this purpose facies associations, or groups of facies, which occur together and are genetically connected, are more important than the geometry or character of an individual lithosome. Fig. 9.5 illustrates how the conventional lithostratigraphic terminology is to be applied during mapping of laterally or vertically occurring facies associations.

Walther's Law

Facies associations, which occur side by side today, and pass laterally into each other, will be preserved vertically above one another in geological record. Conversely, the lateral associations of facies in geologic past can be inferred from what is seen in vertical stratigraphic sections today, using the Law of Correlation of Facies formulated by Walther in 1884. This law states "... the various deposits of the same facies area and, similarly, the sum of the rocks of different facies areas were formed beside each other

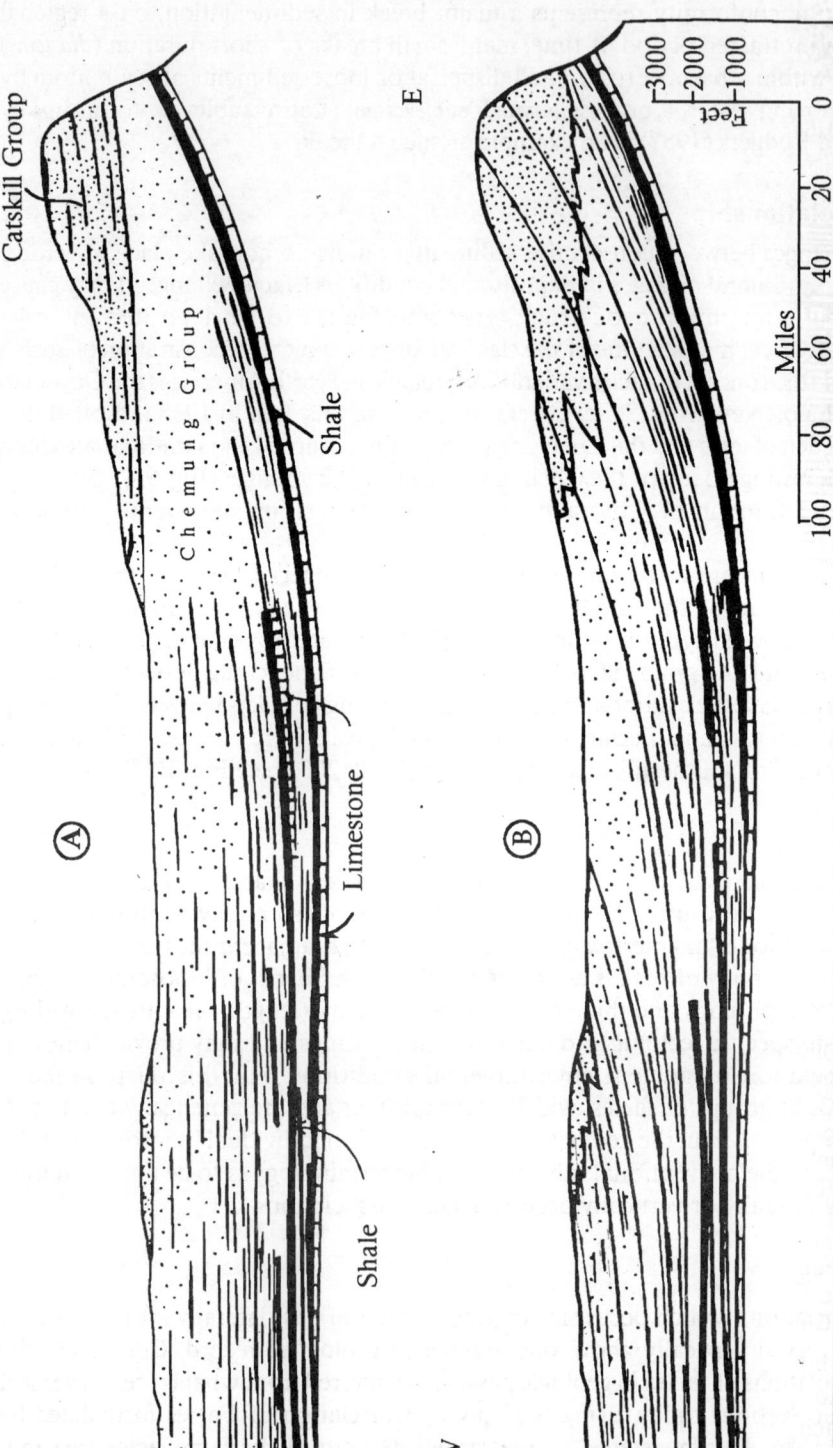

Fig. 9.4: Sections across the Devonian Catskill Delta, New York, showing stratigraphic interpretation : (A) before acceptance of the facies concept and (B) after introduction of the facies concept in the area by Chadwick and Cooper in 1933. The curved lines in B indicate time surfaces (simplified and redrawn after Dunbar and Rodgers 1957).

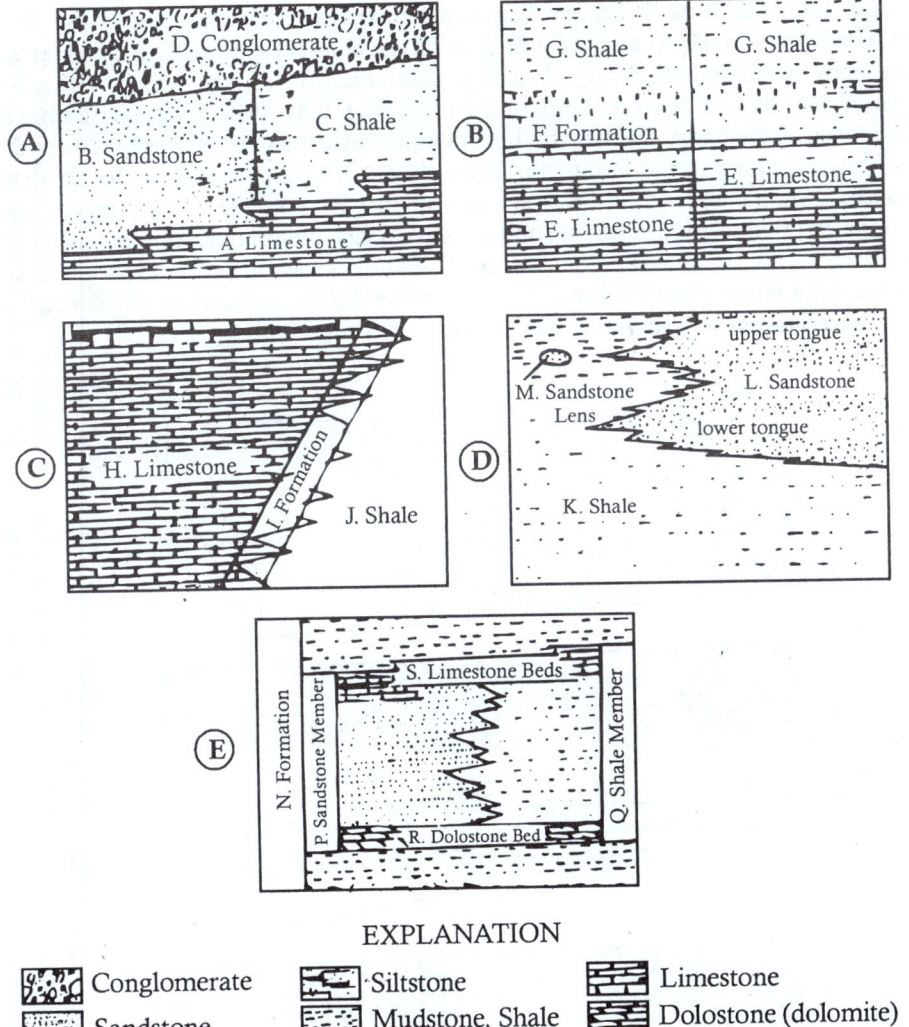

EXPLANATION

Conglomerate	Siltstone	Limestone
Sandstone	Mudstone, Shale	Dolostone (dolomite)

Fig. 9.5: Different types of lithostratigraphic boundaries in areas of facies change (redrawn from the Code of Stratigraphic Nomenclature proposed by the North American Commission 1983). (A) Boundaries at sharp lithological contacts and in laterally gradational sequence; (B) alternative boundaries in a vertically gradational or interlayered sequence; (C) possible boundaries for a laterally intertonguing sequence; (D) possible classification of parts of an intertonguing sequence; (E) key beds (R & S) are used as boundaries to distinguish Q-Shale Member from the other parts of N-Formation. A lateral change in composition between the key beds requires that another name, P-Sandstone Member, be applied. The key beds are part of each member. With permission of American Association of petroleum Geologists.

in space, but in a crustal profile we see them lying on top of each other...... Only those facies and facies areas can be superimposed, without a break, that can be observed beside each other at the present time"(translated by Blatt *et al.* 1972, p.187-188 from the original in German).

The significance of Walther's Law can be explained by referring to the fluvial sedimentation model. At any point of time fluvial channel sands coexist, and laterally pass into, the finer grained floodbasin (marsh swamp) deposits which in turn pass laterally into deltaic deposits bordering the sea (Fig. 9.6). With progress of time and subsidence of the basin, the sand bodies cut into, and are preserved partly on top of silt and clay belonging to the floodplain facies. These in turn partially overlie the deltaic deposits. Complete preservation of such a sequence in geologic record is possible only if the basin of deposition subsides continuously. In fact, continuous subsidence (without any break) is an essential prerequisite for the Walther's Law to be valid. An example of detrital sand lenses preserved in vertical sequence due to continuous sibsidence (with some oscillations) of a fluvial delta complex is illustrated in Fig. 9.7.

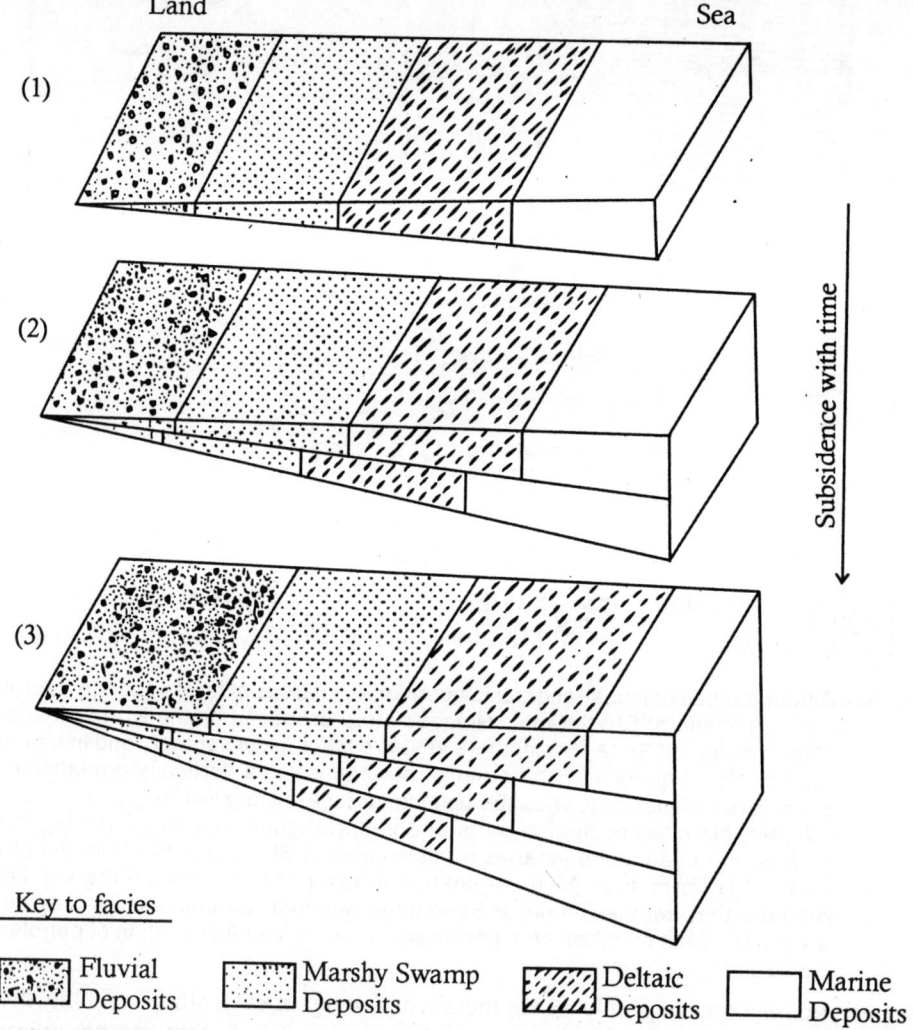

Fig. 9.6: Diagrams illustrating Walther's Law. Facies associations occurring side by side today are preserved vertically above one another in geological record due to continuous subsidence of the basin.

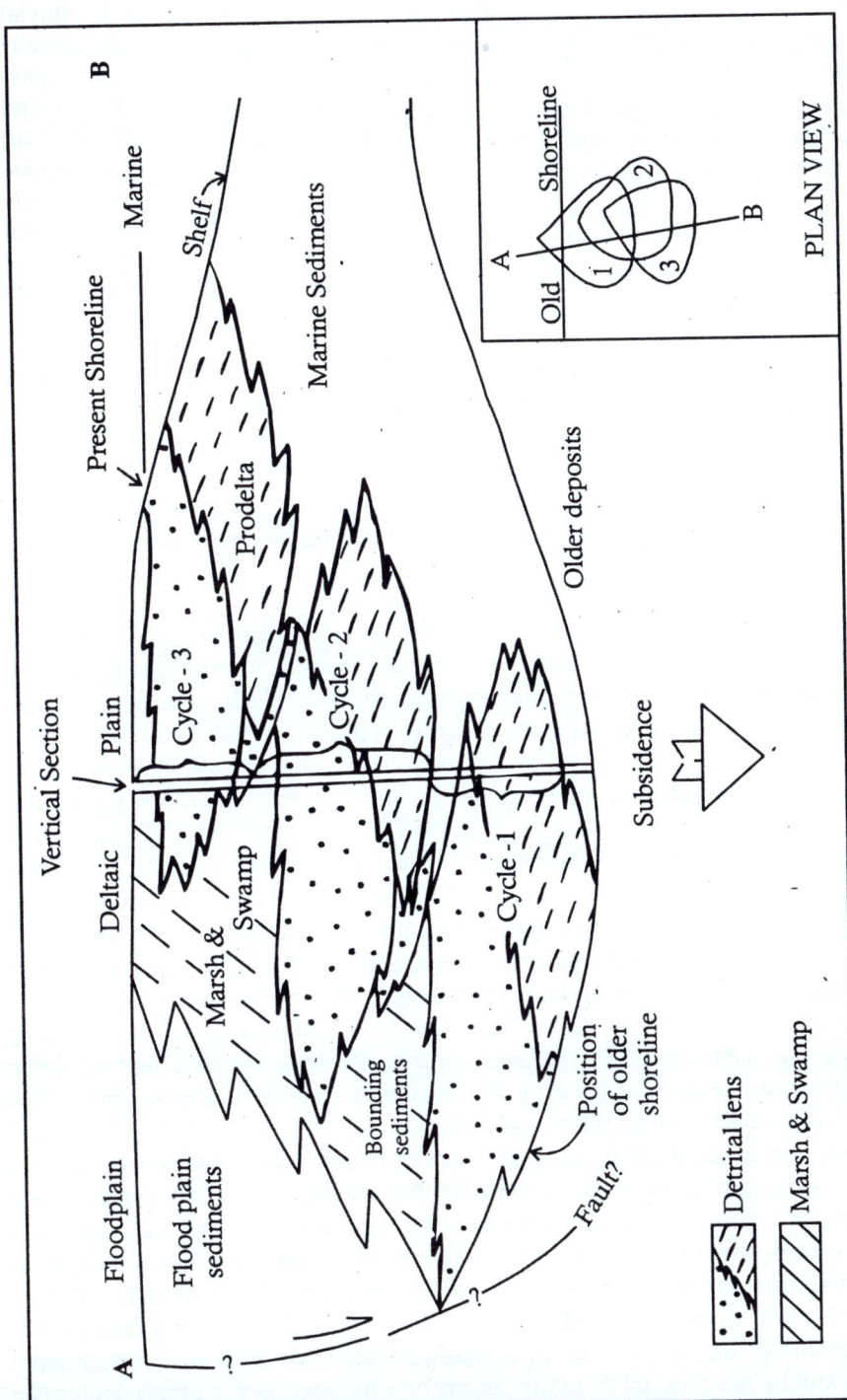

Fig. 9.7: Overlapping cycles produced in a hypothetical delta complex by subsidence (after Coleman and Gagliano 1964, as modified by Weimer 1976 (Inset shows projection of the detrital lenses in plan view). With permission of American Association of Petroleum Geologists.

Cyclicity

In the example cited above cycles of fining-upward sequences are repeated vertically by the process of channel migration. This may take place without a major break during the process of sedimentation. Such repetitions are called autocyclic. Repetitions may also happen on a much larger areal and temporal scale due to, for example, tectonic or base level changes leading to superposition of disconformity bounded deposits. These are termed *allocycles*. The North American Code (1983) provides for classification of these mappable allostratigraphic units into a hierarchical system of decreasing rank: allogroup, alloformation, allomember (Table 9.2).

The boundaries between these *allocycles* are produced by large-scale stratigraphic events related to regional sea-level fluctuations. Allocycles have been discovered in many coal basins of the world. The Pennsylvanian cyclothem of Illinois and the Carboniferous cyclothem of Yorkshire, which are cited as classical examples, have been attributed to sea-level fluctuations, either due to periodic diastrophism or to waxing and waning of land glaciers. Some authors have also attempted to correlate these periodic fluctuations to the orbital cycles of the earth's precession (31,000 year cycles), obliquity (43,000 year cycles) and eccentricity (100,000 year cycles), discovered by Milancovitch in 1941 and subsequently named after him.

During the last two decades earth cycles of various magnitudes have been discovered. Thirtytwo-million year cycles involving changes in the distribution of marine unconformities, patterns of faunal diversity, occurrence of black- and red-clays have been recorded. These have been attributed to cyclic changes in global climatic conditions caused by variation in the rates of volcanic activity at oceanic ridges (Fischer and Arthur 1977). These 32 m.y. cycles are superimposed on much larger cycles of about 200 m.y. duration.

Seismic surveys have led to the discovery of major cyclic erosional breaks of different magnitude during the Phanerozoic (Fig. 9.8). Of these, the major (first-order) breaks have been correlated to global sea-level fluctuations caused by changes in ocean-floor spreading rates (Vail, Mitchum and Thompson III 1977). These have some correspondence to the 200 m.y. cycles of Fischer and Arthur. The rise in sea-level could have been caused by buoying up of sea-water due to heating near the spreading midoceanic ridges (Pitman 1978). The lower order sea-level changes have been attributed to glacio-eustatic changes (Vail *et al.* 1977) or, alternatively, to flexures during basin formation on an elastic crust (Watts *et al.* 1982).

Freezing and locking up of sea-water in the form of land ice would cause a general lowering of the sea-level, but whether this would cause a transgression or a regression of the shoreline would depend on the rates of sediment input, tectonism and isostatic adjustments. A regression may take place even when the sea-level is high, if the sediment input is also high at this stage. Similarly, a transgression may occur even when the sea-level is low, if the rate of subsidence of the land area is fast, either due to tectonic or due to isostatic causes. The local sea-level changes therefore, are the net product of many factors of which eustacy, isostasy, tectonism and sediment input are more important (see Reading 1986 for a more detailed discussion).

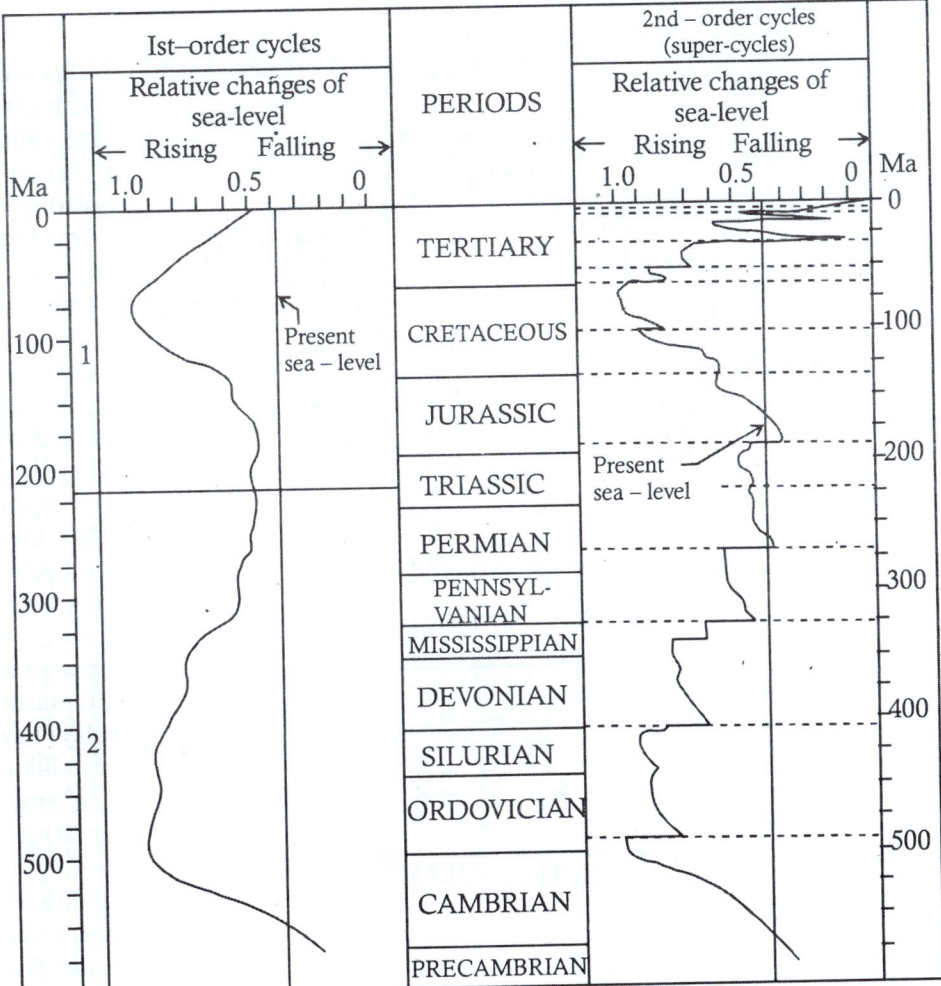

Fig. 9.8: First- and second- order global cycles of sea-level fluctuations during the Phanerozoic Eon (after Vail, Mitchum and Thompson III 1977). With permission of American Association of Petroleum Geologists.

Two orders of *transgressive-regressive cycles* (called T-R cycles), the larger one corresponding to the third-order cycle of Vail *et al.* (1977) and the smaller (named fourth-order cycle) coinciding with the transgressive-regressive maxima of Vail's third-order cycle, have been identified in the coal-bearing Upper Cretaceous Strata of Utah (Reyer 1983). Discovery of short (metre scale) basinward discontinuities within the Helderberg Group or rocks (Devonian) in New York State, led some modern stratigraphers to suggest that the stratigraphic record, barring minor exceptions, consists of thin, shallowing-up cycles bounded by geologically instantaneous rise of base level. The basic assumption behind this model is that the mode of stratigraphic accumulation is episodic rather than gradual, as traditionally believed. Goodwin and Anderson (1985), the advocates of this model, have demonstrated how episodic deposits might be produced by eustatic responses to orbital perturbations superimposed on continuous basin subsidence (Fig. 9.9). They have also worked out in detail how (geologically) instantaneous sea-level rise followed

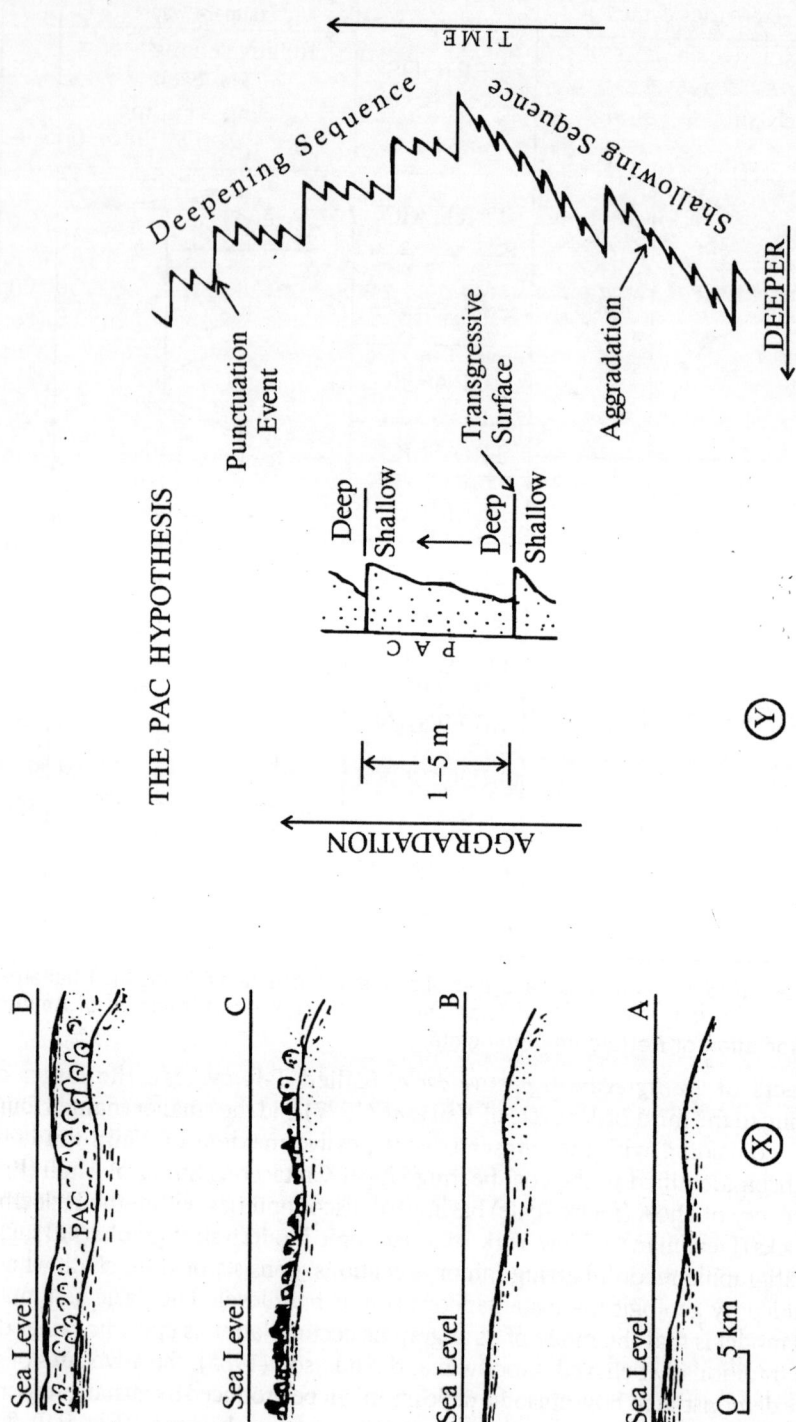

Fig. 9.9: (Y) Sequence of events leading to 'Punctuated Aggradational Cycles' (PAC). Each PAC is bounded by a surface of abrupt change to deeper water facies. (X) Illustrates how this change may take place by sea-level fluctuation (after Goodwin and Anderson 1985). With permission of The University of Chicago Press.

by aggradation might lead to the development of what they termed *Punctuated Aggradational Cycles* (PAC). Such a model of sedimentation, based on inferred, short-lived episodes, precludes the application of Walther's Law except within cycles. The advocates of PAC believe that the PAC boundaries, being short-ranging stratigraphic units, will provide the tool for detailed intrabasinal correlation.

Analysis of Cyclicity

Rock units, arranged apparently randomly in vertical stacks, may contain a hidden regularity. This regularity, if deciphered, may provide clues to the inherent genetic relationship between different facies. In a simple method of studying this relationship, de Raff, Reading and Walker (1965) worked out the frequency of facies which are in contact with each other in a vertical succession. This led to a 'spider diagram' depicting the facies relationship. Nowadays, more sophisticated methods are used for finding the 'hidden motif' within cycles (Miall 1973; Cant and Walker 1976).

The procedure for quantitative analysis of cyclicity, based on Gingerich's (1969) technique for analysis of cyclicity in fluvial sequences, determines the probability with which a certain rock type will follow another in the vertical column. If the probable sequence of transition, worked out in this way, does not give a complete picture, the 'hidden elements' may be detected by a second stage of analysis when the 'difference matrices' are computed. Finally, a test for the Markovian property is carried out in order to know if each depositional event of the system was influenced by the event immediately preceding it, i.e., whether the depositional processes have followed a continuous sequence.

The method outlined above has been shown to be statistically incorrect (Walker 1984a), but the alternative procedure of analysis, claimed to be dependable (Harper, Jr. 1984), is too sophisticated to be discussed here. Experience, however, has shown that the system of analysis proposed by Gingerich (1969) gives fairly dependable clues to genetic interpretation of facies in many cases.

A mechanical procedure of analysis of cyclicity may fail to take into account important geological features such as the nature of contact between facies. The conclusions which prove to be statistically significant, may, in some cases be geologically unimportant. Moreover, every method of mathematical analysis based on summarization of the original observations, runs the risk of losing valuable geological information in the process. The answer to this problem is to return to the raw data at every stage of analysis. Only the geologically meaningful solutions should be accepted (Reading 1986).

Correlation

Traditionally, *correlation* implies demonstration of *time equivalence* of particular stratigraphic units from the type sections to the areas of their geographic extension. The criteria used for correlation should, therefore, be independent of lithology, fossil content or other material basis of stratigraphic division. Some criteria for determination of time which is independent of all these should be used. Although some authorities feel that in stratigraphy this "narrow definition of correlation should be retained" (Rodgers, 1959). More modern texts advocate use of different types of criteria for correlation (Nichols, 1999). These are as follows.

1. Correlation of lithostratigraphic units based on lithologic criteria and stratigraphic position of the concerned rock bodies.
2. Correlation of biostratigraphic units based on fossil content of the units.
3. Correlation of magnetostratigraphic units based on magnetic properties of rock bodies different from those of the underlying and overlying rocks.
4. Correlation of chronostratigraphic units based on time criteria. That is, a body of rocks whose upper and lower boundaries are isochronous surfaces.
5. Correlation of allostratigraphic units based on the position of a unit with respect to unconformities or their relative conformities (see 'sequence stratigraphy').

Techniqnes of Correlation

Heavy Mineral Assemblages

In the early days of geological exploration *heavy mineral* assemblages were often used for stratigraphic correlation. The technique is based on the assumption that heavy minerals are incorporated into layers of sediments by progressive de-roofing of a source rock. When such a process takes place at different geographic locations, but within a short span of time, the layers of sediments may be taken as time equivalent. The method worked well in many cases. For example, the proportions of epidote, hornblende, and staurolite could be successfully used for correlating the Tertiary stratigraphic units of Assam, India. This technique reached such a refinement that local stratigraphic irregularities could be detected by studying the heavy mineral assemblages of only a few odd samples (Evans *et al.* 1934).

The method of correlation with the help of heavy minerals alone has its drawback. When the heavy minerals occur within a stratigraphic column in reverse order of their stability (i.e. when only the more stable minerals are found in the older horizons), one may suspect that the less stable ones have been removed by circulating interstratal fluids. An attempt of stratigraphic correlation by heavy minerals is of little value under such a circumstance (Pettijohn 1975).

Wireline Logging

The best possible way of obtaining subsurface information is by drilling. Well cuttings continuously brought to the surface by circulating drilling fluid ('mud') do not provide precise information on the nature of stratiraphic contacts in the subsurface. To obtain this information core samples must be collected continuously during drilling. As this is an extremely expensive process, a continuous record of the electrical, sonic and radioactive properties of the sections drilled is generally maintained by *wireline logging*. Logging is conducted by introducing into the borehole a *sonde* containing an energy source and a detector. The source is a combined current and potential electrode for electrical logging, an ultrasonic transducer for sonic logging, a neutron emission source for neutron logging, or a gamma ray (Co^{60}, Cs^{137}) source for density logging. Recording is done continuously as the sonde is moved up the borehole (Table 9.3).

The conventional electrical logs record *spontaneous potential* (S.P.) and *resistivity*. The former is a measure of the electrical current generated at the contact of rock layers by interaction of the drilling mud with the formation water. The latter measures the voltage drop across potential electrodes for known currents passed through current electrodes (Fig. 9.10A).

Contrasting salinity of the pore fluid and the drilling mud is essential for the generation of spontaneous potential. As the mud comes into contact with sand or a shale layer filled with salt-water, the electrochemical potentials add up to produce S.P. currents across the sand-shale sequence. The smooth S.P. curve that generates against a shale layer, when it is thick, is called the shale line or *base line*. Against porous sand beds, the S.P. curve shifts to the left. The spontaneous potential generated is related to resistivity of the formation water (R_w) and the resistivity of the mud filtrate (R_{mf}). Hence the S.P. log can be used not only for delineating the bed boundary and the bed thickness, but also for estimating the resistivity of the formation water. Under suitable conditions the S.P. curve can also be used as an indicator of the shale content of a sand layer. The higher the shale concentration, the lower the amplitude of S.P. (Fig. 9.10B).

Table 9.3: Log types, properties measured, and geologic uses (abridged from Cant 1984)

Log	Property measured	Units	Geologic uses
Spontaneous Potential	Natural electric potential (compared to drilling mud)	Millivolts	Lithology (in some cases), correlation,curve shape analysis, identification of porous zones
Resistivity	Resistance to electric current flow	Ohm-metres	Identification of coals, bentonites, fluid evaluation
Gamma-ray	Natural radioactivity of K, Th, U	API units	Lithology (shaliness), correlation, curve shape analysis
Sonic	Velocity of propagation of sound wave	Microseconds /metre	Identification of porous zones, coal, tightly cemented zones
Caliper	Diameter of hole	Centimetres	Evaluate hole conditions and reliability of other logs
Neutron	Concentrations of hydrogen (water and hydrocarbons) in pores	Percent porosity	Identification of porous zones
Density	Bulk density (electron density) includes pore fluid in measurement	Kilograms per cubic metre (kg/m^3)	Identification of hydrite, halite, non-porous carbonates, etc.
Dipmeter	Resistivity changes indicating dips	Degrees (and direction)	Structural and stratigraphic analysis, environmental interpretation.

Since S.P. is generated by the difference in concentration of electrolytes in the mud and the formation water, there will be no S.P. if the salt concentration is of the same order in the mud and the formation in contact. In such a case a gamma-ray log may be used for determination of the sand-shale boundary. Shales containing K^{40} isotope provide a natural source of gamma-ray flux. In fact, a combination of a number of logs (S. P., sonic, density and neutron logs) is used for accurate determination·of the proportion of shale in a sandy shale. The combination of gamma-ray and sonic logs is perhaps the best for the purpose of subsurface correlation (Fig. 9.11). The sonic log is effective in detecting very high velocity rocks (e.g., carbonates or well-cemented calcareous sandstones) and also very low velocity rocks (e.g., coal, porous or fractured rocks with interstitial fluid, unconsolidated sediments). The gamma-ray log, by detecting the shaliness of a rock, may indicate whether a sequence is 'coarsening' or 'fining-upward' although this log does not measure grain-size.

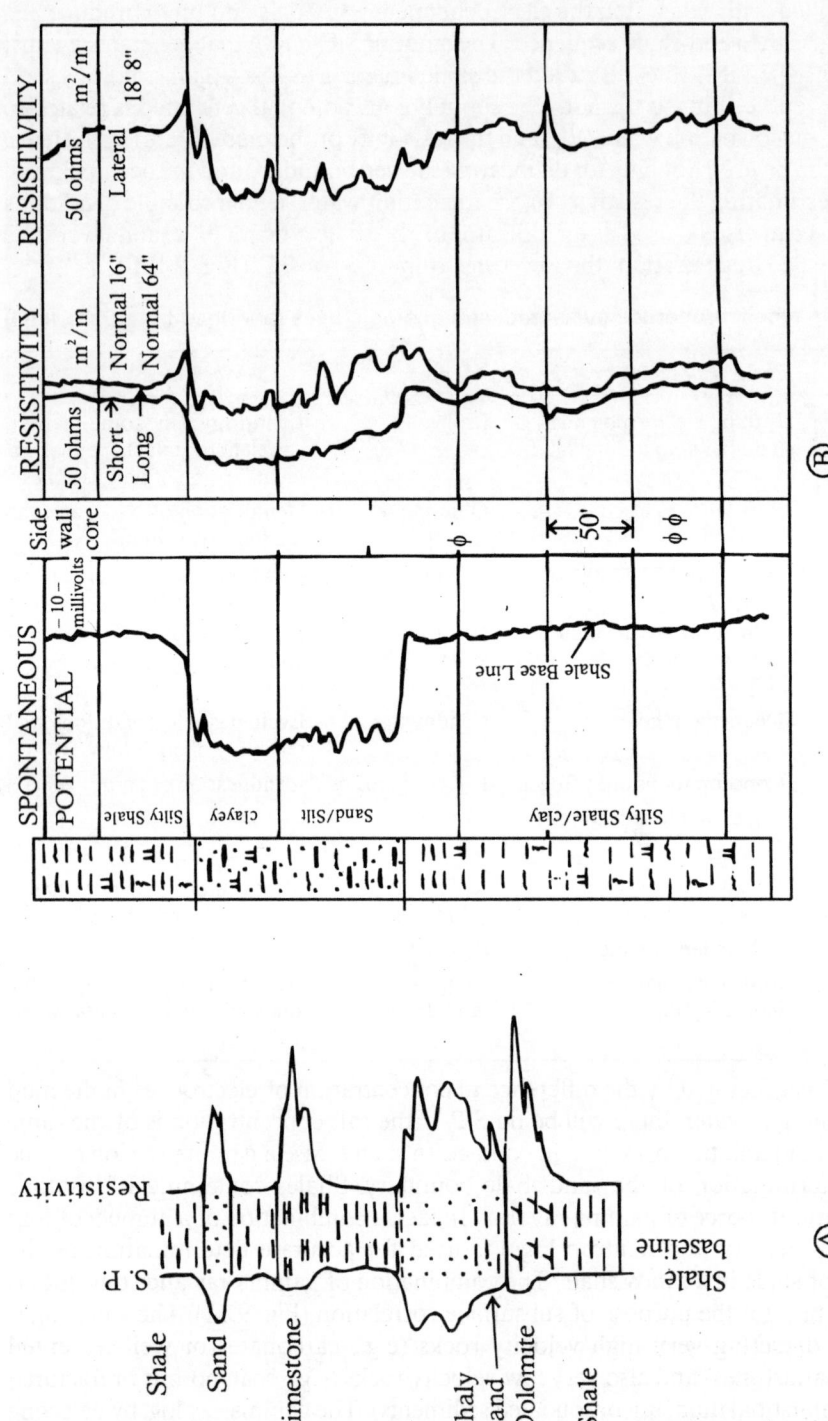

Fig. 9.10: (A). Typical Spontaneous Potential (S.P.) and Resistivity patterns for sand, shale and limestone. (B). An electrolog showing S.P. and Resistivity (Short, Long-Normal, and Lateral) patterns for different lithological units within a borehole. Locations of side wall cores are shown by φ.

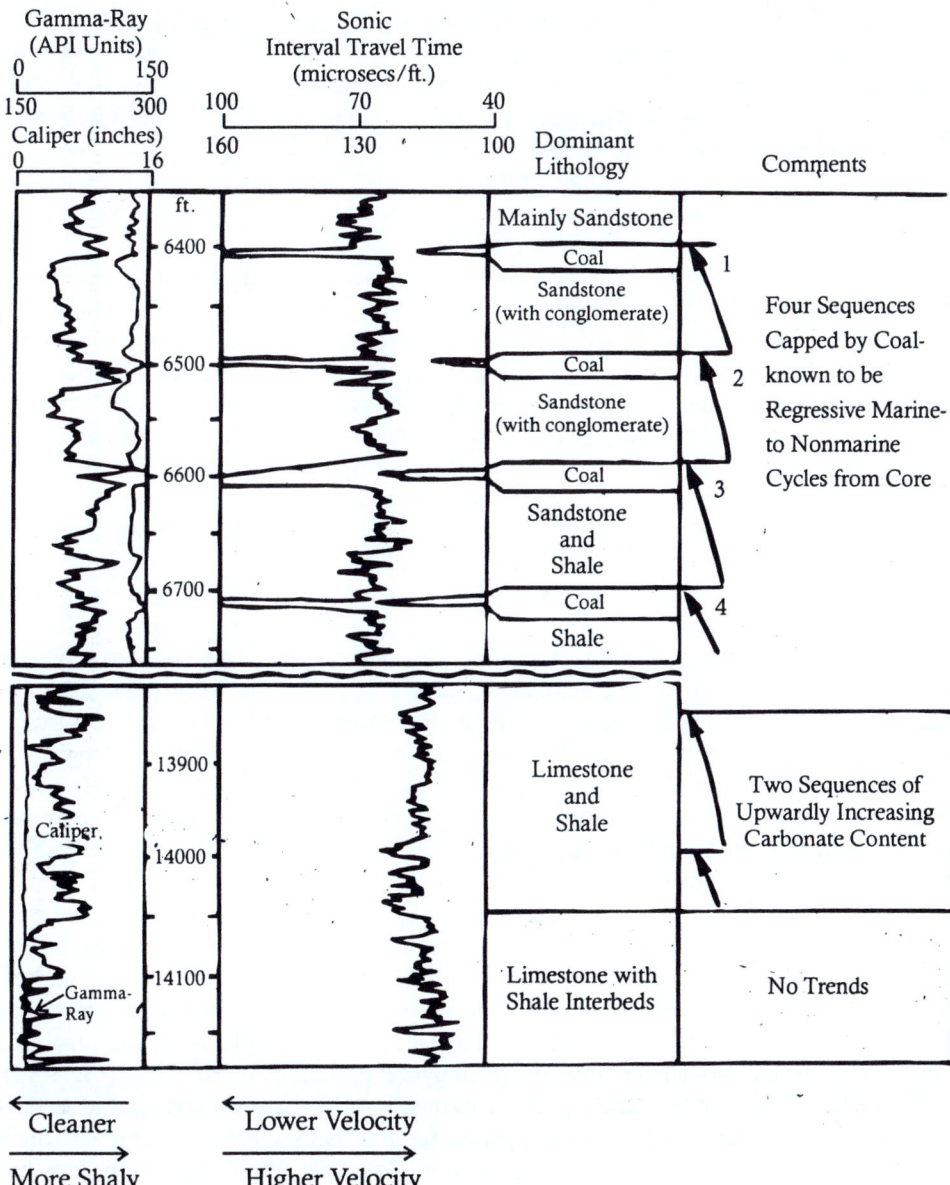

Fig. 9.11: Gamma-ray and sonic logs against coal and some typically siliciclastic and carbonate sections Reprinted from Cant 1984 with permission of Geological Association of Canada.

The equipment for resistivity measurement in boreholes consists of a current and a potential electrode. An electric current is sent through the current electrodes and the potential difference between the two potential electrodes is measured. The greater the spacing of the electrodes, the deeper the penetration of the current into the formation. By using electrodes of different spacings (Normal 16", Long Normal, 64", Lateral, 18'8") resistivities of the mud-invaded zone and also the true formation resistivity beyond the zone of flushing can be measured (Fig 9.10B). Resistivity measurement can aid particularly in correlating the carbonate beds, which have high resistivity and low porosity. A well-cemented calcareous sandstone, which is more resistive than a porous, non-calcareous one, may be detected by resistivity logging. Both sonic and resistivity logs may be used for detecting the fluid content of a rock.

Resistivity measurements using direct current work only when the drilling mud is conducting. In empty wells, as also in wells filled with oil-based drilling fluids or gas, resistivity is measured by transmitting alternating current into the formation. The amount of current induced into the system is a measure of the formation resistivity. This is the method of *induction logging*.

Resistivity logs may also be used for measurement of the dip of a formation within a borehole. For this purpose microresistivity is measured simultaneously by four (five in the case of high resolution measurements) identical microresistivity electrodes mounted on four equally spaced pads of the dipmeter tool inserted into the borehole. A dipping bed produces a resistivity record of comparable magnitude at different depths on different walls of the borehole. Mathematical correlation methods are then applied for measurement of these displacements. Individual features are matched together for computation of the dip and azimuth of the bedding. These computations, as well as correction for the effect of deviation of the borehole, are done by a microcomputer built into the system.

The *dipmeter* output is in the form of an arrow plot. The combined dipmeter-inclinometer plot, which shows dip as well as azimuth is called a 'tadpole plot.' Vertical variations in dips recorded in a dipmeter log help in long- and short-interval correlation. Structural and stratigraphic features can also be detected in the subsurface by dipmeter logging.

The methods commonly used for correlation of borehole logs are: tracing of marker beds, sequence analysis and slicing (Cant 1984). A bed having a distinctive log character may be used as a marker bed for the purpose of correlation when it has an appreciable areal extent. The deposits in low-energy environments, such as lakes and oceans, which generally extend over large areas, make good markers. Those in high-energy environments, such as rivers and sea beaches, are impersistent and lack distinctive characters. Such beds may be correlated for short distances with reference to the markers that may occur immediately above or below. In the Nahorkatiya Oilfield in Assam, for example, a number of multistory sand bodies, each of small areal extent, were mapped between two well-established marker horizons by S.P. and Short Normal logs (Fig. 9.12).

Sequence analysis in wireline logs involves matching of readily recognizable vertical patterns such as coarsening or fining-upward sequences by their log characters. Such sequences, called 'events', may match palaeontologically established time lines but they often cut across lithological or facies boundaries. When all other techniques of correlation fail, a log interval may be subdivided arbitrarily into slices for studying the nature of variation within small intervals. Thick deltaic deposits, for example, may be correlated over large areas with the help of the overlying and underlying markers but the intermediate thick pile of sediment is 'sliced' to study the nature of facies variation within short vertical distances.

Interpretation of facies or depositional environment from the shape of log signatures alone is risky because no log pattern is uniquely related to any particular environment. Fig. 9.13 gives the idealized gamma-ray or S.P. patterns found in some of the common depositional environments. Each log pattern shown in this diagram can generate in

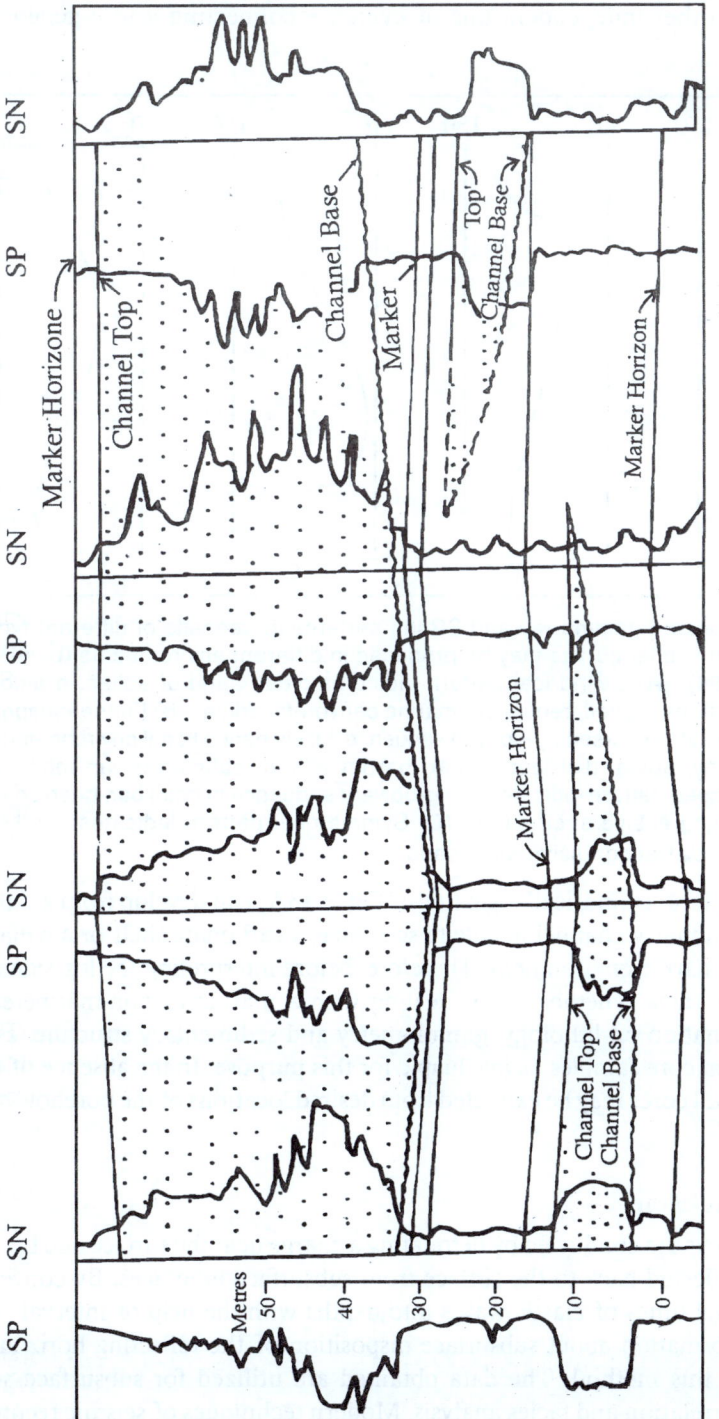

Fig. 9.12: Correlation of fluvial channel sands in the subsurface of Nahorkatiya, India by Spontaneous Potential (SP) and Short Normal (SN) resistivity logs (based on Azad *et al.* 1971). With permission of Schlumberger Technical Services, Inc.

any of the depositional conditions mentioned. In choosing between them one must utilize some other independent line of evidence connecting it to a genetic (facies) model.

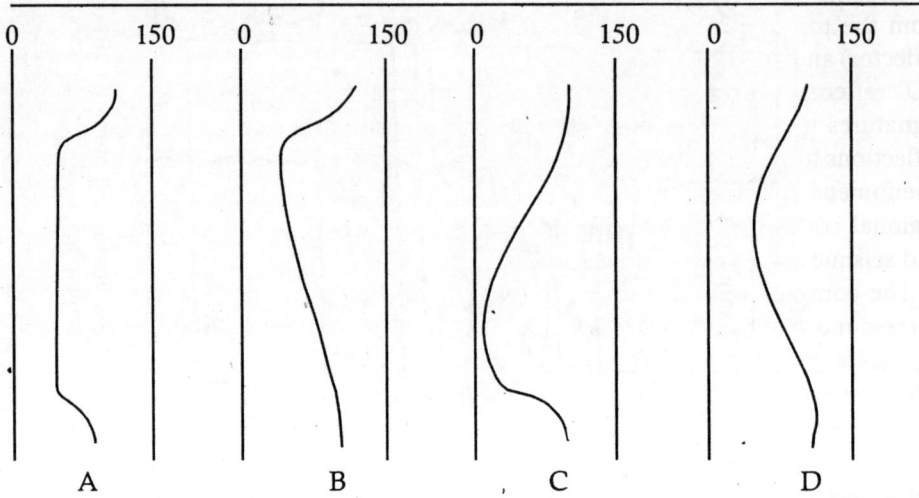

Fig. 9.13: Idealised gamma-ray and SP log patterns in deposits of different types. Note that similar curves may be produced in different environments (based on Cant 1984). (A) Cylindrical pattern indicates clean sand of eolian, braided fluvial, carbonate shelf, reef or submarine canyon-fill origin. (B) Funnel-shaped pattern indicates coarsening-upward sequence. Likely depositional environment: crevasse splay, fluvial distributary bar, barrier island, submarine fan lobe. (C) Bell-shaped pattern indicates fining-upward sequence in point bar, deep sea channel, transgressive shelf sand. (D) Symmetrical pattern indicates sandy offshore bar or transgressive shelf sand.

For example, a 'cylindrical' log pattern (Fig. 9.13A) can develop from a fluvial sand within a distributary channel in a delta sequence, a carbonate shelf or a reefal growth in a shallow marine environment. Therefore, before interpreting the log signature it is important to know whether one is dealing with a marine or non-marine sequence. Direct information on lithology, granulometry and sedimentary structure, as they be obtained from core samples, is invaluable for this purpose. In the absence of borehole cores, side-wall cores may be extracted from desired locations of the borehole by special techniques.

Seismic Correlation

Seismic surveys are conducted by detonating a charge near the surface and by detecting the waves reflected back to the surface from subsurface interfaces. By converting the two-way travel times of elastic waves into depths with the help of interval velocities, accurate information about subsurface disposition of the reflecting horizons can be obtained by this method. The data obtained are utilized for subsurface structural mapping, correlation and facies analysis. Modern techniques of seismic recording and display allow visualization of the subsurface picture over large areas.

For the purpose of geological interpretation composite cross-sections running across the basin under study are constructed using seismic reflections. One such composite profile prepared for the western part of the Bengal Basin by means of seven seismic reflectors is shown in Fig. 8.22. Two of these reflectors, one emanating from the top of an extensive basalt flow underlying the Tertiary sediments ('R'-reflector) and the other corresponding to the interface of a limestone and a shale bed ('D' reflector), were used as markers for geological interpretation. The seismic signatures of both the reflectors could be correlated over large areas. Five other reflections from shallower horizons terminated at different distances basinward. This phenomenon was interpreted as a basinward change of lithofacies. The type or regional correlation achieved in the Bengal Basin by co-ordinated electro-logging and seismic survey is illustrated in Fig. 9.14.

The conventional technique of seismic interpretation expects the reflections to correspond to lithological interfaces characterized by sharp contrasts in *acoustic impedance* (product of velocity and density). Marker reflections generated from such interfaces of regional extent are characterized by their coherence over a number of seismograms, The 'D'-and the 'R'-reflectors in the Bengal Basin are two examples of such reflections. Apart from the marker reflections a number of minor ones, apparently not connected with strong lithological interfaces, are also found in every sedimentary basin. Such a situation is depicted in Fig. 9.15. Seismic reflections represented by thick black lines in this diagram follow stratal surfaces which are ancient depositional planes, corresponding to time lines. These reflections cross rock formation (lithofacies) boundaries.

Seismic techniques have their limitations. Subsurface rock bodies having thickness less than the wavelength of seismic waves go undetected. Thus the units seen in seismic profiles represent packages of beds. Moreover, the wave propagation velocities for shales, sandstones and limestones overlap at depth, thereby making the seismic method unsuitable for detection of their interfaces (Gardner, Gardner and Gregory 1974).

Unconformities are often the best seismic reflectors. Angular unconformities involving sharp changes in dip are the easiest to detect by seismic surveys. Even when angular changes are not involved, an unconformity may be detected by the contrast in acoustic impedance created by the presence of weathered zone at the plane of erosion. In fact, unconformities correlatable on a global scale by seismic reflections, and interpreted to be the effect of global sea-level fluctuations, have given rise to a new concept in stratigraphy called *sequence stratigraphy*.

SEQUENCE STRATIGRAPHY

History

In 1941, L.L. Sloss identified unconformity-bounded packages of sediments of craton-wise extension in Montana, USA. He named these packages 'sequences' in lithostratigraphic sense (see Sloss 1988 for a review of these developments). Recognition of unconformities of regional extent and identification of eustacy as the driving mechanism for generation of unconformity-bounded sequences, provided the ground

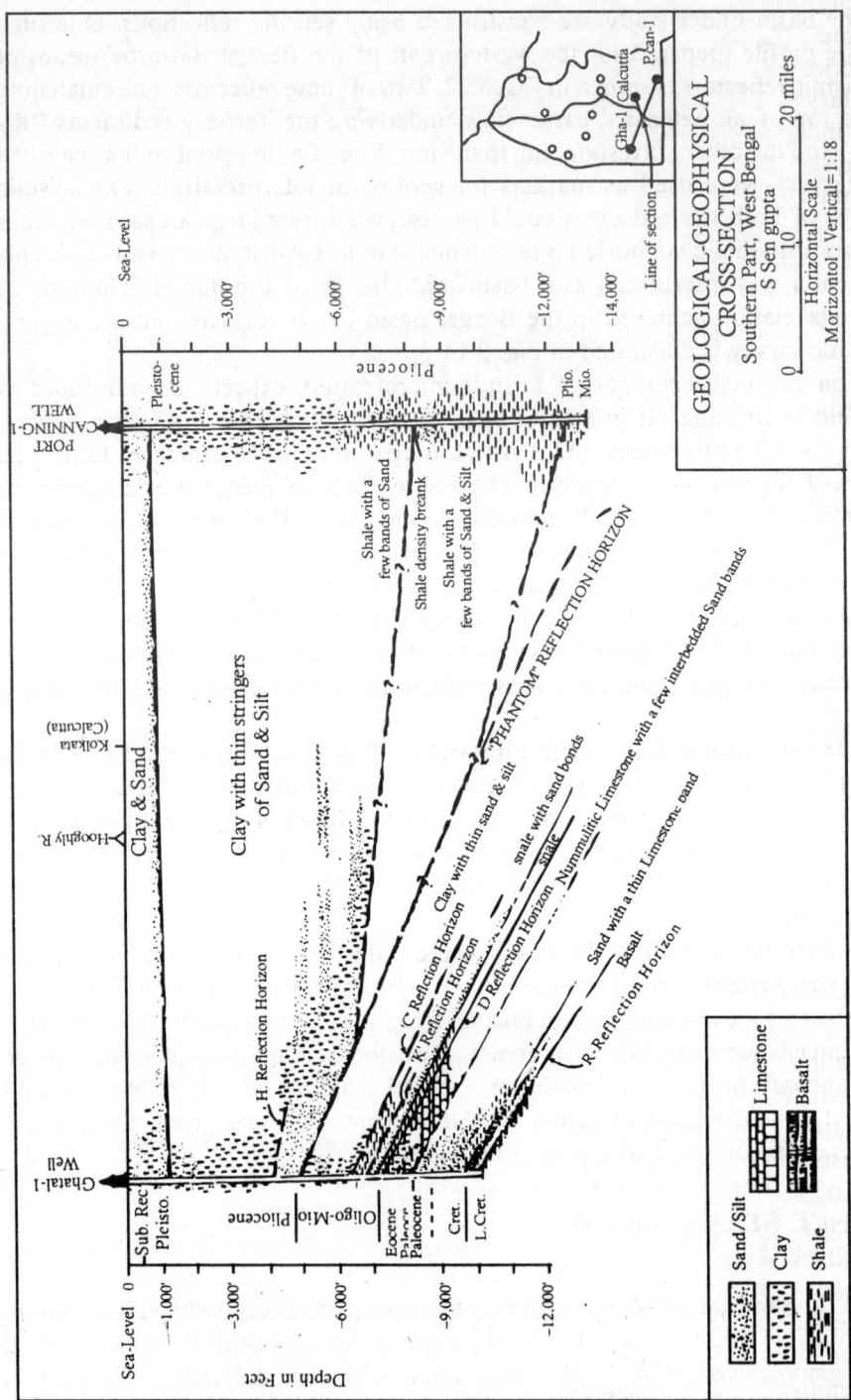

Fig. 9.14: Cross-section across southern part of West Bengal showing correlation of subsurface rock untis by seismic reflections and electrical logs (after Sengupta 1966a). With permission of American Association of Petroleum Geologists.

for revival of the concept of sequence stratigraphy (Vail 1987; Haq *et al.*1987). At the same time development of high-resolution seismic techniques at the EXXON Laboratories at Houston provided the tools for identification of unconformity-bounded sedimentary sequences in the subsurface. Thus, a global chronostratigraphic correlation scheme, independent of formation boundaries and rock types, could be established from physical stratigraphic analysis. This led to the foundation of the sequence stratigraphic technique, as we know it today (Sloss 1988).

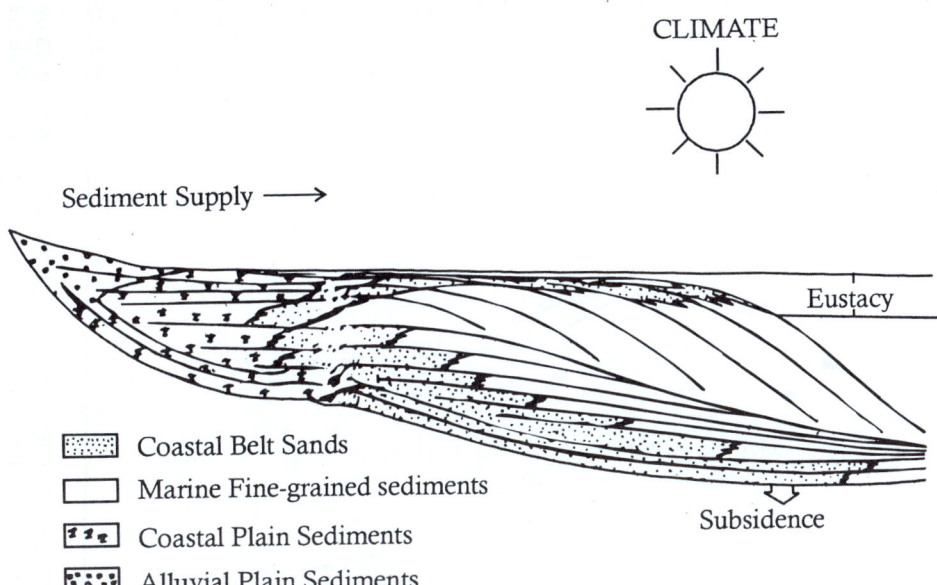

CLIMATE

Sediment Supply ⟶

Eustacy

▨ Coastal Belt Sands

☐ Marine Fine-grained sediments

Coastal Plain Sediments

Alluvial Plain Sediments

Subsidence

Fig. 9.15: Diagram illustrating development of stratigraphy as a funciton of sediment supply, subsidence, sea-level change (eustacy) and climate. Thick black lines indicate seismic reflections (based on Vail, 1987). With permission of American Association of Petroleum Geologists.

Basic Concepts

To have a grasp of the sequence stratigraphic concept it is important to remember that preservation of a sedimentary sequence in nature depends largely on two factors, the rate at which sediments are available, and the rate at which new space is created for preservation of these sediments. The sediments include the clastics as well as the non-clastics. The space for accommodation of these sediments is created either by subsidence of the basin of deposition, or by a rise in sea-level due to eustatic changes. Subsidence of land due to tectonic movements and transgression of sea on land due to eustatic changes will have the same effect on sedimentation. The following three conditions may arise out of the combination of subsidence and sea-level changes.
1. When the sediment supply equals the space available for its accommodation, the deposits build up vertically. The phenomenon is called *aggradation* (Fig. 9.16A).
2. When the space available for accommodation of the sediments grows faster than the rate at which sediments are available, the sedimentary sequence shows *retrogradation* (Fig. 9.16B). This happens during landward movement of the shoreline

(*transgression*). 3. When the volume of sediments exceeds the space available for its accommodation, an upward change to a shallower facies takes place within the basin. This phenomenon, called *progradation*, takes place during *regression*, when the shoreline moves away from land (Fig. 9.16 C). Many of the stratigraphic features seen in regional seismic profiles are the results of fluctuation of sea-level. Of these the following are more important (Fig. 9.17). Steeply inclined surfaces called *clinoforms* are produced when progradational sediment packages build layers into deeper water. *Offlaps* are stratal packages building upwards and outwards into the basin. *Toplaps* are created when a succession of packages of sediments prograde basinward without aggradation. *Downlaps* are inclined surfaces terminating abruptly on horizontal surfaces lying below. These structures are the results of sedimentation under marine condition, but some structures of non-marine origin may resemble these features. For example, Gilbert-type deltas, and also some forereefs may show downlaps. Differential compaction of mudstone and sandstone at the basin edge may appear as downlaps in regional seismic profiles because thin units of mudstones may go undetected.

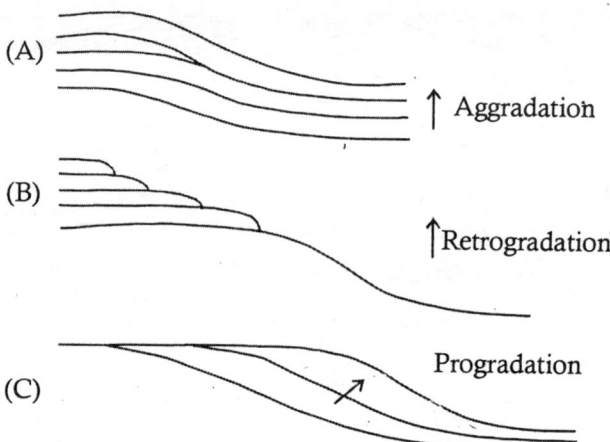

Fig. 9.16: Situations arising out of various combinations of rate of sediment supply and rate of creation of the space available for their accommodation.

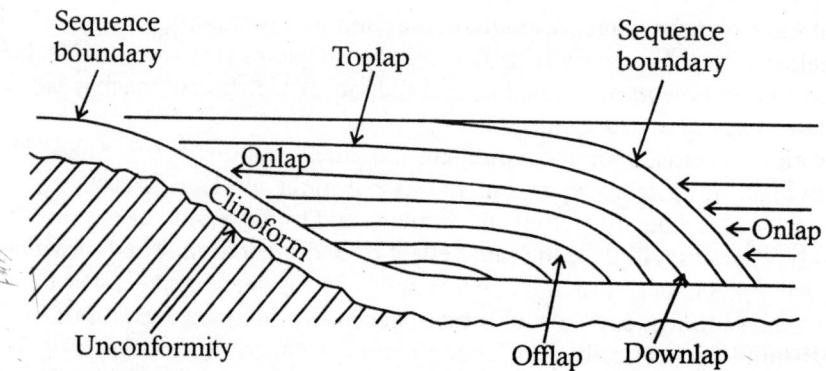

Fig. 9.17: Stratigraphic features recognizable in regional seismic profiles. (Based on Nichols 1999). With permission of Blackwell Science Ltd., a Blackwell Publishing Company.

Terminology

The EXXON Group coined a new set of terms to designate the sedimentary sequences generated out of subsidence and sea-level fluctuations. Following is a list of some of these terms and their definitions (Mitchum *et al.* 1977 & 1977a).

A *depositional sequence* is defined as a "relatively conformable, genetically related succession of strata bounded by unconformities or their correlative conformities". This sequence represents the sedimentary unit deposited between two episodes of significant sea-level fall. A fall in the sea-level exposes the beach and foreshore, subjecting them to subaerial erosion. The unconformity thus developed may cover whole of the shelf area if the fall in the sea-level is significant enough. This unconformity, recognized in seismic profiles by termination of a series of continuous reflectors against a surface at high angles, is reckoned as a *sequence boundary*. In the outer shelf, where there is no erosion, the surface equivalent to this unconformity is the '*correlative conformity*' in Mitchum's definition.

While the sequence boundaries are created by events of major sea-level fall, bodies of sediments may be deposited during phases of intermittent sea-level rise between two major events of fall. Several types of *systems tracts* may develop out of these movements. Systems tract, the fundamental mapping unit in sequence stratigraphy, is defined as a three dimensional assemblage of several lithofacies which are the products of same processes or environments. This means that they have to be *genetically linked*.

Like the Code of Stratigraphic Nomenclature, the Sequence Stratigraphic scheme proposes its own hierarchy of stratal units. The fundamental unit, *parasequence*, is defined as "a relatively conformable succession of genetically related succession of beds bounded by marine flooding surfaces and their correlative surfaces" (Van Vagoner *et al.* 1990). A parasequence set is "a succession of genetically related parasequences that form a distinctive stacking pattern, bounded, in many cases by major marine flooding surfaces and their correlative surfaces. Parasequence and parasequence set boundaries form in response to an increase in water depth. Under certain depositional conditions parasequence and parasequence set boundaries may coincide with sequence boundaries" (Van Wagonar *et al.* 1990).

Parasequences are composed of bedsets, beds, laminasets, and laminae. "Each stratal unit, with the exception of the lamina, is a genetically related succession of strata bounded by chronostratigraphically significant surfaces, and each surface is a single, physical boundary that everywhere separates all of the strata above from all of the strata below over the extent of the surface... Because of these properties, bounding surfaces that are correlated using well logs, cores, or outcrops, provide a high-resolution chronostratigraphic framework for facies analysis" (Van Wagoner *et al.* 1990).

Time range, thickness and areal extent of the stratal units in sequence hierarchy are shown in Fig. 9.18. The characteristics of lamina, laminaset, bed, and bedset are given in Table 9.4. It may be noted that some of these terms duplicate similar terms in the Code of Stratigraphic Nomenclature but carry different meanings, thus leaving a scope for confusion.

Depositional Sequences

The sequences of beds referred to above, occur in cycles because they are related to the cyclic patterns of sea-level fluctuation. A complex pattern of sinusoidal curve is generated by superimposition of the short-term sea-level fluctuations on longer-term

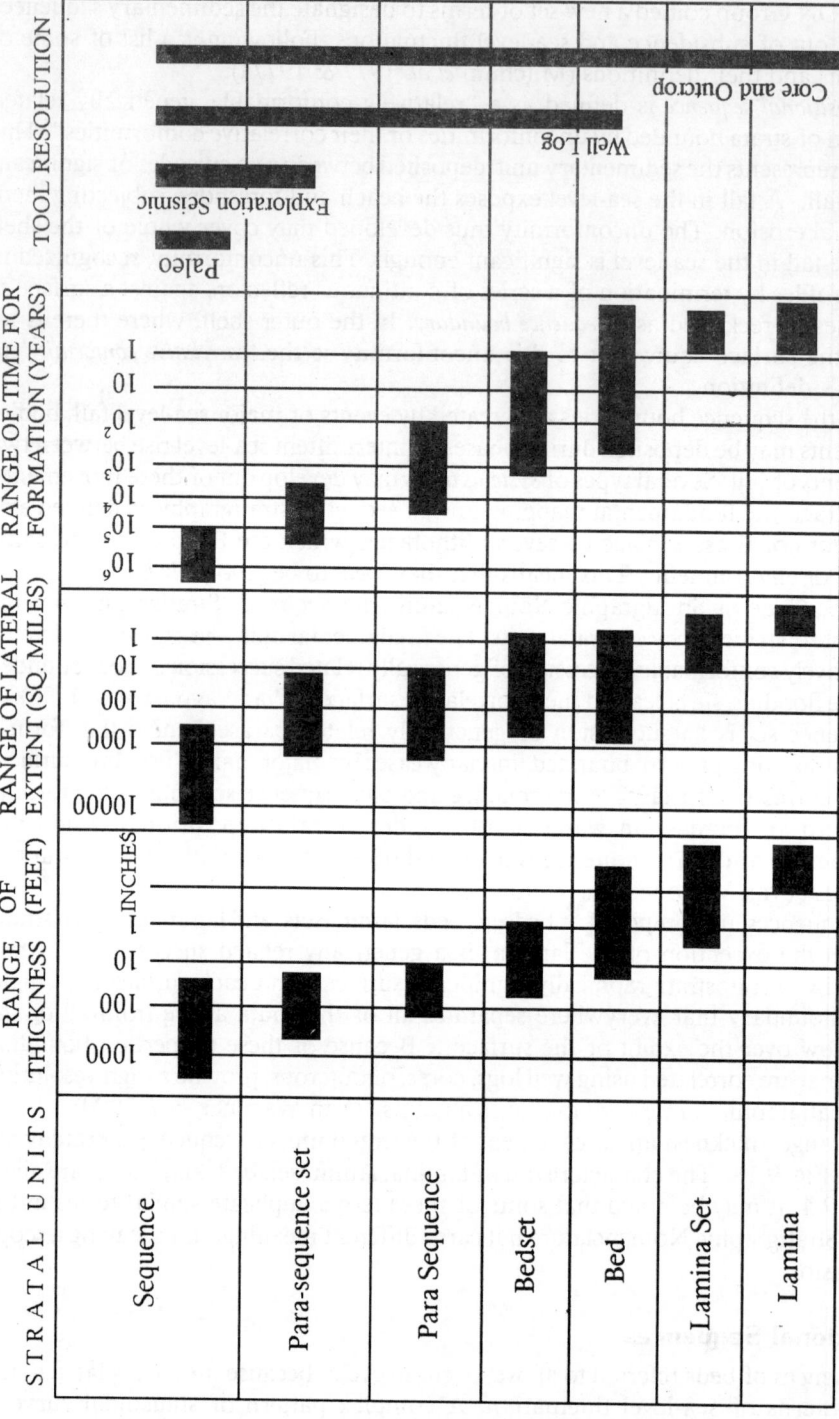

Fig. 9.18: Hierarchy of stratal units in sequence stratigrphy (after Van Wagoner *et al.* 1990). With permission of American Association of Petroleum Geologists.

Table 9.4: Detailed Characteristics of lamina, laminaset, bed and bedset (from Campbell 1967)

Stratal Unit	Definition	Characteristics of constituent stratal units	Depositional processes	Characteristics of bounding surfaces
BEDSET	A relatively conformable succession of genetically related beds bounded by surfaces (called bedset surfaces) of erosion, non-deposition, or their correlative conformities	Beds above and below bedsets always differ in composition, texture, or sedimentary structure from those composing the bedset	Episodic or periodic (same as bed below)	(Same as Bed below) plus • Bedsets and bedset surfaces form over a longer period of time than beds • Commonly have a greater lateral extent than bedding surfaces
BED	A relatively conformable succession of genetically related laminae or laminasets bounded by surfaces (called bedding surfaces) of erosion, non-deposition or their correlative conformities	Not all beds contain laminasets	Episodic or periodic Episodic deposition includes deposition from storms, floods, debris flows, tubidity currents Periodic deposition includes deposition from seasonal or climatic changes	• Form rapidly, minutes to years • Separate all younger strata from all older strata over the extent of the surfaces • Facies changes are bounded by bedding surfaces • Useful for chronostratigraphy under certain circumstances • Time represented by bedding surfaces probably greater than time represented by beds • Areal extents vary widely from square feet to 1000s square miles
LAMI-NASET	A relatively conformable succession of genetically related laminae bounded by surfaces (called laminaset surface) of erosion, non-deposition or their correlative conformities	Consists of a group or set of conformable laminae that compose distinctive structures in a bed	Episodic, commonly found in wave- or current-rippled beds, turbidites, wave rippled intervals in hummocky bedsets, or cross-beds as reverse flow ripples or rippled toes of foresets	• Form rapidly, minutes to days • Smaller areal extent than encompassing bed
LAMINA	The smallest megascopic layer	Uniform in composition/texture. Never internally layered	Episodic	• Form very rapidly, minutes to hours • Smaller areal extent than encompassing bed

cycles (Fig. 9.19). High- and Low-stand systems tracts occur at the crest and trough positions of these cycles respectively. Parasequences are deposited during shorter-term sea-level fluctuations within the longer term ones. The bounding planes of these parasequences are the marine flooding surfaces. A depositional sequence consisting of a highstand and a lowstand systems tract generally includes a number of parasequences.

Facies Associations

Three of the major situations related to falling sea-level, rising sea-level, and high sea-level are discussed below.

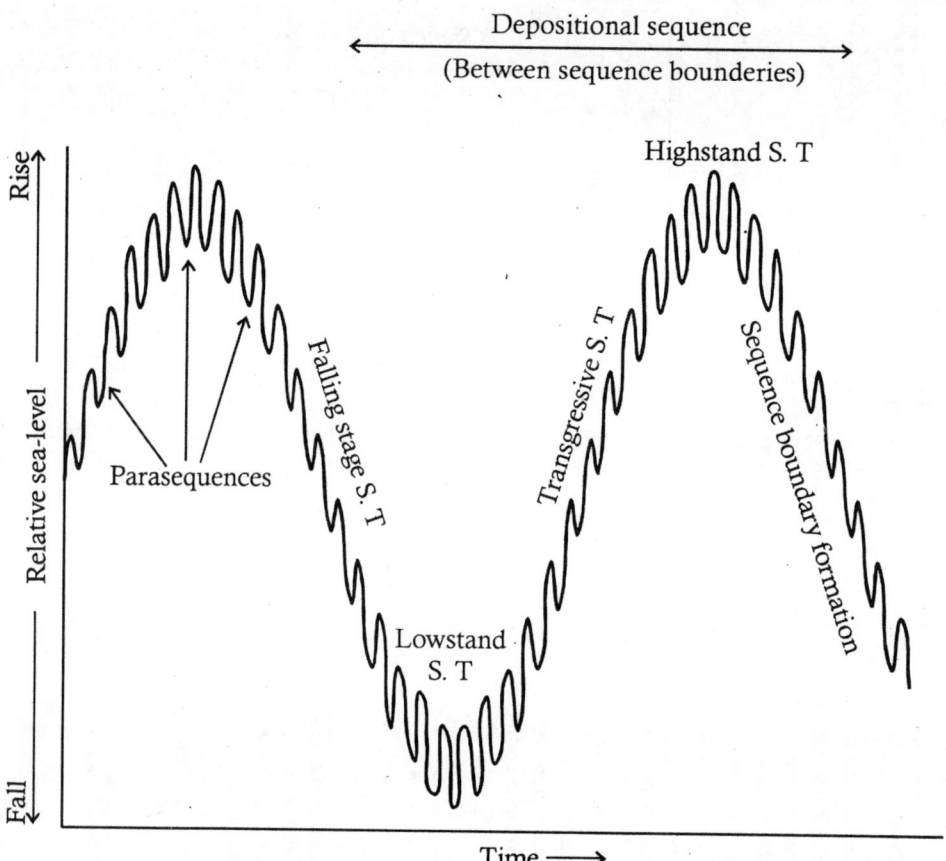

Fig. 9.19: Diagram illustrating development of high- and lowstand systems tracts (S.T.) and parasequences consequent to relative sea-level rise and fall (based on Nichols 1999). With permission of Blackwell Science Ltd., a Blackwell Publishing Co.

Case 1. Falling Sea-Level: Lowstand Systems Tract

Deposits laid down immediately after a relative fall in sea-level but before the next phase of sea-level rise represent lowstand systems tracts. A falling sea-level lowers the base level thus affecting the river systems at the shelf edge. With lowering of the base level rivers incise into their floodplains and carry the products of erosion into the sea where they are redistributed by turbidity currents. As a result the submarine fan lobes grow fast and change vertically to shallower facies. Progradation of coarser turbidites mark the sequence boundary in the deep sea (Fig. 9.20A). The products of erosion of the hinterland develop a blanket over the canyons on the continental slope.

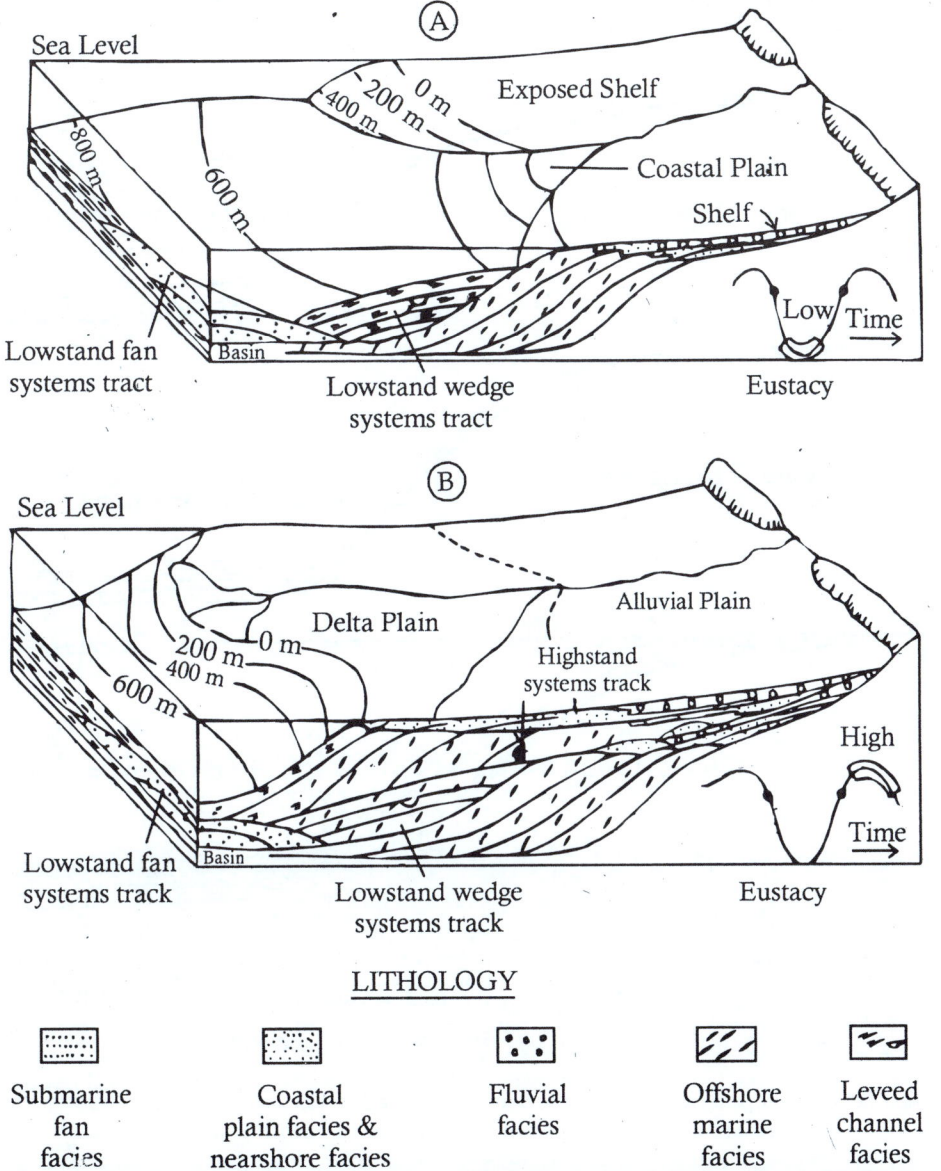

LITHOLOGY

| Submarine fan facies | Coastal plain facies & nearshore facies | Fluvial facies | Offshore marine facies | Leveed channel facies |

Fig. 9.20: (A) Lowstand Systems Tract (LST) and (B) Highstand Systems Tract (HST) produced respectively by sea-level fall and rise. Note the timing of the model with respect to the sea-level sine curve (based on Posamentier *et al.* 1988).

In areas showing ramp geometry (a gentle slope without a sharp break in profile) the zone of erosion is limited but the facies associations show a distinct shift towards the basin. Under this condition, a sequence boundary is defined as the plane above the falling stage systems tract (Fig. 9.21A). A stage of falling sea-level is usually followed by a phase of slow rise in the relative sea-level. Topsets and clinoforms are produced during this phase. In limestone bearing areas, a fall in sea-level leads to karst formation on land when the climate is humid. In arid climate, it leads to hypersaline conditions developing sabkhas in coastal areas.

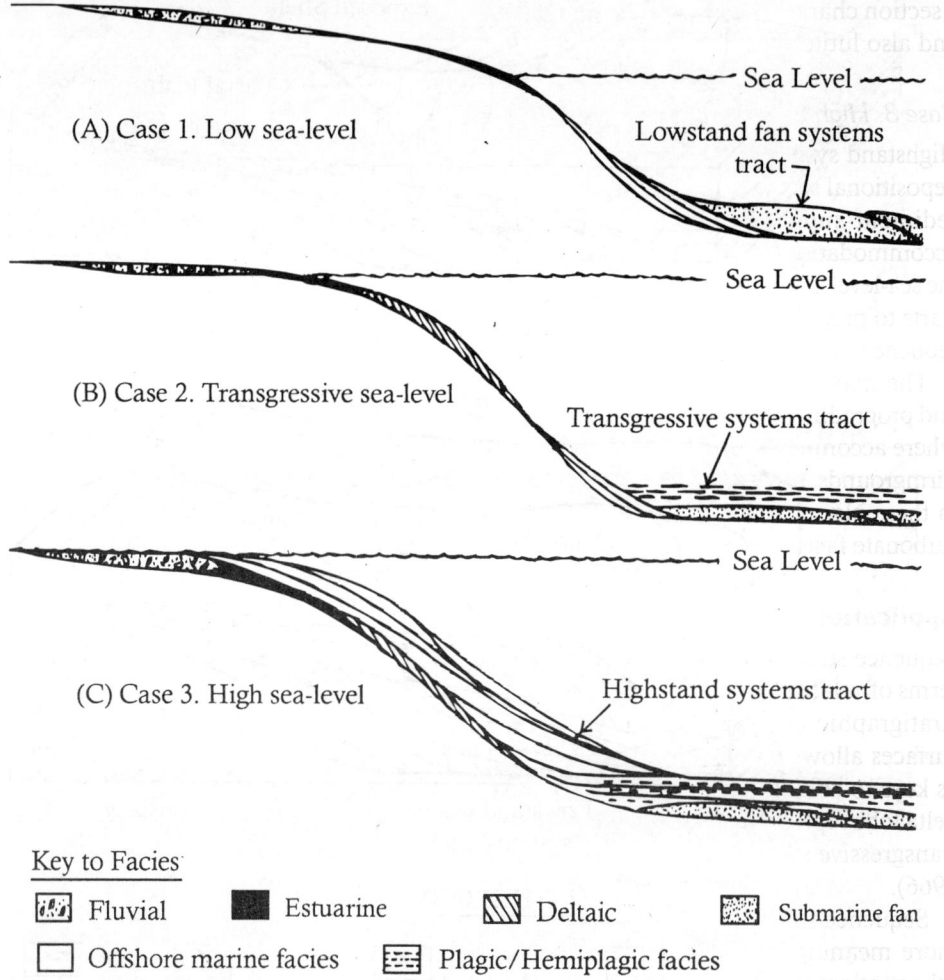

Key to Facies

▨ Fluvial   ■ Estuarine   ⧄ Deltaic   ▨ Submarine fan

☐ Offshore marine facies   ▦ Plagic/Hemiplagic facies

Fig. 9.21: Development of systems tracts at shelf edge following sea-level rise and fall (based on illustrations by Van Wagoner *et al.* 1990).

Case 2. Rising Sea-Level: Transgressive Sysems Tract

A transgressive systems tract is produced when the sea-level rises and the coastal plains are inundated. The sequence boundaries, under such a condition, may be recognized by incision of fluvial deposits into the shelves. Estuarine sedimentation takes place on the drowned river valleys. Deltaic complexes are the first deposits to form above the preexisting lowstand sequence boundary. These deposits are essentially retrogradational, being confined to topsets and clinoforms. Coarse sediments are confined to the areas near the shore. Not much sediment is transported to the deeper basin areas because there is very little subaerial erosion. Starved of sediments the submarine clastic deposits grow at a slow rate. On the other hand, production of carbonates, pelagic and hemipelagic sediments is enhanced in the deep sea areas (Fig. 9.21B). These mark the top of the transgressive systems tract. The transgressive systems tract pass distally into a section characterized by a low rate of sedimentation. Glauconitic, phosphatic shales, and also lutites rich in organic content are produced under such condition.

Case 3. High Sea-Level: Highstand Systems Tract
Highstand systems tracts are produced when the sea-level is high. The exact nature of depositional sequence at this stage is controlled by the relative rates of sea-level rise and sediment supply. When the rate of sea-level rise is low, the rate of creation of accommodation space is balanced by the sediment supply This causes aggradation. When the sea-level is at its peak very little new accommodation space is created and the sequence starts to prograde. The succession showing a sequence of aggrading beds followed by a sequence of prograding beds mark the highstand systems tract (Fig. 9.21C).

The maximum flooding surface is marked by the plane demarcating the aggrading and prograding sequences. During transgression deposits accumulate in the basinal areas where accommodation space is available. The distal areas remain starved of sediments. Firmgrounds, hardgrounds, and extensive erosional surfaces mark the sequence boundary in these places. A rapid rise in sea-level may result in an abrupt change from shallow carbonate facies to pelagic limestones on both ramp and rimmed shelves.

Application of Sequence Stratigraphy

Sequence stratigraphic concept advocates interpretation of sedimentary sequences in terms of relative changes in sea-level. This approach has opened up new dimensions to stratigraphic correlation and facies analysis. Identification of maximum flooding surfaces allows large-scale correlation of the stratigraphic units. Parasequences serve as keys to correlation of the smaller units. In the Mississippi delta, for example, each delta lobe represents a parasequence (Fig. 7.37), The first three of these lobes represent transgressive systems tracts. Others are highstand systems tracts (Kolb and van Lopik 1966).

Sequence stratigraphic technique prefers genetic criteria for correlation. Hence it is more meaningful than the conventional technique which uses purely descriptive, lithostratigraphic criteria for correlation.

Observations on closely spaced outcrop and subcrop sections (regional reflection seismic profiles followed by boreholes) are essential for sequence stratigraphic correlation. For temporal correlation, additional, bio-stratigraphic or magneto-stratigraphic information is necessary.

The sequence stratigraphic technique, initially intended to deal with marine deposits only, is applicable to sediments of fluvial origin also (see for example, Banerjee, *et al.*

1996; Friedman and Sanders 2000). This is so because base level changes due to sea level fluctuations, are eventually reflected in the fluvial sequences. When the base level is lowered due to a fall in sea-level, a large part of the land surface is exposed, allowing extensive development of fluvial channels. Conversely, during phases of marine transgression and sea-level highstands, the amount of overbank accretion is enhanced. Palaeosols and soils bearing marks of oxidation are produced on land during phases of marine lowstand. Highstand conditions lead to waterlogging and organic growth. All these features serve as markers for stratigraphic correlation.

In the Upper Carboniferous Breathitt Group of eastern Kentucky stacked fluvial bodies are incised into open marine and delta plain deposits. Systems tracts there could be identified from facies associations, sedimentary body geometry and sequential position of the deposits (Aitken and Flint, 1995). In the Triassic Ivishak Formation of the Prudhoe Bay Field on the Northern slope of Alaska, recognition of sequence boundaries, parasequence set boundaries and their up-dip equivalents aided in long distance correlation and classification (Emery and Mayers 1996, pp.130-133). While the Ivishak study is based on continuous well logs, the technique of field logging of the outcrops at centimetre scale was used for the Breathitt Group study. The marker beds identified on the logs were walked out along individual outcrops and correlated.

Examples of the application of sequence stratigraphy in fluvial systems can be found in Wright and Marriott (1993); Lindsay, Holliday and Hulbert (1991). Posamentier, Allen and James (1992) applied high-resolution sequence technique for the study of Coulee Delta, Alberta.

In India, sequence stratigraphic technique has been applied in exploration for petroleum in western offshore (Kutty and Agarwal, 1993). In the Bengal Basin, the stratigraphic architecture during phases of Oligocene lowstand has been conceptualized by sequence stratigraphic technique (Das, Chakrabarti and Dutta, 1998). In the late Cretaceous Bagh carbonate sequences of the Narmada Valley Tripathi and Lahiri (2000) identified three hardground surfaces and interpreted them as representatives of marine oscillation events.

10

BASIN ANALYSIS: A SYNTHESIS

INTRODUCTION

Basin analysis, the ultimate goal of a sedimentological study, unravels the complete depositional history of a sedimentary basin. This involves synthesis of observations on the framework of the basin, its sediment fill, and also the biota associated with the sediments. The stages of basin evolution and the patterns of sediment dispersal are also to be worked out in detail. In short, basin analysis calls for complete knowledge of all aspects covering the processes and products of sedimentation within the sedimentary prism which constitutes a basin. This also demands a command over all the techniques and methods of sedimentology and related disciplines on the part of the analyser.

Until about the fifties of the 20th century basin analysis was a relatively simple affair. It involved, broadly speaking, erection of local stratigraphic columns within the sedimentary basin and correlation of these units with the components of the conventional geosynclinal model which was popular at that time. Basin analysis today is far more complicated. Mapping the basement configuration of a sedimentary basin now calls for intensive application of various geophysical techniques. The geometry of the sediment fill has to be worked out in three dimensions using subsurface data, the nature of the fill has to be interpreted in terms of facies models, and the dispersal patterns have to be determined in quantitative terms, using the modern techniques of palaeocurrent analysis. To complete the story of basin evolution, one would like to have radiometric dates not only for the layers of sediments, but also for the intrusives occurring within the basin.

Finally, today's students of sedimentology would like to interpret the stages of basin evolution in terms of lithospheric plate movements, following the currently popular model of plate tectonics. Constant feedback of information from field to laboratory and *vice versa* is essential. The ultimate aim is to interpret the products of sedimentation in terms of the processes involved. Co-ordinated effort on the part of specialists in various branches of earth science is required to achieve this goal. In most of the exploratory organizations this is done by a team designated as 'the basin study group.'

A simple flow chart for basin analysis is presented in Table 10.1. The procedure to be followed is divided into three broad categories: geological, geophysical and geochemical. The sequence of analysis to be followed is briefly discussed below. For details of the analytical techniques the reader should refer to the chapters cited.

The study of a virgin sedimentary basin commences with mapping the depositional boundary and obtaining as much information as possible on the basin architecture.

Table 10.1: Flow sheet for basin analysis

Nature of Investigation		Work to be undertaken	Phase I Field	Phase I Lab.	Phase II Field	Phase II Lab.	Phase III Field	Phase III Lab.
Geological	Field studies	Mapping of the basin boundary and the lithic fill. Establishing lithostratigraphic units, unconformities etc.	I					
		Preparation of measured stratigraphic columns from surface and subsurface data. Study of sedimentary structures, biota.			II			
	Laboratory	Thin section petrography, modal, textural and heavy mineral analysis,		I				
		Clay mineral studies by XRD, SEM etc. Palaeontological, biometric and palynological studies.				II		
Geochemical		Determination of major and trace elements, organic content, maturation etc. Radiometric dating.						III
		Magnetostratigraphy						III
Geophysical Field		Aeromagnetic and gravity surveys.	I					
		Reflection seismic survey, DSS.			II			
		Drilling and borehole logging.			II		III	
Purpose		Determination of:	Basin configuration, nature of sediment fill, facies variation.		Basin floor pattern, palaeocurrent, palaeohydraulics, palaeoecology, depositional and diagenetic conditions.		Seismic facies, and sequences. Regional correlation. Pattern of sedimentation in relation to tectonics. Maturation of sediments.	
Conclusions			Nature of the basin, its lithic fill and provenance of sediments.		Basin architecture. Sediment dispersal pattern, depositional enviornment and diagenetic trends.		Basin evolution with respect to geodynamics and global sea-level changes. Hyrocarbon prospects.	

The approach is necessarily different for relatively small, intracratonic basins (e.g. pockets of Gondwana deposits within the Indian craton) and large sedimentary basins of the type of the Bengal Basin, which run across cratonic or continental margins and continue on the oceanic crust. While conventional geological mapping followed by exploratory drilling provides answers to most of the questions in the former case, a well co-ordinated geological-geophysical approach is essential for investigation of basins of the latter type.

An *aeromagnetic survey*, which provides the quickest method for obtaining the depth of the basin-floor by detecting the boundary between the magnetically susceptible, presedimentary, often metamorphosed basement, and the non-magnetic sediments overlying it, is useful in detecting the gross thickness of sediments. The overall direction of thickening and thinning can also be determined by the same method. An aeromagnetic survey is generally followed by ground *gravitymeter survey* for obtaining information on density anomalies in the subsurface. The data obtained are sometimes useful although the inherent ambiguity in the interpretation of gravity data may not always provide a unique solution to problems.

BASIN AND ITS LITHIC FILL

Mapping, Petrography & Sedimentary Strcuctures

Geological mapping of the lithic fill should be started simultaneously with the aeromagnetic survey. The purpose of mapping is to establish the litho-stratigraphic units following the Code of Stratigraphic Nomenclature. Mapping begins at the level of formations. The lower and the higher ranking units in the hierarchy (member and group respectively) can be demarcated within an area once the formations have been established. The nature of unconformity below and within the stratigraphic sequence should also be worked out in the course of mapping (Chapter 9). Three-dimensional disposition of the rock units within a basin is represented by a variety of composite maps prepared from measured stratigraphic sections. In the absence of suitable exposures, information on rock units is obtained either from boreholes or from two-way seismic travel times from reflecting interfaces in the subsurface. The reflection times, converted to depths by means of interval velocities, are used for preparation of structural maps, isopachs, palaeogeologic and palaeotopographic maps.

Structural maps show relief of the surface of interest with respect to a datum which is commonly the mean sea-level. These are comparable to the topographic maps prepared at the surface. Palinspastic adjustments for folding, faulting and regional tilt may be incorporated into these maps. *Isopach maps* depict variation in thickness of the stratigraphic unit of interest. This variation is represented by contours; zero contour, indicative of total merger of the upper and lower surfaces of the unit, defines the limit of pinchout. The shape of the unit thus defined often helps to decipher the origin and depositional environment of the concerned rock unit.

Lithofacies maps depict aerial variations in rock types (e.g., sand, shale, carbonates), their proportions, volumes and ratios (e.g. sand/shale, shale/carbonate) within particular stratigraphic units. *Isolith maps* show distributions as well as thickness variation of individual facies. Aerial distributions of cumulative thickness of particular rock types is shown by net rock-type maps.

Having obtained a broad picture of the basin architecture and the overall nature of the basin fill, one should undertake a petrographic study of the rock samples collected from the field. *Modal analysis* of sandstone samples by the point-counting method provides quantitative information on the mineral proportions. The sandstones are broadly classified into arenites and wackes on the basis of matrix content. The proportions of quartz, fieldspars and rock fragments are then plotted on a QFRF diagram for further subdivision of the sandstone type (see Fig. 3.3). Proportions of undulose, non-undulose, and polycrystalline varieties of quartz, plotted on the Indiana diamond diagram (see Fig. 2.4), give clues to the nature of the non-sedimentary source rock. The heavy mineral assemblage (see Table 2.2), degree of sorting and rounding (see Fig. 4.9) of the clastic particles provide insight not only into the possible provenance, but also into the transportational history of the sediments. Grain-size data may be utilized in a limited way for interpreting palaeoenvironment and palaeohydraulics (Chapter 4).

The carbonate rocks may be classified either by their depositional texture, as proposed by Dunham (1962), or by identification of the allo- and orthochemical components under the microscope (Folk 1959). The type of carbonate (calcite/aragonite/dolomite) is deciphered by staining (Friedman 1959). Replacement textures, if any, may also be identified at this stage. Identification of mineralogy of the carbonates and the fine-grained argillaceous sediments, by more sophisticated methods, are undertaken at an advanced stage of investigation. The mineralogical assemblage within terrigenous or chemical sediments are often indicative of the chemical conditions (Eh and pH) prevailing at the site of deposition (see Fig. 3.19).

In the second phase of fieldwork the primary sedimentary structures are studied in detail. These can aid in stratigtraphic studies in many different ways. Graded bedding and ripple marks, for example, provide clues to the attitude (top and bottom) of the strata. For this purpose, wave ripples are of greater help than current ripples. Sharp crests and indented troughs of wave ripples can be easily identified even when they occur in reversed position at the sole of beds (see Fig. 5.24). In areas of structural complexity, when the original sequence is inverted, the depositional top and bottom of the sequence can be determined from the wave ripples. The external morphology of a current ripples is of little help in the determination of the depositional top of the sequence but the internal cross-lamination might be useful in such a case (see Fig. 5.24). Some of the common scour marks, such as flutes and grooves, have been traditionally used for determination of the depositional top. Such interpretations are based on the belief that the groove and the flute marks are the counterparts of the original casts formed as scours on the bed surface. However, beds having positive reliefs resembling flutes have been reported from the upper surface of non-cohesive sediment beds also (Friedman and Sanders 1974). Thus the technique of determination of the depositional top from scour marks may not always provide dependable results.

Palaeoslope and Palaeocurrent

The initial dip or slope of a basin-floor at the time of sedimentation is called the palaeoslope. The term is related to certain subaqueous deposits only. Eolian deposits, controlled wholly by wind movement, have very little relationship with the slope of the depositional surface. In fluvial environment, where the flow is essentially unidirectional, palaeoslope can be obtained by averaging the cross-bedding foreset azimuths over an area. This is not necessarily so for marine or tidal environments where the palaeoslope may not be very consistent. Before attempting an interpretation of the palaeoslope, therefore, the nature of the depositing medium (water or wind) and the environment of sedimentation must be ascertained by criteria independent of palaeocurrent direction.

Palaeocurrent analysis involves the study of ancient sediment dispersal patterns. Sediment properties indicating palaeocurrent are of two types: directional and scalar. Table 10.2 lists the more important of these properties. Following is a brief description of them.

Table 10.2: Properties of Sediments Indicating Palaeocurrent
(Modified after Pettijohn 1962)

Directional Properties

Planar Structures
    Cross-bedding
        Planar-tabular
        Trough-lenticular
        Low angle (<10°)
        High angle (>10°)
Linear Structures
    On the base of bed ("sole markings")
        Striation and groove casts
        Flute casts
    Internal structures
        Clast lineation
        Fossil lineation
        Parting lineation
    On the top of bed
        Ripple marks
            Symmetrical
            Asymmetrical
        Rib-and-furrow
        Current crescents

Complex Structures
    Convolute bedding

Scalar Properties

    Pebble and grain-size
    Pebble and grain roundness
    Thickness of bedding and cross-bedding

Directional Properties

The directional elements in rocks indicate palaeocurrent direction at the point of observation. These directions are often allotted weights and treated as vectors. The directional elements again, may be of two types: planar and linear (see Table 10.2).

Planar Directional Elements

Cross-beddings of various shapes and sizes are the commonest planar elements used for palaeocurrent analysis. The direction of inclination of the cross-bedding foreset (the azimuth of cross-bedding) indicates the direction of sediment transport at the spot of observation. High and Picard (1974) discussed the reliability of different types of cross-bedding for palaeocurrent interpretation. For a planar-tabular cross-bedding, palaeocurrent is indicated by the direction of inclination of the foreset. Random measurement of dips on the flanks, in case of a trough type cross-bedding however, might give misleading results. Such observations must be corrected, taking into consideration the plunge of the trough axis, to obtain the trough axial azimuth (Slingerland and Williams 1979).

When the area of study is structurally tilted or folded, a correction has to be applied to restore the bed containing the directional elements to its original position. In the case of cross-bedding this is done by rotating the plane of the cross-bedded unit around the strike of the true bedding through the dip angle of the bedding. Potter and Pettijohn (1977) have discussed this technique. Empirical studies show that tilt correction is negligible when the structural dip is less than 25°, and also when the difference between the strike directions of bedding and cross-bedding is small (<10°). In all other cases tilt correction is necessary. Computer programs for tilt correction are now available (Saha and Chakraborty 1990).

Conclusions on regional palaeocurrent trends should be based, as far as possible, on a large number of observations on cross-bedding of the same type, because they are likely to be indicative of the same depositional environment. Certain procedural problems are faced in the collection and summation of directional data because the conventional technique of arithmetic averaging may not be applicable in this work. When the data are spread on either side of the point of origin, as in a compass dial, the arithmetic mean may give absurd results (for example, the arithmetic mean of 340° and 20° is 180°). The usual technique of obtaining standard deviation can not also be used. The procedure for obtaining the resultant direction in such cases is discussed in the appendix to this chapter.

Summation and representation of palaeocurrent directions also pose special problems. The current directions are generally represented graphically, in the form of a circular histogram ('rose diagram') (see Fig. 10.10). Circular histograms plotted on linear scales may give a wrong impression about the nature of distribution because the area of the 'pie' in such a case is not proportional to the frequency ('density'). Circular histograms with areas proportional to density may be constructed by using the specially designed, non-linear frequency net devised by Nemec (1988).

When the geological formation under study contains a large number of directional features of a particular type, it is necessary to know the minimum number of observations that would give the mean direction for the whole formation with a specified precision. This is a sampling problem. The optimum hierarchical sampling procedure developed by Rao and Sengupta (1970, 1972), who reviewed earlier practices, takes into account the special requirements of the radially distributed data. This technique answers the following questions: 1) the minimum sample size required for estimating, with a desired precision, the mean palaeocurrent direction of a formation and 2) the optimum allocation of samples between and within the outcrops that would allow efficient sampling at minimum cost. A summary of this sampling procedure is given in the appendix to this chapter.

Linear Directional Elements

Ripple marks are the commonest linear structures used in palaeocurrent study. Crest lines of asymmetrical ripples are usually normal to the direction of sediment transport. Orientation of flute marks, grooves, skip marks and other similar features, generally preserved on the undersurface of sediment layers as sole marks, also indicate palaeocurrent. Some other similar features like parting lineation; orientation of shell fragments may also be used in a similar way. Current crescents splay out in the direction of flow and often serve as good indicators of palaeocurrent (see Chapter 5). Orientation and imbrication of sand grains and pebbles indicate palaeocurrent. Pebbles are more commonly used for this purpose because they are easier to measure. In such studies particular care should be taken while measuring orientation of pebbles associated with a cross-bedded unit because pebble orientations are likely to change, depending on their locations on topset, bottomset, or foreset of the cross-bedded unit. Elongate pebbles tend to be transverse to the flow direction on the topset and bottomset of a cross-bedded unit. On the foreset they are longitudinal to the flow direction (Sengupta 1966, Johansson 1976).

Presentation of Directional Data

For convenience of presentation the area of study is arbitrarily divided into a number of sectors of equal size. The resultant vector of all the directional observations within each sector is computed following the procedure given in the appendix. This resultant direction is plotted at the centre of the grid. The length of the arrow is made proportional to the vector magnitude in each case (Fig. 10.1).

While analysing the trend of palaeocurrent direction in a large area it is desirable to emphasise the regional trend of sediment transport. Several techniques are available for this purpose. In the relatively simple, 'moving average' method the area of study is arbitrarily divided into a number of sectors of equal size. The resultant direction of all the data within this area is then moved systematically over the map with the help of a grid. The averages are then recomputed and plotted at the grid coordinates (Fig. 10.2). More sophisticated methods for isolating the local from the regional are available, but the results obtained from the simple technique of 'moving average' compare well with those obtained by these techniques (Pettijohn 1962).

In the sophisticated technique of trend surface analysis, different trends, linear, quadratic or higher are fitted to the observed data. The 'residual' values are then worked out from the difference between the computed and the observed values at each point (Miller 1956; Krumbein 1959). Computer programs for working out the trend surface are also available (Fox 1967).

Scalar Proerties

Several rock properties like grain-size, grain roundness and bed thickness show gradual variation in the direction of sediment transport. Unlike the directional features described above none of these features provide any clue to the flow direction at the point of

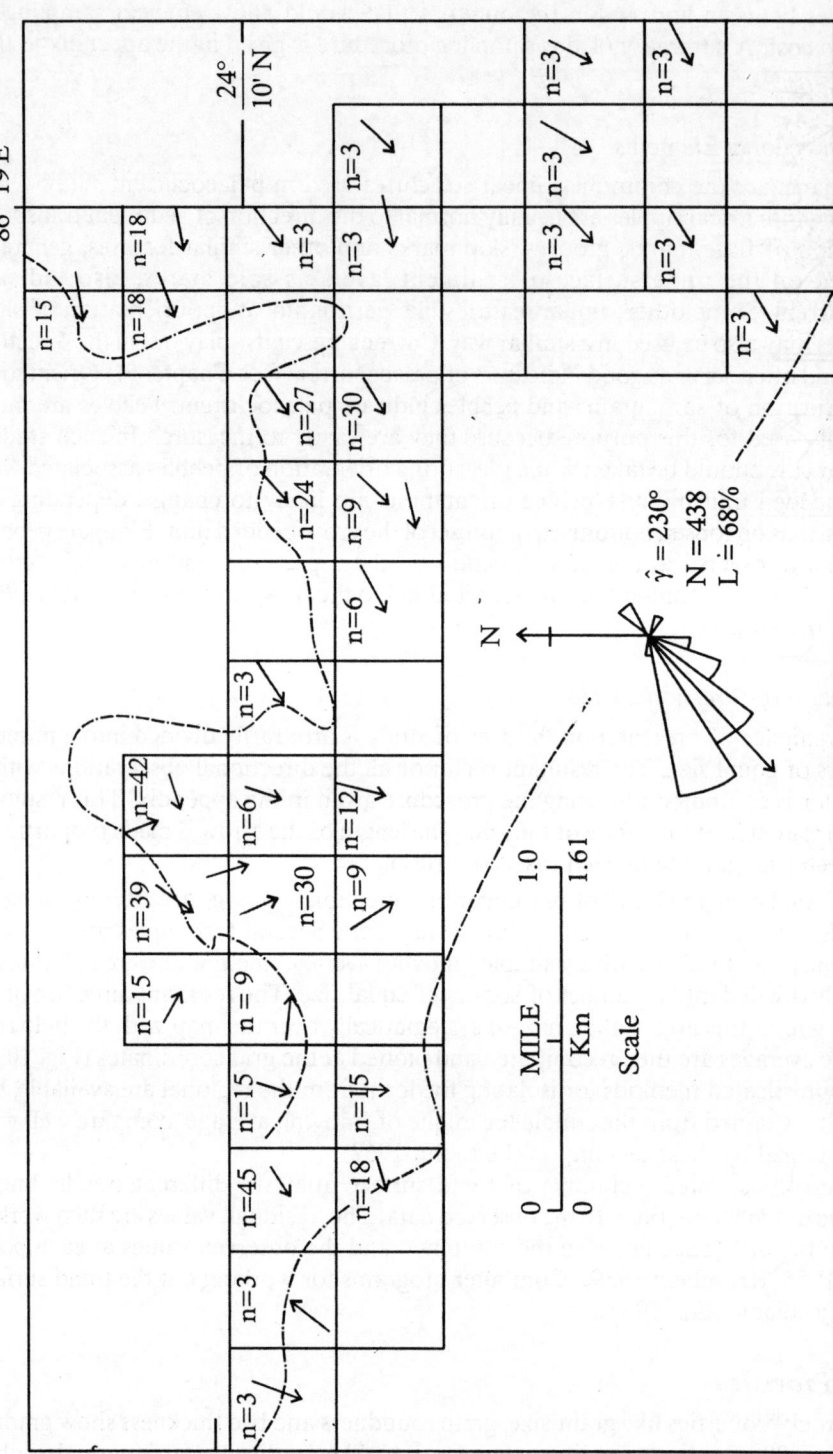

Fig. 10.1: Palaeocurrent directions within Barakar Formation worked out from cross-bedding forest azimuths in the Giridih basin, India (from Sengupta, *et al.* 1988). With permission of Geological, Mining & Metallurgical Society of India.

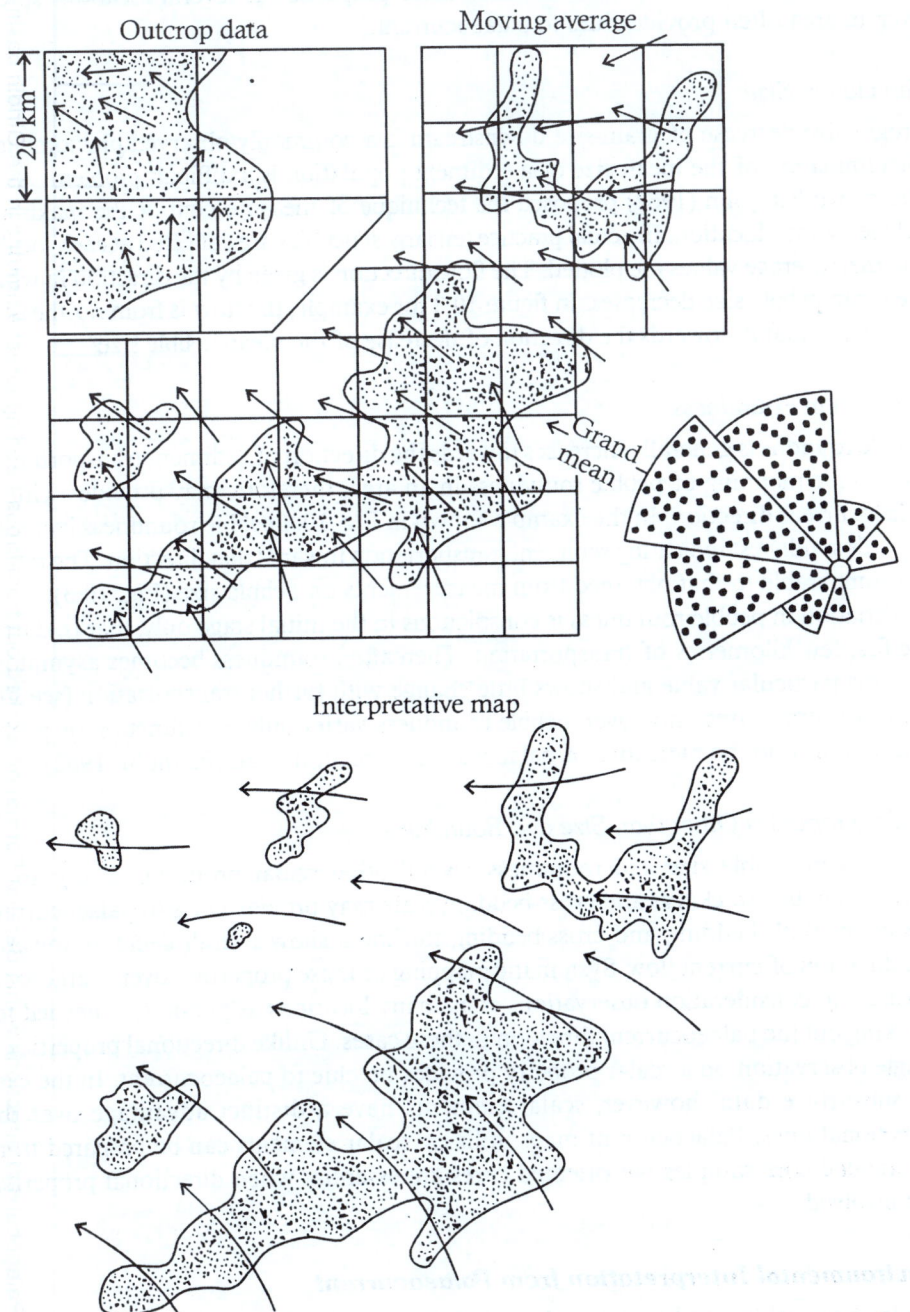

Fig. 10.2: The technique of interpretation of regional palaeocurrent trends from directional data (from Potter and Pettijohn 1963). With permission of Springer Science and Business Media.

observation. However, measurements of these properties at several locations spread over an area often provide clues to palaeocurrent.

Variation in Size

Progressive decrease of grain-size downstream is a commonly observed phenomenon. Determination of the mean size of a sediment population is a difficult problem. As an alternative Pettijohn (1962) proposed the technique of measurement of the maximum pebble size at a location. In actual practice ten largest pebbles are collected at each locality and their average values are plotted. The flow direction is given by the direction in which the mean pebble size decreases. In figure 10.3 for example, the flow is from northeast to southwest, that is, towards the direction of decrease of the mean pebble size.

Variation in Roundness

Pebble roundness generally increases towards the direction of sediment transportation. Systematic mapping of pebble roundness in an area, therefore, may provide a clue to palaeocurrent direction. In the example shown in Fig. 10.4 pebble roundness increases towards southwest indicating sediment transportation towards that direction. The results are comparable to that obtained from measurements on pebble size (Fig. 10.3).

Variation in pebble roundness is conspicuous in the initial stage only, that is, during the first few kilometres of transportation. Thereafter, roundness becomes asymptotic to some particular value and shows little change with further transportation (see Fig. 4.10). In some cases, moreover, pebble roundness varies only as a function of pebble size and may not be indicative of palaeocurrent direction at all (Pettijohn 1962).

Scalar Properties Other than Size and Roundness

Apart from pebble size and roundness several other scalar properties like gradual variation in bed thickness and cross-bedding scale may provide clues to palaeocurrent direction. Both bedding and cross-bedding thickness show overall decrease towards the direction of current flow. Systematic mapping of these properties, over a large area, taking into consideration observations at as many locations as possible, is needed for working out the palaeocurrent direction in these cases. Unlike directional properties, a single observation on a scalar property provides no clue to palaeocurrent. In the case of subsurface data, however, scalar elements have a distinct advantage over the directional ones. Palaeocurrent maps utilizing scalar elements can be prepared from unoriented core samples but oriented cores are essential when directional properties are involved.

Environmental Interpretation from Palaeocurrent

Under favourable conditions, as discussed in the following section, the pattern of palaeocurrent dispersion provides clues to the depositional environment. In fluvial, deltaic, and some turbidite sands the palaeocurrent is generally unimodal, pointing downslope. Eolian palaeocurrent is unrelated to palaeoslope and may have any kind of dispersion depending on the direction of wind flow. In the tidal zones bimodal

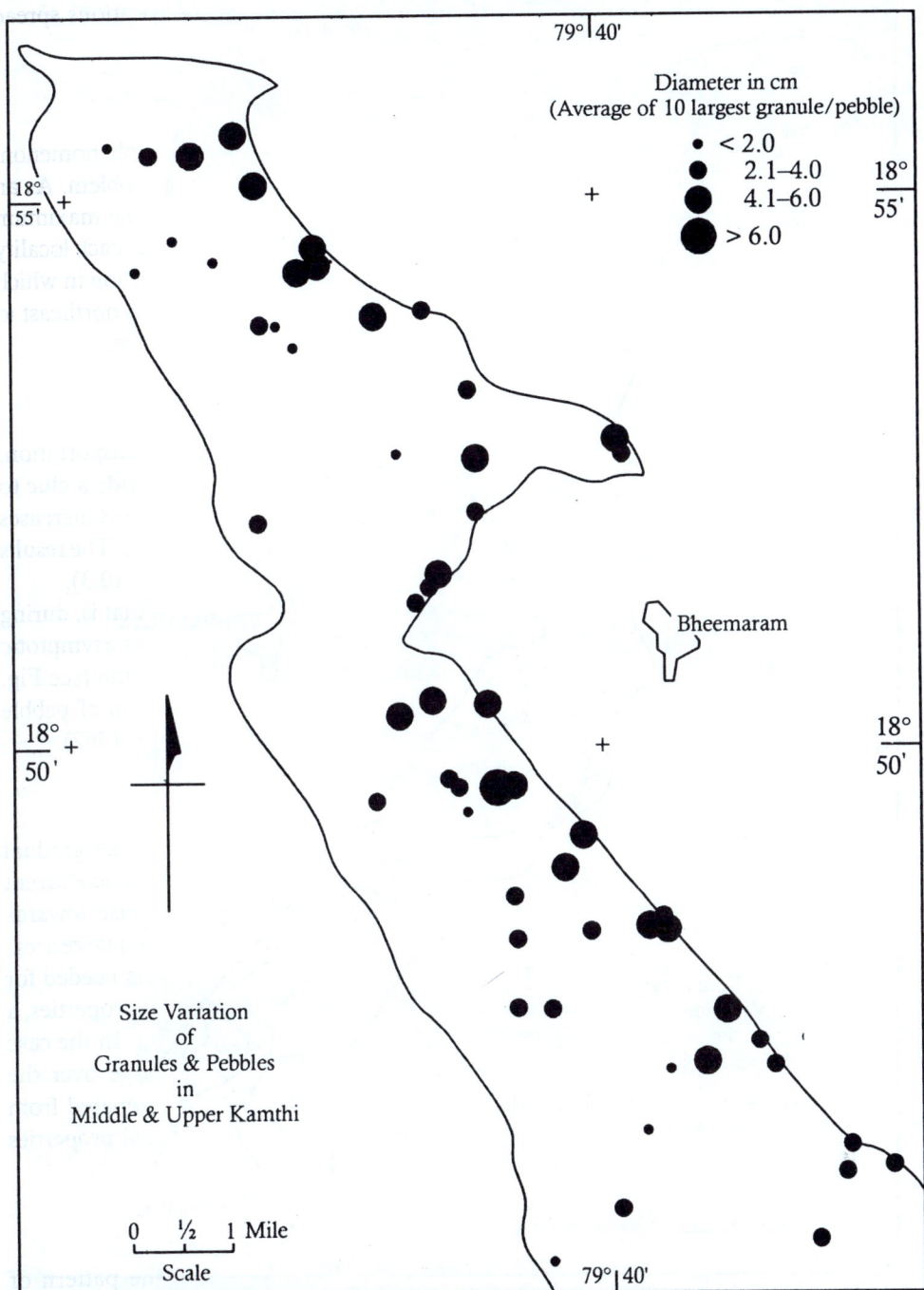

Fig. 10.3: Pictorial representation of size variations of granules and pebbles in the Middle and Upper Kamthi near Bheemaram India. (from Sengupta 1970). With permission of SEMP (Society for Sedimentary Geology).

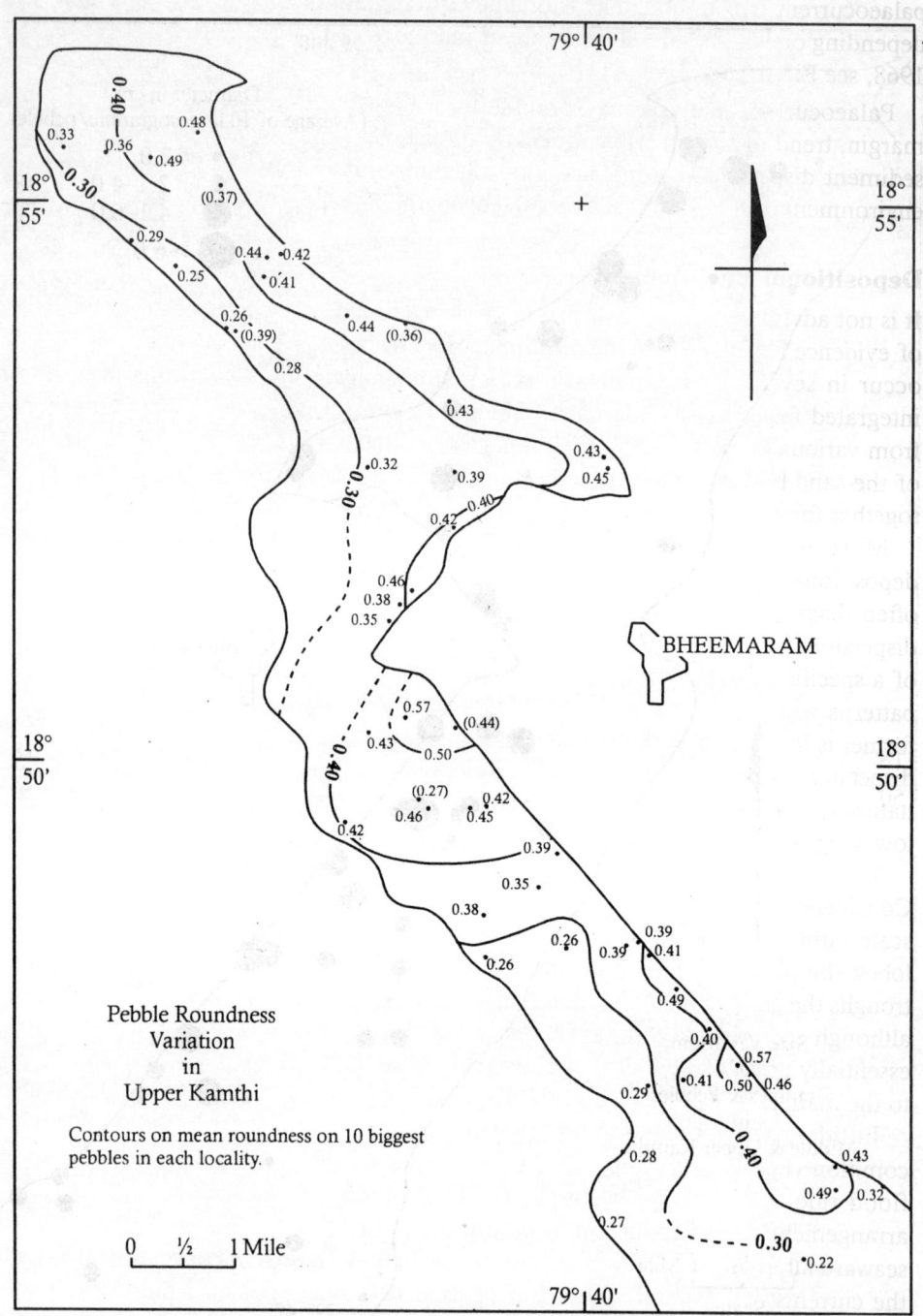

Fig. 10.4: Map of pebble roundness variation in the Upper Kamthi near Bheemaram, India (contours on mean roundness of 10 biggest pebbles in each locality) (form Sengupta 1970). With permission of SEPM (Society for Sedimentary Geology).

palaeocurrent is common. The shoreline current patterns may be of different types, depending on whether the currents are of marine or continental (fluvial) origin (Selley 1968, see Fig. 10.5).

Palaeocurrent analysis may provide information on configuration of the basin margin, trend of the shoreline, direction of sediment source (provenance), pattern of sediment dispersal. This information, together with knowledge of palaeoslope and environment of deposition allows palaeogeographic reconstruction.

Depositional Environment

It is not advisable to draw conclusion on depositional environment from a single line of evidence because a particular sedimentary structure or palaeocurrent pattern may occur in several environments in which similar hydraulic conditions prevail. An integrated facies model approach, taking into consideration the information drawn from various lines of evidence—lithofacies, sedimentary structure, texture, geometry of the sand bodies, nature of coarsening or fining-upward sequences - should be put together for drawing a meaningful conclusion.

Most of the common sedimentary structures may originate in more than one depositional environment but particular combinations of sedimentary structures are often diagnostic of certain depositional conditions (see Chapter 7 for details). The dispersion pattern of the palaeocurrent may, under certain circumstances, be indicative of a specific depositional environment (Selley 1968). For example, the palaeocurrent patterns within fluvial systems may be of two types—radiating and converging. The former is indicative of alluvial fans and the latter is characteristic of coal basins. The dispersion of current directions within piedmont fans and braided stream channels is naturally high. For meandering channels on other hand, the dispersion is particularly low (Fig. 10.5, a-d).

The palaeocurrent patterns for deep-sea fans are unimodal, as in fluvial systems. Considered on a regional scale, however, two different patterns are possible. For small-scale turbidity currents, as in the outlet of submarine canyons or on slopes of delta lobes, the pattern is unimodal with small dispersion (Fig. 10.5 a). In turbidity-filled troughs the general flow direction remains longitudinal and parallel to the trough axis' although sedimentation may take place laterally from the margins. This leads to an essentially unimodal pattern, with two subordinate modes lying nearly at right angles to the main flow direction (Fig. 10.5 f).

In tidal flats bimodal and bipolar current directions, normal to the shoreline, are common (Fig. 10.5 h). Bipolarity develops due to landward flow of current during flood tide and flow in the opposite direction during ebb tide. A perpendicular arrangement of current directions may also be produced when fluvial currents flowing seaward alternate with longshore drifts (Fig. 10.5 i). In tidal inlets, on the other hand, the currents are unimodal, with small dispersion, being guided either by ebb or by flood (Fig. 10.5 e). Sea beaches may show a variety of palaeocurrent patterns, depending on whether the agencies responsible for sediment transportation are wholly marine or have both marine and tidal influence. Deltas produced by fluvial systems debouching into sea or lake may produce radiating palaeocurrent patterns (Fig. 10.5 g).

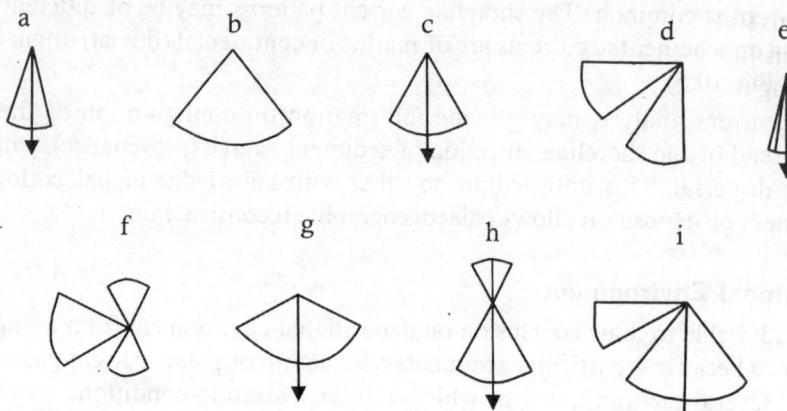

Fig. 10.5: Idealized palaeocurrent dispersion patterns in various depositional environments, (a) meandering stream channels or turbidity flows, (b) alluvial or submarine fans, (c) braided streams, (d) converging fluvial channels as in coal basins, (e) tidal inlets, (f) turbidite filled troughs, (g) radiating distributaries in deltas, (h) tidal flats, (i) longshore drifts superposed on fluvial currents (based on Selley 1968, Allen 1968a).

Subtle changes within the sedimentary structures may also be used as indicators of variation in physical condition within a depositional environment. Cross-strata sets, for example, are sometimes interrupted by sloping surfaces called *reactivation surfaces* (Fig. 10.6B). These may be caused by changes in flow stages (a lowering of water level followed by rising) of the depositing medium. In the case of reversal of flow direction due to tidal effect, the cross-strata above and below the reactivation surface may dip in opposite directions, giving the impression of *herringbone structures* (Fig. 10.6 A).

The interpretation of depositional environment is aided also by the logs of vertical stratigraphic columns, generally prepared during an advanced phase of field work. These may be constructed either from the exposed sections or, preferably, from borehole data when these are available. The field (or borehole) data may be utilized for studying the vertical variation in lithofacies and also for preparing maps. Correlation of borehole data and preparation of statigraphic sections also provide clues to the three-dimensional geometry of sand bodies, which are useful for environmental interpretation. The choice of an appropriate datum is important for this purpose. For example, the same sand body may look like a distributary channel sand having a convex bottom or a sand bar with a convex top, depending on whether the upper or the lower surface is used as the datum (Conybeare 1979).

Wireline logs, run within the boreholes, aid not only in regional correlation, but also in defining the facies boundaries. A bed having a distinctive log character may be used as a marker bed for the purpose of correlation. To some extent the log characters are indicative of depositional environments. The deposits in low-energy environments, as in lakes for example, are good markers but those in high-energy environments, such as rivers, are impersistent. No log pattern, however, is uniquely related to any particular depositional environment. For example, the gamma-ray or

S.P. patterns shown in Fig. 9.13 can be found in any of the depositional environments listed in this figure. A line of evidence, independent of the log pattern, must be used for choosing the most probable depositional environment.

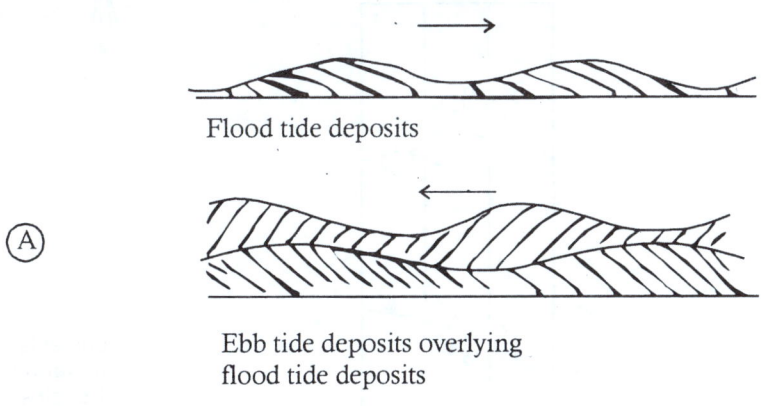

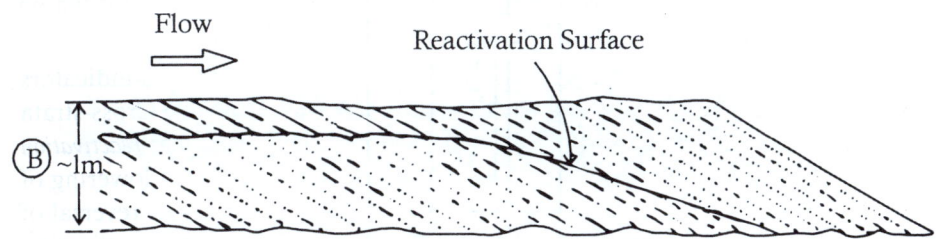

Fig. 10.6: (A) Herringbone structures produced by reversal of current directions in tidal environment (based on Fritz and Moore 1988). (B) Reactivation surface produced by changes in flow stage (after Harms *et al.* 1975).

Under suitable conditions dipmeter logs may be also used for interpreting the depositional environment. Four different dipmeter log patterns are generally recognized (Fig. 10.7): 1) the 'green pattern' showing vertically constant dips, 2) the 'blue pattern' showing downward decrease of dip, 3) the 'red pattern' showing downward increase of dip and 4) the 'random pattern' showing no systematic variation of dip. Although each of these four patterns can generate in a number of depositional environments, with some prior information on the type of environment to be expected, one can make an intelligent guess about the type of environment involved. In the case of an alluvial channel, for example, the thickest and the steepest cross-beds are expected near the basal scour ('the red pattern', see Fig. 7.22), but in a prograding delta the dips will decrease downwards ('the blue pattern', see Fig. 9.1B). the Schlumberger (1970) dipmeter manual discusses in detail the technique of interpretation.

Other uses of stratigraphic logs include analysis of the nature of cyclicity, i.e., identification of auto-and allocycles. The vertically fining-or coarsening-upward cycles may also be worked out following the techniques suggested in Chapter 9. These also have important bearings on interpretation of depositional environment.

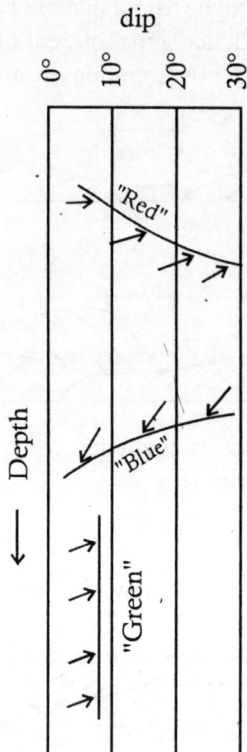

Fig. 10.7: Typical dipmeter log patterns.

Palaeohydraulic Interpretations in Fluvial Channels

The dimensions and types of sedimentary structures are intimately related to the hydraulic conditions responsible for their origin. Sedimentary structures may thus be used as clues to the geometry and hydraulics of ancient flow channels. In one such attempt Bridge (1978, p.725) used the following relationship:

$$V_m = \sqrt{(8gy_m \cdot S_m)/f}$$

where V_m is the mean flow velocity for a cross-secton at the position of mean depth (y_m), S_m is longitudinal water-surface slope along the locus of mean depth (~channel centre line) of a curved channel, and f is the Darcy-Weisbach friction coefficient. This factor is estimated to be 0.02 for horizontal lamination in upper flow regime plane beds, 0.04 to 0.06 for large-scale trough cross-stratification in fully developed dunes and 0.03 for small-scale (ripple) cross-stratification, or tabular cross-stratification from diminished dunes (Bridge 1978, p.736).

Allen (1968) obtained the following empirical relationship between height of subaqueous sand dune (H) and water depth (d):

$$H = 0.086 \ d^{1.19}$$

This relationship, according to Allen (1968, 1970), is valid for sand dunes having heights ranging between 0.1 and 10m. A new set of relations connecting water depth (d) and height (H) of sand ripples and small dunes (H = 0.01-0.09 m) was obtained experimentally by Das and Sengupta (1998): $H = 0.096.d^{1.02}$. Field tests show that the error between computed and observed water depths obtained by this equation is small. Moreover, the influence of grain-size and flow velocity on H vs.d relationship is negligible when small sand ripples (H < 2.0 cm) are involved, and the flow ranges between 30 and 50 cm/s.

Using a series of hydraulic relationships valid for meandering stream channels, Casshyap and Khan (1982) worked out the palaeohydraulics of the Permian streams responsible for Gondwana sedimentation in the Bokaro basin of India. Using similar techniques, Gardner (1983) worked out the palaeohydrology and palaeomorphology of a Carboniferous meandering stream in eastern Kentucky. That such interpretations can provide a consistent picture of the regional palaeohydraulic changes was shown by contouring the palaeohydraulic parameters of the Karharbari and the Barakar Formations (Permian) in the Giridih basin, India (Sengupta *et al.* 1988, see Fig.10.8). Starting from Allen's (1968) equation, several well-known hydraulic relationships were used in the following sequence for obtaining these results.

$$H = 0.086 \, d^{1.19} \quad \text{(Allen 1968)}$$

$$W = 42 \, d^{1.11} \quad \text{(Allen 1968)}$$

$$L_m = 10.9 \, W^{1.01} * \quad \text{(Leopold, Wolman and Miller 1964)}$$

$$L_m = 106.1 \, \overline{Q}^{0.46} * \quad \text{(Carlston 1965)}$$

$$V = \overline{Q}_{ma}/A \quad \text{(Schumm 1972)}$$

*These equations, where the units are in f.p.s. system, are used only as intermediate steps for estimating the water discharge. Leopold *et al.* equation serves as an intermediate step for estimation of mean annual discharge. For practical purposes mean annual flood ($\overline{Q}_{ma}$) may be taken to be approximately the same as bankfull discharge (Schumm 1972, p.103).

Explanation of Notations

H = mean dune height in metres

d = bankfull depth in metres

W = channel width in metres

L_m = meander wave length in feet

$\overline{Q}$ = mean annual discharge in cubic feet/second

A = channel cross-sectional area in square metres

V = flow velocity in metres/second

$\overline{Q}_{ma}$ = mean annual flood in cubic m/sec = bankfull discharge (Q)

This technique of palaeohydraulic interpretation uses dune height as the starting point for computations. The full height of the original dune is seldom preserved in geological record, hence a correction is often necessary. Jopling (1980) suggested a method of correction on the assumption that about four-fifths of the actual dune heights are preserved in the cross-beddings. The true dune height may be estimated from this assumption.

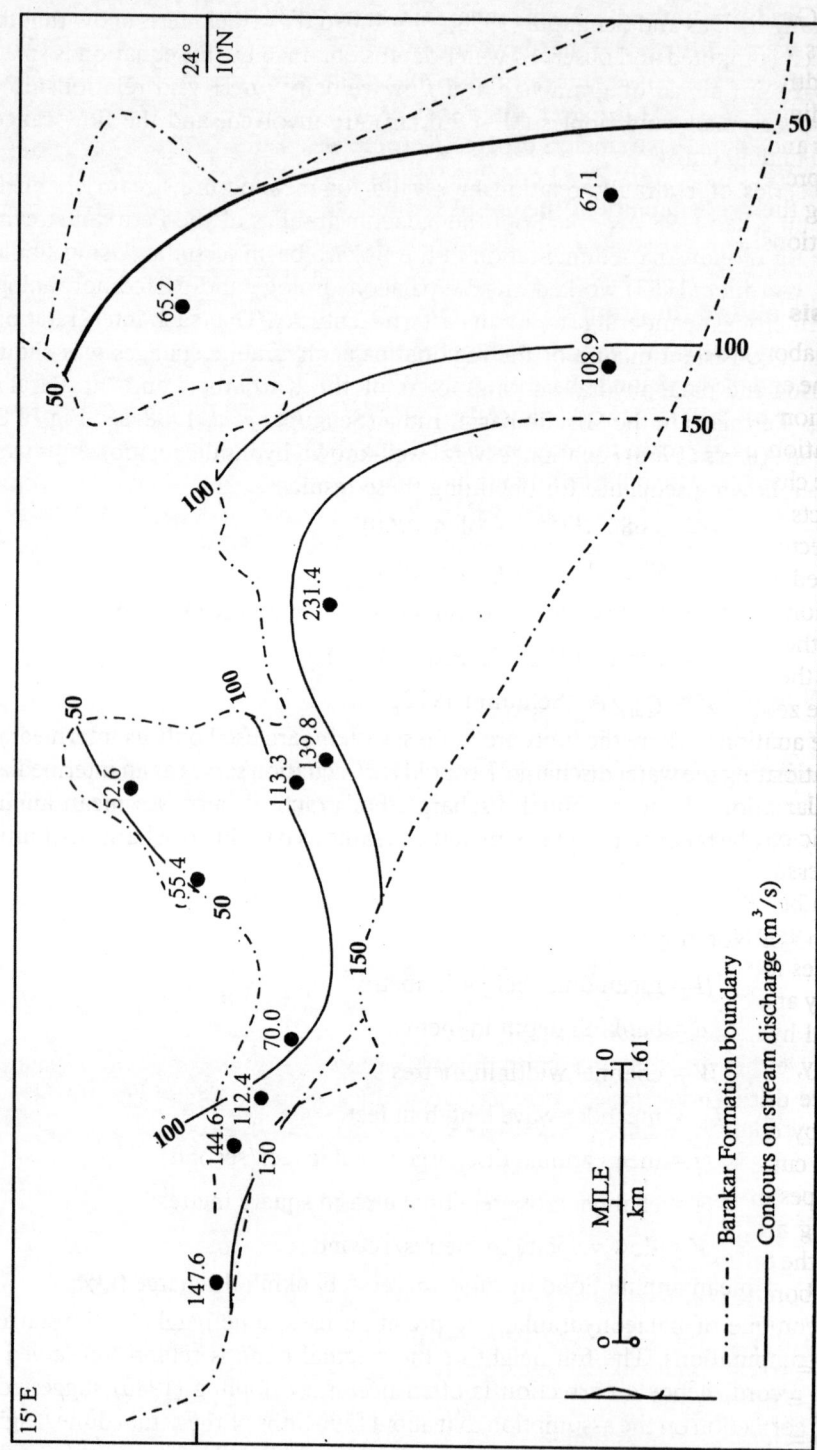

Fig. 10.8: Barakar (Permian) stream discharge (m³/s) in the Giridih basin worked out from magnitudes of bedforms using hydraulic equations (from Sengupta et. al. 1988). With permission of Geological, Mining & Metallurgical Society of India.

In the Giridih Gondwana basin, where channel geometry and palaeohydraulic parameters were estimated by the above mentioned method, the contours of channel depth, width and discharge run normal to the palaeoflow direction obtained from cross-bedding azimuths. Moreover, the trends of width, depth and discharge parallel each other and all these parameters progressively increase in the downstream direction, as in the present day rivers (see Leopold, Wolman and Miller 1964, p.240), thereby confirming the dependability of the use of sedimentary structures in palaeohydraulic interpretations.

Diagenesis and Maturation

Detailed laboratory investigation of the core and surface samples is recommended towards the end of the second phase of study. While optical microscopy is adequate for identification of the framework minerals, XRD or SEM techniques are needed for determination of the clay mineral species. Such studies often provide clues to the diagenetic changes, which the sediments have undergone. The pattern of diagenesis in turn reflects the burial, thermal and fluid circulation events to which the sediments were subjected after deposition. Under suitable conditions the trend of diagenesis can also be used as a clue to geodynamics and sedimentary basin evolution. For example, precipitation of high-temperature zeolite cements within sandstones in the back-arc basins of the Western Pacific was interpreted by Lee and Klein (1986) as the effect of upwelling thermal convective circulation followed by a lower temperature downwelling phase. The zeolite cement was preserved because the sediments underwent slow burial during the active rifting phase of basin formation.

Maturation of the organic material within sediments during burial in deep basins is of particular interest to petroleum geologists. During progressive burial of sediments the organic components are converted to liquid or gaseous hydrocarbons, depending on the pressure and temperature to which the sediment has been subjected. Liquid petroleum being preserved only within a narrow range of pressure and temperature, called the 'liquid window', it is of utmost importance for petroleum geologists to obtain clues to the burial history of the sediments.

Porosity and permeability of the reservoir rocks, factors of importance in prospecting for natural hydrocarbon, are affected during diagenesis and cementation. Reduction of porosity, either due to compaction during burial or due to chemical precipitation, takes place during diagenesis. Although some idea of the packing density may be obtained by counting under a microscope the number of grain-to-grain contacts in randomly cut rock sections, *in situ* determination of porosity is best done by using various types of wireline logs. Combined SP-resistivity log, sonic log, neutron log and density log are particularly helpful for *in situ* determination of porosity. Sonic log measures the time taken by acoustic waves to travel through one foot of rock formation along the borehole. Porosity (ϕ_s) is then determined by the formula :

$$\phi_s = (\Delta t - \Delta t_{ma})/(\Delta t_f - d\Delta_{ma})$$

where Δt_f = transit time through the pore fluid,

Δt_{ma} = transit time through the rock matrix,

Δt = interval transit time

(in modern logging manuals Δt is often replaced by t).

The sonic log, which responds to primary porosity, is valid only for clean compact formations with intergranular pores filled by liquids. The presence of hydrocarbon in a liquid or gaseous state upsets measurements.

Sediment Chemistry

Analysis of the major and trace elements of the sediment samples are undertaken at an advanced stage of basin analysis. Ratios of various oxides, $(Fe_2O_3 + MgO)/TiO_2$, Al_2O_3/SiO_2, K_2O/Na_2O, $Al_2O_3/(CaO + NaO)$, may help decipher the tectonic setting of the sedimentary basin (see Table 8.2). A similar analysis may also be attempted by plotting the framework mineralogy of sandstones on Dickinson-Suczek diagrams with QFL or QmFL as end members (see Fig. 8.10). Analysis of the immobile trace elements may also be used for working out the tectonic settings of the basin. A progressive increase in the proportions of light rare earth elements and decrease in U and Sc are expected from oceanic island arcs to passive margin settings passing through continental island arcs and active continental margins (Chapter 8).

Basin Evolution and Tectonics

Most of the large sedimentary basins are defined by a series of step faults bordering the cratonic margin, a sharp increase in sediment thickness at the basin centre, and highly deformed strata along the borderland. While the mobile belt at the borderland can be studied by conventional geological techniques, geophysical methods have to be applied for investigating the first two features because they generally lie concealed in the subsurface.

Subsidence of the basin margin concomitant with sedimentation allows a thick pile of sediment to accumulate. Simultaneously, a series of *growth faults* develop in the areas of rapid subsidence (see Dailly 1976 for details). The beds located above these growth faults at shallower levels show flexures. Another zone of faulting generally marks the belt of adjustment between the stable shelf of the basin and the deeper trough of relatively rapid subsidence. This was formerly designated as the *'hinge zone'* of the basin (Weeks 1952). In the Bengal Basin a conspicuous flexure in the Nummulitic (Eocene) limestone horizon in the subsurface detected by seismic survey, was designated as the 'hinge zone' (Sengupta 1966a). This is also the area where the seismic reflections from the shallower sections above the 'hinge' terminate sharply. The zone is also marked by an abrupt increase in thickness of sediments downdip (see Fig. 8.22). This zone, therefore, is not a hinge in the tectonic sense but possibly a stratigraphic feature resulting from increase in thickness of the sediment section and consequent compaction of layers in the lower part (see Conybeare 1979). Such an interpretation suggests that the so-called hinge line of the Bengal Basin is in reality the depositional limit of the shelf sediments. On the other hand, later discovery of continuation of the Nummulitic limestone into the deeper horizons in the form of a flexure across the 'hinge line' suggests that the limestone hinge may be the shallower expression of a deep-seated system of fault (Kaila *et al.* 1992). The challenge of basin analysis lies in searching for the most plausible answer from a host of such possible solutions.

A sedimentary basin occupies only a thin layer in the outermost part of the earth's crust. Thus our knowledge of a sedimentary basin remains incomplete without information of the lithospheric plate below the basin. While conventional reflection seismic survey can provide information on most of the major tectonic features within the basin, deep seismic surveys (DSS) are required for obtaining information from deeper parts of the lithosphere. The importance of detection of thinning of the crustal layer (e.g., the North Sea Graben) below a basin, or of a shear zone traversing the whole lithosphere plate (as in the Rocky Mountains) in tracing the stages of evolution

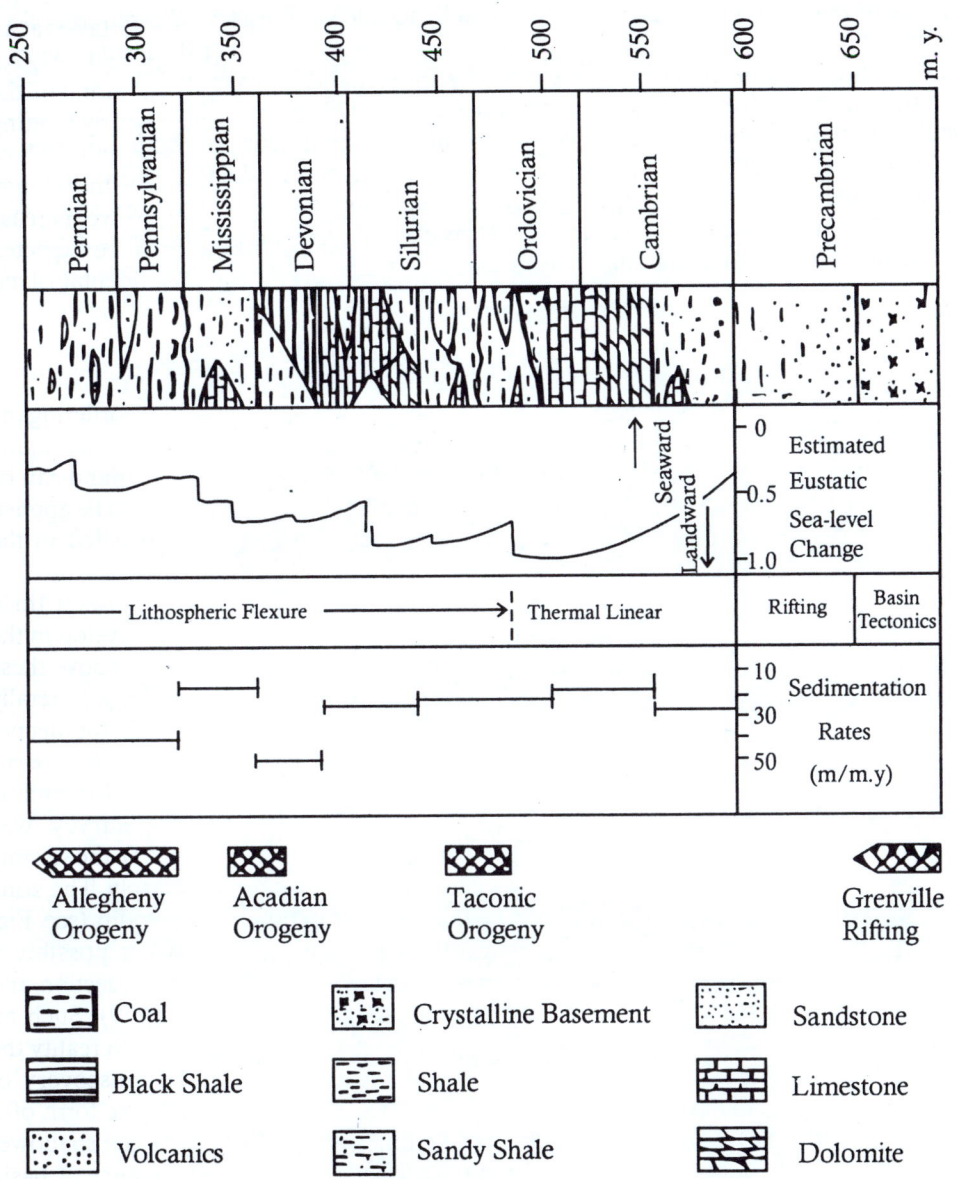

Fig. 10.9: Correlation diagram of the Appalachian Basin combining information on sediment accumulation rates, sea-level changes, orogenic events and stages of basin evolution against time scale. The ultimate aim of basin analysis is to prepare such correlation diagrams for sedimentary basins (redrawn after de V. Klein 1987). With permission of Elsevier Science Publishers B.V.

of a sedimentary basin, can hardly be overemphasized. A basin can be classified according to its genesis only when its location with respect to the plate, the type of the underlying crust (continental or oceanic) and the major structural movements which have affected it are known (see Chapter 8 and Table 8.6). Well co-ordinated geological, geophysical and geocheimcal explorations are needed for this purpose.

The ultimate aim of basin analysis should be to trace the history of evolution of a sedimentary basin in the context of global dynamics. 'Correlation of the local sections into a composite time scheme', as suggested by Dunbar and Rodgers (1957), is essential for this purpose. Measurement of the time span involved in basin evolution and sedimentation by radiometric, time-stratigraphic or other similar methods is needed. All these are represented in composite correlation diagrams of the type shown in Fig. 10.9. Recognition of stratigraphic 'sequences' and 'systems tracts' and correlation of the local events with global phenomena such as plate movements and sea-level changes are taken to be the goal of basin analysis today.

APPENDIX

Techniques of Palaeocurrent Analysis: Treatment of Directional Data

Computation of Resultant Directions

The average palaeocurrent direction within an area is represented by the resultant vector ($\hat{\gamma}$) of all observations. This direction is computed as follows:

Ungrouped Data

Each directional observation (cross-bedding azimuth, ripple orientation, and the like) is treated as a unit vector with components $\sin \alpha_i$ and $\cos \alpha_i$. The sums of sines and cosines for the sample α_i.............. α_n are computed as:

$$V = \sum_1^n \cos \alpha_i, \ W = \sum_1^n \sin \alpha_i$$

The vector resultant (sample mean direction) is given as:

$$\hat{\gamma} = \tan^{-1}(W/V), \text{ if } V > 0,$$
$$\hat{\gamma} = 180° + \tan^{-1}(W/V), \text{ if } V < 0$$
$$\hat{\gamma} = 90° + \tan^{-1}(W/V), \text{ if } V = 0, \text{ and } W > 0$$
$$\hat{\gamma} = 270° + \tan^{-1}(W/V), \text{ if } V = 0, \text{ and } W < 0$$
$$\hat{\gamma} \text{ is undetermined, if } V = W = 0.$$

Grouped Data

Suppose the data are grouped in the usual way into k classes with x_i denoting the classmark of the ith class and the ith class has n_i observations for: $i = 1, \ldots, k.$

$$Let \ V' = \sum_{i=1}^k n_i \cos x_i, \text{ and } W' = \sum_{i=1}^k n_i \sin x_i$$

Then the mean direction is obtained as in the case of ungrouped data, with W and V replaced by W' and V' respectively.

Variability

A measure of variability of the observations within an area is given by $1 - \overline{R}$, where $R = (W^2 + V^2)^{1/2}$, and $\overline{R} = R/n$, n being the number of observations, and R is the concentration. In palaeocurrent maps concentration of observations is often represented by the notation L called *consistency ratio* where $L = R/n \times 100$.

Palaeocurrent Analysis

Example: The following cross-bedding foreset azimuths have been measured within the Barakar sandstone outcrops in a Gondwana coal basin. Work out : I) the mean palaeocurrent direction (vector resultant), ii) dispersion (consistency ratio) of these directions.

The directions measured are: 5°, 25°, 45°, 48°, 50°, 315°, 321°, 322°, 330°, 332°, 335° 336°, 337°, 338°, 342°, 352°.

Computation of Mean Flow Direction

The data are grouped and sines and cosines are worked out in the following way.

Table 10.3: Directional data of Barakar sandstone

Interval	Midpoint	ni	Sin φ	cosφ	$W = n_I\, Sin\phi$	$V = n_I\, Cos\phi$
0°-.19°	9.5°	1	0.17	0.99	0.17	0.99
20°-39°	29.5°	1	0.49	0.87	0.49	0.87
40°- 59°	49.5°	3	0.76	0.65	2.28	1.95
300°-319°	309.5°	1	-0.77	0.64	-0.77	0.64
320°-339°	329.5°	8	-0.50	0.86	-4.06	6.89
340°-359°	349.5°	2	-0.18	0.98	-0.36	1.97
Total		16			-2.25	13.31

Mean flow direction is given by the resultant vector $(\hat{\gamma}) = \tan^{-1} W/V$

where $W = \Sigma n_i . \sin\phi; \quad V = \Sigma n_i . \cos\phi$

$\hat{\gamma} = \tan^{-1}(-2.25/13.31) = -9.5949$ (or 10°), $360° - 10° = 350°$

$R = \sqrt{W^2 + V^2} = 13.50$

Concentration is given by the consistency ratio, $L = R/n = 13.50 / 16 = 0.844$ or 84.37%. A graphical representation (circular histogram) of these directional data is shown in Fig. 10.10.

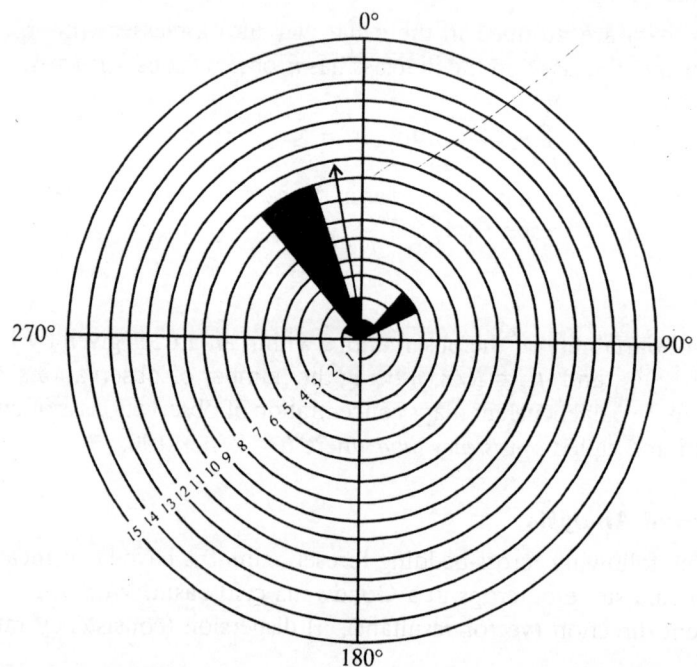

Fig. 10.10: Circular histogram ('rose diagram') of the directional data shown in table 10.3. The arrow indicates direction of the resultant vector

Sampling of Directional Data

The procedure for sampling of directional data (like cross-bedding azimuths) is summarized below. The reader may refer to Rao and Sengupta (1970, 1972) for further details.

Pilot Survey

Collection of Samples: Before the actual sampling of the directional data like cross-bedding azimuths is undertaken, it is necessary to conduct a pilot survey in the area of study with a small number of representative samples. An equal number of observations from each outcrop facilitates computation but this is not essential. Supposing the number of outcrops visited is n, and the number of observations taken from each of them is m, the total sample size $N = n.m$. Let ϕ_{ij} denote the jth observation from the ith outcrop ($j = 1,...m$; $i = 1, ..., n$).

Analysis of Samples:

1. Sine and cosine values are computed for each direction (f_{ij}) measured during the pilot survey. The length of resultant direction for each outcrop is obtained as follows:

$$R_i^2 = \left(\sum_{j=1}^{m} \cos \phi_{ij}\right)^2 + \left(\sum_{i=1}^{m} \sin \phi_{ij}\right)^2 = C_i^2 + S_i^2$$

where R_i is the outcrop resultant for the ith outcrop, and m is the number of observations within each outcrop.

2. The overall resultant R for all outcrops is given by

$$R^2 = \left(\sum_{i=1}^{n} C_i\right)^2 + \left(\sum_{i=1}^{n} S_i\right)^2$$

where n is the number of outcrops surveyed, and C_i and S_i are as defined above.

3. The analysis of variance (ANOVA) for the directional data is performed using Table 10.4. $\hat{\omega}$ and $\hat{\beta}$, the estimates for within outcrop and between outcrop concentration parameters respectively, are obtained by equating columns (4) and (5) of Table 10.4.

Table 10.4: ANOVA Table for Angular Data
(Rao and Sengupta, 1970, p. 536)

Source of Variations (1)	df (2)	SS (3)	MS (4)	E (MS) (5)
Between outcrops....	$n-1$	$\sum_{1}^{n} R_i - R$	$(\sum R_i - R)/(n-1)$	$\dfrac{1}{2}\left(\dfrac{1}{\omega} + \dfrac{\overline{m}}{\beta}\right)$
Within outcrops......	$N-n$	$N - \sum_{1}^{n} R_i$	$(N - \sum R_i)/(N-n)$	$\dfrac{1}{2\omega}$
Total	$N-1$	$N-R$	....	....

Note: $\overline{m}$ is a weighted average of the m_i's. If $m_i = m$ for all outcrops then $\overline{m} = m$.

4. An efficient sampling scheme should allow sampling at minimum cost. The optimum number of observations $m*$ to be taken at an outcrop is obtained from the relation: $m* = \sqrt{C_1 \hat{\beta} / C_2 \cdot \hat{\omega}}$.

where C_1 and C_2 are the costs for reaching an outcrop, and taking an observation within the outcrop, respectively. An approximate idea of the relative cost (C_1/C_2), say 10 : 1, or 20 : 1 is needed for this purpose.

5. The optimum number of outcrops to be sampled, $n*$, is given by the equation:

$$n* = \kappa_0 \left[\left(1 / \hat{\beta}\right) + \left(1 / m \cdot \hat{\omega}\right) \right]$$

where κ_0 is the concentration required for a desired confidence level $(1 - \alpha)$ and the semiangle of confidence (ψ_0). κ_0 is obtained from Table 10.5. The amount of precision required and the desired confidence level determine κ_0.

Table 10.5: Value of Concentration Parameter κ_0 Required for $\hat{\gamma}_N$ to Attain a Given Precision (Rao and Sengupta, 1970, p. 537)

ψ_0 Semiangle of Confidence	$(1-\alpha)$ Confidence Level	κ_0 Concentration Value Required
5^0	0.90	354.6292
	0.95	504.0610
	0.99	870.6882
10^0	0.90	88.7598
	0.95	126.1125
	0.99	217.9236
20^0	0.90	22.1772
	0.95	31.5221
	0.99	54.4496

Example: Sampling of cross-bedding azimuths from the Upper Kamthi member around Bhimaram (Bheemaram), India.

1. Fourteen outcrops of Upper Kamthi around Bhimaram were visited during the pilot survey . From each of these outcrops 10 observations on cross-bedding azimuths were recorded i.e. $n_u = 14$, $m_u = 10$ for Upper Kamthi. The total number of observations for the 14 outcrops $Nu = 14 \times 10 = 140$.

2. The overall resultant based on all the 140 observations is $R_u = (82.5159, -22.5615)$, with a vector length $R_u = \sqrt{(82.5159)^2 + (-22.5615)^2} = 85.5447$. The sum of the lengths of individual resultants is $\sum_{i=1}^{14} R_{iu} = 91.8074$.

3. Table 10.6 shows the ANOVA for the Upper Kamthi data. The value of F-statistic ($F = 1.2593$) being insignificant at 5% level (the tabulated critical value being 1.83 with 13 and 126 degrees of freedom), the variation between outcrops was not taken to be significant. In other words, all the Upper Kamthi outcrops were

taken to have the same mean direction, which is also the Upper Kamthi Formation direction. The reasons for drawing this conclusion are discussed in detail in Rao and Sengupta (1970, p. 539).

Table 10.6: ANOVA Table for Upper Kamthi Data
(Rao and Sengupta, 1970, p. 539)

Source	df	SS	MS	E (MS)	F
Between outcrops....	13	6.2627	0.4817	$\frac{1}{2}\left(\frac{1}{\omega_u} + \frac{10}{\beta_u}\right)$	F = 1.2593
Within outcrops........	126	48.1926	0.3825	$\frac{1}{2\omega u}$	....
Total	139	54.4553	....	....	....

4. The estimate of ω, say $\hat{\omega}_{\hat{u}}^*$, is obtained by pooling the between and within sum of the squares and the degrees of freedom. This gives $\hat{\omega}_{\hat{u}}^* = 2.5526$.

5. The equation for κ now reduces to $\kappa = N\hat{\omega}_{\hat{u}}^*$. Therefore, the number of observations N which one should take in order to attain a semiangle of say, $10°$ with 95% confidence (the corresponding value of $\kappa_0 = 126.1125$) is $N = \kappa_0/\hat{\omega}_{\hat{u}}^*$ 126.1125/2.5526 = 49.40 or roughly, 50 observations. This means that for Upper Kamthi one can estimate the mean direction to within $\pm10°$ with 95% confidence based on 50 observations.

REFERENCES

(Figures within parentheses indicate page numbers of citations)

Adams, J. E. and Rhodes, M. L. 1960. Dolomitization by seepage refluxion. *American Association Petroleum Geologists Bulletin*, vol. 44, pp. 1912-1920. (45)

Adhikari, A. K., Roy, J. and Sengupta, S. 1981. Estimates of parameters of grain-size distribution from weight frequencies. *Mathematical Geology*, vol. 12, pp. 247-265. (65)

Ahlbrandt, T. S., and Fryberger, S. G. 1982. Introduction to eolian deposits. *In* P. A. Scholle and D. Spearing (eds.), *Sandstone Depositional Environments*, pp. 11-47. American Association of Petroleum Geologists, Tulsa, OK., 410 pp. (171)

Aissaoui, D. M., Buigues, D. and Purser, B. H. 1986. Model of reef diagenesis: Mururoa Atoll, French Polynesia, pp. 27-52. *In* J. H. Schroeder and B. H. Purser (eds.) *Reef Diagenesis*. Springer-Verlag, Berlin, 455 pp. (37)

Aitken, J. D. 1967. Classification and environmental significance of cryptalgal limestones and dolomites with illustrations from the Cambrian and Ordovician of southwestern Alberta. *Journal Sedimentary Petrology*, vol. 37, pp.1163-1178. (135)

Aitken, J. F., and Flint, S.S., 1995. The application of high-resolution sequence stratigraphy to fluvial systems: a case study from the Upper Carboniferous Breathitt Group, eastern Kentucky, USA. *Sedimentology*, vol. 42, pp.3-30. (278)

Albritton, C. C. Jr. (ed.), 1963. *The Fabric of Geology*, Addisson-Wesley Publication Co. Inc., Reading, 372 pp. (5)

Allen, J. R. L. 1963. A classification of cross-stratified units with notes on their origin. *Sedimentology*, vol. 2, pp. 93-114. (109, 110, 112)

Allen, J. R. L. 1964. Studies in fluviatile sedimentation : six cyclothems from the Lower Old Red Sandstone, Anglo-Welsh basin. *Sedimentology*, vol. 3, pp. 163-198. (188)

Allen, J. R. L. 1965. The sedimentation and palaeogeography of the Old Red Sandstone of Anglesey, North Wales. *Proceedings Yorkshire Geological Society*, vol. 35, pp. 139-184. (131)

Allen, J. R. L. 1965a. A review of the origin and characteristics of Recent alluvial sediments. *Sedimentology* (special issue), vol. 5, 2 : pp. 89-191. (179)

Allen, J. R. L. 1965b. Upper Old Red Sandstone palaeogeography (Farlovian) in South Wales and the Welsh Borderland. *Journal Sedimentary Petrology*, vol. 35, pp. 167-195. (188)

Allen, J. R. L. 1968. The nature and origin of bedform hierarchies. *Sedimentology*, vol. 10, pp. 161-182. (104, 294)

Allen, J. R. L. 1968a. *Current Ripples*. North Holland, Amsterdam, 433 pp. (105, 123, 292, 295)

Allen, J. R. L. 1970. *Physical Processes of Sedimentation. An Introduction*. George Allen and Unwin Ltd., London, 248 pp. (150)

Allen, J. R. L. 1970a. Studies in fluviatile sedimentation : a comparison of fining-upwards cyclothems, with special reference to coarse-member composition and interpretation. *Journal Sedimentary Petrology*, vol. 40, pp. 298-323. (187)

Allen, J. R. L. 1973. Development of flute mark assemblages, I. Evolution of pairs of defects. *Sedimentary Geology*, vol. 10, pp. 157-177. (121)

Allen, J. R. L. 1975. Development of flute mark assemblages, 2. Evolution of trios of defects. *Sedimentary Geology*, vol. 13, pp. 1-26. (121)

Allen, J. R. L. 1977. The possible mechanism of convolute lamination in graded sand beds. *Journal Geological Society London*, vol. 134, pp. 19-31. (127)

Arya, B. C. and Rao, C. N. 1979. Bioturbation structures from the Middle Proterozoic Narji Formation, Kurnool Group, Andhra Pradesh, India. *Sedimentary Geology*, vol.22, pp.127-134. (144, 145)

Ashley, G. M. 1990. Classification of large-scale subaqueous bedforms : a new look at an old problem (report of the SEPM bedforms and bedding structures symposium). *Journal Sedimentary Petrology*, vol. 60, pp. 160-172. (105, 106)

Aubouin, J. 1965. *Geosynclines: Developments in Geotectonics*. Elsevier, Amsterdam, 335 pp. (208)

Auden, J. B. 1933. Vindhyan sedimentation in the Son Valley, Mirzapur district. *Memoir, Geological Survey India*, 62 (2), pp. 141-256. (53, 139, 194)

Awramik , S. M., 1971. Precambrian columnar stromatolite diversity: reflection of metazoan appearance. *Science*, vol. 174, pp. 825-827. (138)

Awramik, S. M. 1984. Ancient stromatolites and microbial mats, pp.1-22. *In* Y. Cohen, R. W. Castenholz, and H. O. Halvorson (eds.), *Stromatolites*, A. R. Liss Inc. New York. (138)

Azad, J., Bhattacharyya, S., Datta, B. D. and Stevens, T. E. 1971. Hydrocarbon accumulation in the Nahorkatiya oilfield. Assam. *Proceedings VIII World Petroleum Congress, Moscow*, vol. II, pp. 1-10. Applied Science Publishers Ltd., UK. (*In* P. Berger *et. al.* 1983). (265)

Baas, J. H. 1994. A flume study on the development and equilibrium morphology of current ripples in very fine sand. *Sedimentology*, vol. 41, pp. 185-209. (104)

Baas-Becking, L. G. M., Kaplan, I. R. and Moore, D. 1960. Limits of natural environment in terms of pH and oxidation-reduction potentials. *Journal Geology*, vol. 68, pp. 243-284. (58)

Bagati, T. N. 1979. Geochemistry of carbonate rocks of the Krol Formation of the Simla Himalaya, India. *Himalayan Geology*, vol. 9, pt. II, pp. 462-497. (34, 35)

Bagnold, R. A. 1941. *The Physics of Blown Sand and Desert Dunes*. Methuen, London (reprinted 1954), 265 pp. (4)

Bagnold, R. A. 1954. Experiments on gravity-free dispersion of large solid spheres in a Newtonian fluid under shear, *Philosophical Transactions Royal Society, London* (A), 225, pp. 49-63. (118)

Bagnold, R. A. 1956. The flow of cohesionless grains in fluids. *Philosophical Transactions Royal Society, London* (A), vol. 249, pp. 235-297. (21)

Baksi, A. K., Ray Barman, T., Paul D. K. and Ferrad, Edward. 1987. Widespread Early Cretaceous flood basalt volcanism in eastern India: Geochemical data from Rajmahal-Bengal-Sylhet traps. *Chemical Geology*, vol. 63, pp. 133-141. (244)

Banerjee, D. M. 1986. Proterozoic and Cambrian phosphorites-regional review: Indian subcontinent. *In* P. J. Cook and J. H. Shergold (eds.), *Phosphatic Deposits of the World*, vol. I. *Proterozoic and Cambrian Phosphorites*, pp. 70-90, Cambridge University Press, Cambridge. (46)

Banerjee, Indranil. 1960. Stratigraphy and sedimentation in the South Karanpura Coalfield, Bihar. *Quarterly Journal Geological, Mining & Metallurgical Society India*, vol. 32, pp. 189-203. (188, 189, 251)

Banerjee, Indranil. 1964. Size-roundness relation in the Barakar sandstones of the South Karanpura Coalfield, India. *Sedimentology*, vol. 3, pp. 22-28. (76)

Banerjee, Indranil. 1966. Turbidites in a glacial sequence: a study from the Talchir Formation, Raniganj Coalfield, India. *Journal Geology*, vol. 74, pp. 593-606. (168)

Banerjee, Indranil. 1973. Sedimentology of Pleistocene glacial varves in Ontario, Canada. *Geological Survey Canada Bulletin* 226. (166)

Banerjee, Indranil. 1974. Barrier coastline sedimentation model and the Vindhyan example. *Geological, Minning & Metallurgical Society India. Golden Jubilee Volume*, (vol. 46), pp. 101-127. (193, 194, 227)

Banerjee, Indranil. 1980. The Vindhyan tidal sea. *In* K. S. Valdiya, *et al.* (eds.) *Geology of Vindhyanchal*, pp. 80-87, Hindusthan Publishing Corporation (India), New Delhi, 231, pp. (194)

Banerjee, Indranil. 1996. Populations, trends and cycles in combined-flow bedforms. *Journal Sedimentary Research*, vol. 66, pp. 868-874. (117)

Banerjee, I., Kalkreuth, W., and Davies, E. H. 1996. Coal seam splits and transgressive-regressive coal couplets: A key to stratigraphy of high-frequency sequences. *Geology*, vol. 24, pp.1001-1004. (277, 278)

Banerjee, Indranil and McDonald, B. C. 1975. Nature of esker sedimentation. *In* A. V. Jopling and B. C. McDonald (eds.), *Glaciofluvial and Glaciolacustrine Sedimentation*, pp. 132-154. Society of Economic Paleontologists and Mineralogists, Spl. Pub. No. 23. (165)

Banerjee, Indranil and Sengupta, Supriya. 1963. The Vindhyan Basin—a regional reconnaissance of the eastern part. *Quarterly Journal Geological*, Mining & Metallurgical Society India, vol. 35, pp. 141-149. (177, 227)

Banerji, A. K. 1977. On the Precambrian Banded Iron Formations and the manganese ores of the Singhbhum region, Eastern India. *Economic Geology*, vol. 72, pp. 90-98. (51)

Basu, Abhijit, 1976. Petrology of the Holocene fluvial sand derived from plutonic source rocks: implications to paleoclimatic interpretations. *Journal Sedimentary Petrology*, vol. 46, pp. 694-709.

Basu, Abhijit, 1977. Provenance and AlSi order-disorder of detrital alkali feldspars. *Journal Geological Society India*, vol. 18, pp. 477-492. (14)

Basu, Abhijit, 1985. Reading provenance from detrital quartz. *In* G. G. Zuffa (ed.), *Provenance of Arenites*, pp. 231-247. NATO-ASI. V. C-148. Reidel, Holland. (14)

Basu, Abhijit and Suttner. L. J. 1975. Use of structural state of alkali feldspars in provenance interpretations. *Proceedings IXth International Sedimentological Congress*, Nice, Theme 3, pp. 1-9. (14, 15)

Basu, Abhijit, Young, S. W., Suttner, L. J., James, W. C. and Mack, G. H. 1975. Re-evaluation of the use of undulatory extinction and polycrystallinity in detrial quartz for provenance interpretation. *Journal Sedimentary Petrology*, vol. 45, pp. 873-882. (15)

Basu, P. K., Biswas, S. and Acharyya, S. K. 1987. Late Quaternary ash beds from the Son and Narmada Basins, Madhya Pradesh. *Indian Minerals*, vol. 41, pp. 66-72. (53)

Basumallick, Sudhir, 1962. Petrology and zonation of the Rewa sandstone in Maihar, Madhya Pradesh, *Quarterly Journal Geological, Mining & Metallurgical Society India*, vol. 34, pp. 11-28. (24)

Bates, C. C. 1953. Rational theory of delta formation. *American Association Petroleum Geologists Bulletin*, vol. 37, pp. 2119-2162. (202)

Bathurst, R. G. C. 1975. *Carbonate Sediments and Their Diagenesis*. Elsevier, Amsterdam, 658 pp. (131)

Berger, P. *et al.* 1983. Schlumberger Well Evaluation Conference, India. Schlumberger Technical Services, Inc., 263 pp. (226, 231, 235)

Berner, R. A. 1981. A new geochemical classification of sedimentary environments. *Journal Sedimentary Petrology*, vol. 51, pp. 359-365. (49)

Bertrand-Sarfati, J., and Trompette, R. 1976. Use of Stromatolites for intrabasinal correlation: example from the Late Proterozoic of the northwestern margin of the Taoudenni basin, pp. 517-534, *In* M. R. Walter (ed.), *Stromatolites. Developments in Sedimentology*, vol. 20, Elsevier Scientific Publishing Co., Amsterdam, 790 p. (138)

Best, Jim. 1992. On the entrainment of sediment and initiation of bed defects: insights from recent developments within turbulent boundary layer research. *Sedimentology*, vol. 39, pp. 797-811. (91, 99)

Bhatia, M. R. 1983. Plate tectonics and geochemical composition of sandstones. *Journal Geology*, vol. 91, pp. 611-627. (220)

Bhatia, M. R. and Crook, A. W. 1986. Trace element characteristics of graywackes and tectonic setting discrimination of sedimentary basins. *Contribution to Mineralogy & Petrology*, vol. 92, pp. 181-193. (223)

Bhattacharji, S. and Singh, R. N. 1984. Thermo-mechanical structure at the southern part of the Indian shield and its relevance to Precambrian basin evolution. *Tectonophysics*, vol. 105, pp. 103-120. (234)

Bhattacharyya, A. and Morad, S. 1993. Proterozoic braided ephemeral fluvial deposits: an example from the Dhandraul Sandstone Formation of the Kaimur Group. Son Valley, Central India. *Sedimentary Geology*, vol. 84, pp. 101-114. (194)

Biswas, B. 1963. Results of exploration for petroleum in the western part of the Bengal basin, India. *Proceedings 2nd Symposium Development Petroleum Resources*. ECAFE, Mineral Resources Development Series, No. 18, pt. 1, pp. 241-250. (206)

Biswas, S. K. 1981. Basin framework, Palaeo-environment and depositional history of the Mesozoic sediments of Kutch basin, Western India. *Quarterly Journal Geological, Mining & Metallurgical Society India*, vol. 53, pp. 56-85. (239)

Biswas, S. K. 1982. Rift basins in western margin of India and their hydrocarbon prospects with special reference to Kutch basin. *American Association Petroleum Geologists Bulletin*, vol. 66, pp. 1497-1513. (239, 240)

Biswas, S. K. and Deshpande, S. V. 1983. Geology and hydrocarbon prospects of Kutch, Saurashtra and Narmada basins. *In* L. L. Bhandari et al. (eds.), *Petroliferous Basins of India*, vol. I, pp. 111-126. *Petroleum Asia Journal*, V-VI, No. 4, 189 pp. (239)

Blatt, H. and Christie, J. M. 1963. Undulatory extinction in quartz of igneous and metamorphic rocks and its significance in provenance studies of sedimentary rocks. *Journal Sedimentary Petrology*, vol. 33, pp. 559-579. (14)

Blatt, H., Middleton, G. and Murray, R. 1972. *Origin of Sedimentary Rocks*. Prentice-Hall, Englewood Cliffs, N. J., 634 pp. (6, 29, 101, 253)

Boersma, J. R. 1967. Remarkable types of mega cross-stratification in the fluviatile sequence of a subrecent distributary of the Rhine Amerongen, the Netherlands. *Geol. Mijnbouw*, vol. 46, pp. 217-235. (112)

Borchert, H. 1960. Genesis of marine sedimentary iron ores. *Trans. Inst. Min. Met.*, vol. 69, pp. 261-279. (11)

Bose, Abhijit and Sengupta, Supriya. 1993. 'Infra-Kamthi' of the central Godavari Valley-petrological evidence of marine influence during the Permian. *Proceedings National Acad. Sci. India*, vol. 63, Sec. A Pt. 1 (Spl. Pub. on Petrology), pp. 149-166. (52)

Bose, P. K. and Das, N. G. 1986. A transgressive storm- and fair-weather wave-dominated shelf sequence: Cretaceous Nimar Formation, Chakrud, Madhya Pradesh, India. *Sedimentary Geology*, vol. 46, pp. 147-167. (163)

Bose, P. K., Chaudhuri, A. K. and Seth, Ashoke. 1988. Facies, flow and bedform patterns across a storm-dominated inner continental shelf. Proterozoic Kaimur Formation, Rajasthan, India. *Sedimentary Geology*, vol. 59, pp. 275-293. (163)

Bouma, A. H. 1962. Sedimentology of Some Flysch Deposits: *A Graphic Approach to Facies Interpretation*. Elsevier, Amsterdam, 168 pp. (157)

Bouma, A. H. 1969. *Methods for the Study of Sedimentary Structures*. Wiley Interscience, New York, 458 pp. (6)

Bridge, J. S. 1978. Palaeohydraulic interpretation using mathematical models of contemporary flow and sedimentation in meandering channels. *In* A. D. Miall (ed.), *Fluvial Sedimentology*, pp. 723-742. Canadian Society of Petroleum Geologists Memoir 5, Calgary, Alberta, 859 pp. (294)

Bridge, J. S. and Diemer, J. A. 1983. Quantitative interpretation of an evolving ancient river system. *Sedimentology*, vol.30, pp.599-623. (181)

Brink, A. H. 1974. Petroleum Geology of Gabon basin. *American Assoc. Petroleum Geologists Bulletin*, vol. 58, pp. 216-235. (238)

Brookfield, M. E. 1984. Eolian Facies. *In* R. G. Walker (ed.), *Facies Models*, 2nd ed., pp. 91-104. Geological Association of Canada, Toronto, Ont. 317 pp. (169, 170, 171, 172)

Burger, Heinz and Skala, Wolfdietrich. 1976. Comparison of sieve and thin-section technique by Monte-Carlo model. *Computers and Geoscience*, vol. 2, pp. 123-139. (66)

Burke, K., Dessauvagie, T. F. J. and Whiteman, A. J. 1972. Geological history of the Benue valley and adjacent areas. *In* T. F. J. Dessauvagie and A. J. Whiteman (eds.), *African Geology*, Ibadan, Nigeria, pp. 187-205. (233)

Burley, S. D., Kantorowicz, J. D. and Waugh, B. 1985. Clastic diagenesis. *In* P. J. Brenchley and B. P. J. Williams (eds.), *Sedimentology, Recent Developments and Applied Aspects*, pp. 189-226. Published for the Geological Society by Blackwell Scientific Publications, Oxford. (54)

Busch, D. A. 1974. *Stratigraphic Traps in Sandstones-Exploration Techniques*. Memoir 21, American Association of Petroleum Geologists, Tulsa, Oklahoma, 174 pp. (202)

Campbell, C. V. 1967. Lamina, lamina set, bed, and bedset. *Sedimentology*, vol. 8, pp. 7-26. (273)

Cant, D. J. 1984. Subsurface facies analysis. *In* R. G. Walker (Ed.), *Facies Models*, 2nd ed. pp. 277-310. Geol. Assoc. of Canada, Toronto, Ont. 317 pp. (261, 263, 264, 266)

Cant. D. J. 1989. Zuni Sequence: the foreland basin. Lower Zuni sequence: Middle Jurassic to Middle Cretaceous. Chapter 11. *In* B. D. Rickets (ed.) *Western Canada Sedimentary Basin*: A Case History. pp. 251-254. Canadian Society of Petroleum Geologists. Calgary. (234)

Cant. D. J. and Walker, R. G. 1976. Development of a braided-fluvial facies model for the Devonian Battery Point Sandstone, Quebec. *Canadian Jour. Earth Sc.*, vol. 13, pp. 102-119. (259)

Carlston, C. W. 1965. The relation of free meander geometry to stream discharge and its geomorphic implications. *American Journal Science*, vol. 263 pp. 864-885. (295)

Carozzi, A. 1960. *Microscopic Sedimentary Petrography*. John Wiley, New York, 485 pp. (6)

Carver, R. W. (ed.). 1971. *Procedures in Sedimentary Petrology*. John Wiley, New York, 458 pp. (6)

Casshyap, S. M. 1970. Sedimentary cycles and environment of deposition of the Barakar Coal Measures of Lower Gondwana, India. *Journal Sedimentary Petrology*, vol. 40, pp. 1302-1317. (188, 189)

Casshyap, S. M. 1973. Paleocurrents and paleogeographic reconstruction in the Barakar (Lower Gondwana) sandstones of Peninsular India. *Sedimentary Geology*, vol. 9, pp. 283-303. (188)

Casshyap, S. M. and Khan, Z. A. 1982. Paleohydrology of Permian Gondwana seams in Bokaro Basin, Bihar, *Journal Geological Society India*, vol. 23, pp. 419-430. (295)

Casshyap, S. M. and Tewari, R. C. 1982. Facies analysis and paleogeographic implications of a late Paleozoic glacial outwash deposit, Bihar, India. *Journal Sedimentary Petrology*, vol. 52, pp. 1243-1256. (19, 167)

Chakrabarti, Amitabha. 1977. Upward flow and convolute laminations. *Senckenbergiana Marit.*, vol. 9, pp. 285-305. (127)

Chakrabarti, A. 1990. Traces and dubiotraces: examples from the so-called Late Proterozoic siliciclastic rocks of the Vindhyan Supergroup around Maihar, India. *Precambrian Research*, vol.47, pp.142-153. (145)

Chakraborty, Tapan. 1991. Sedimentology of a Proterozoic erg.: the Venkatpur Sandstone, Pranhita-Godavari Valley, South India. *Sedimentology*, vol. 38, pp. 301-322. (173)

Chanda, S. K. 1963. Cementation and diagenesis of the Lameta Beds, Lametaghat, M.P. India. *Journal Sedimentary Petrology*, vol. 33, pp. 728-738. (58)

Chanda, S. K. and Bhattacharyya, Ajit. 1974. Ripple-drift cross-lamination in tidal deposits: example from the Precambrian Bhander Formation of Maihar, Satna District, Madhya Pradesh, India. *Geological Society America Bulletin*, vol. 85, pp. 1117-1122. (194)

Chaudhuri, Asru. 1970. Precambrian stromatolites in the Pranhita-Godavari Valley (South India). *Palaeogeography, Palaeoclimatology, Palaeoecology*, vol. 7, pp. 309-340. (140)

Chow, Ven Te. 1959. *Open-Channel Hydraulics* (International Student Edition). McGraw-Hill Kogakusha Ltd., Tokyo, 680 pp. (89, 94)

Christiansen, C., Blaesild, P. and Dalsgaard, K. 1984. Re-interpreting 'segmented' grain-size curves. *Geological Magazine*, vol. 121, pp. 47-51. (71)

Clifford, N. J. 1993. Formation of riffle-pool sequences: field evidences for an autogenic process. *Sedimentary Geology*, vol.85, pp.39-51. (182)

Colby, B. R. and Hembree, C. H. 1955. Compositions of total sediment discharge Niobrare River near Cody, Nebraska, *U.S. Geological Survey Water Supply* Paper No. 1357. 187 pp. (97)

Coleman, J. M. 1969. Brahmaputra River, channel processes and sedimentation. *Sedimentary Geology*, vol. 3, pp. 131-239. (113)

Coleman, J. M. and Gagliano, S. M. 1964. Cyclic sedimentation in the Mississippi river deltaic plain. *Trans. Gulf Coast Association of Geological Societies*, vol. 14, pp. 67-80. (255)

Conybeare, C. E. B. 1979. *Lithostratigraphic Analysis of Sedimentary Basins*. Academic Press, New York, 555 pp. (292, 298)

Cook, H. E., Field, M. E. and Gardner, J. V. 1982. Characteristics of sediments on modern and ancient continental slopes. *In* P. A. Scholle and O. Spearing (eds.), *Sandstone Depositional Environments*, pp. 329-364. AAPG Memoir No. 31, 410 pp. (148)

Crook, K. A. W. 1974. Lithogenesis and geotectonics: the significance of compositional variation in flysch arenites (graywackes). *In* R. H. Dott and R. H. Shaver (eds.), *Modern and Ancient Geosynclinal Sedimentation*, pp. 304-310. Society of Economic Paleontologists and Mineralogists, Spl. Pub. No. 19, 380 pp. (222)

Curray, J. R. and Moore, D. G. 1971. Growth of the Bengal deep-sea fan and denudation in the Himalayas. *Geological Society America Bulletin*, vol. 82, pp. 563-572. (212, 236, 244)

Curray, J. R., *et al.* 1979. Tectonics of the Andaman Sea and Burma. *In* J. S. Watkins, L. Montadert and P. W. Dickerson (eds.), *Geological and Geophysical Investigations of Continental Margins*. pp. 189-198, American Association of Petroleum Geologists, Memoir No. 29, Tulsa, OK. (237)

Curtis, C. D. 1976. Stability of minerals in surface weathering reactions: a thermochemical approach: *Earth Surf. Proc.*, vol. 1, pp. 63-70. (13)

Dailly, G. S. 1976. A possible mechanism regarding progradation, growth faulting, clay diapirism and overthrusting in a regressive sequence of sediments. *Bulletin, Canadian Petroleum Geology*, vol. 24, pp. 92-116. (298)

Das, S. K., Chakrabarti, P. K. and Dutta, T. K. 1998. Sequence stratigraphic analysis of Oligocene sequence with special reference to lowstand systems tract in Ichapur-Jaguli area, Bengal basin. *ONGC Bulletin*, vol.35, pp.213-225. (278)

Das, S. S. and Rao, C. N. 1992, Micro-burrows from the Chamuria Formation, Madhya Pradesh, India. *Palaios*, vol.7, pp. 548-552. (145)

Das, S. S. and Sengupta, S. 1998. Sand ripples in palaeohydraulic analysis. *Indian Journal of Geology*, vol. 70, pp. 163-170. (295)

Dasgupta, H. C. and Batabyal, A. K. 1988. Genesis of the manganese and iron ore deposits of Noamundi basin, Eastern India: a dilemma (abs.). *Intl. Conf. on Metallogeny Related to Tectonics of the Proterozoic Mobile Belts*. Calcutta. (51)

Datta, Basudeb, Sarkar, Soumen and Chaudhuri, Asru K. 1999. Swaley cross-stratification in medium to coarse sandstone produced by oscillatory and combined flows: examples from the Proterozoic Kansapathar Formation, Chattisgarh Basin, M.P, India. *Sedimentary Geology*, vol. 129, pp. 51-70. (117)

Demaison, G. J. and Moore, G. T. 1980. Anoxic environments and oil source bed genesis. *American Association Petroleum Geologists Bulletin*, vol. 64, pp. 1179-1209. (176)

de V. Klein, G. 1987. Current aspects of basin analysis. *Sedimentary Geology*, vol. 50, pp. 95-118. (223, 299)

Dickinson, K. A., Berryhill, H. L. Jr. and Holmes, C. W. 1972. Criteria for recognizing ancient barrier coastlines. *In* J. K. Rigby and W. K. Hamblin (eds.), *Recognition of Ancient Sedimentary Environments*, pp. 192-214. Society of Economic Paleontologists and Mineralogists, Spl. Pub. No. 16, 340 pp. (190)

Dickinson, K. A. and Suczek, C. A. 1979. Plate tectonics and sandstone composition. *American Association Petroleum Geologists Bulletin*, vol. 63, pp. 2164-2182. (16, 220, 221, 298)

Dickinson, W. R. 1974. Plate tectonics and sedimentation. *In* W. R. Dickinson (ed.), *Tectonics and Sedimentation*, pp. 1-27. Society of Economic Paleontologists and Mineralogists, Spl. Pub.No. 22, 204 pp. (207, 213)

Dietz, R. S. 1963. Collapsing continental rises: an actualistic concept of geosynclines and mountain building. *Journal Geology*, vol. 71, pp. 314-333. (212)

Dietz, R. S. and Holden, J. C. 1974. Collapsing continental rises: actualistic concept of geosynclines-a review. *In* R. H. Dott and R. H. Shaver (eds.), *Modern and Ancient Geosynclinal Sedimentation*, pp. 14-25. Society of Economic Paleontogists and Mineralogists, Spl. Pub No. 19, 380 pp. (212)

Dobkins, J. E. Jr. and Folk, R. L. 1970. Shape development on Tahiti-Nui. *Journal Sedimentary Petrology*, vol. 40, pp. 1167-1203. (79)

Donahue, J. 1969. Genesis of oolite and pisolite grains: An energy index. *Journal of Sedimentary Petrology*, vol. 39, pp. 1399-1411. (36)

Dott, R. H. Jr. 1964. Wacke, graywacke and matrix—what approach to immature sandstone classification? *Journal Sedimentary Petrology*, vol. 34, pp. 625-632. (22)

Dott, R. H. Jr. and Bourgeois, Joanne. 1982. Hummocky stratification: significance of its variable bedding sequences. *Geological Society America Bulletin*, vol. 93, pp. 663-680. (117)

Drake, C. L., Ewing, M. and Sutton, G. H. 1959. Continental margin and geosynclines: the east coast of North America, North of Cape Hatteras. *In* L. H. Ahrens et al. (eds.), *Physics and Chemistry of the Earth*, vol. 3, pp. 110-198, Pergamon Press, London. (212)

Dunbar, C. O., Rodgers, John, 1957. *Principles of Stratigraphy*. John Wiley & Sons. Inc., New York, 356 pp. (5, 6, 32, 202, 248, 249, 250, 251, 252, 300)

Dunham, R. J. 1962. Classification of carbonate rocks according to depositional texture. *In* W. E. Ham (ed.), *Classification of Carbonate Rocks*, pp. 108-121. AAPG Memoir No. 1, American Association of Petroleum Geologists, Tulsa, Okalahoma. (38, 43, 282)

Dunham, R. J. 1969. Vadose pisolite in the Capitan reef (Permian), Mexico and Texas. pp.182-191 *In* G. M. Friedman (ed.) *Deposional Environments in Carbonate Rocks, a Symposium*. SEPM Spl. Pub. No. 14, 209 pp. (36, 37)

Dzulynski, S. and Walton, E. K. 1965. *Sedimentary Features of Flysch and Ggraywackes. Developments in Sedimentrology*, vol. 7, Elsevier, Amsterdam, 274 pp. (124)

Edwards, M. 1986. Glacial environments. *In* R. G. Reading (ed.), *Sedimentary Environments and Facies*, pp. 445-470. Blackwell Scientific Publications, Oxford, 2nd ed. 615 pp. (167)

Einstein, H. A. 1950. The bed-load function for sediment transportation in open channel flows. United States Department of Agriculture, Soil Conservation Service, Washington, D. C., Technical Bulletin No. 1026, p. 1-71. (96)

Ekdale, A. A. 1988. Pitfalls of paleobathymetric interpretations based on trace fossil assemblages. *Palaios*, vol. 3, pp. 464-472. (144)

Evans, P. 1932. Tertiary succession in Assam. *Mining & Geological Institute India*, Transactions, vol. 27 (3). (235)

Evans, P., Hayman, R. J. and Majeed, M. A. 1934. The graphical representation of heavy mineral analyses. *Quarterly Journal Geological, Mining & Metallurgical Society India*, vol. 6, pp. 27-48. (260)

Eyles, N. and Miall, A. D. 1984. Glacial facies. *In* R. G. Walker (ed.), *Facies Models*, 2nd ed. pp. 15-38. Geological Association of Canada, Toronto, Ont. 317 pp. (164, 166)

Feo-Codecido, Gustavo, 1956. Heavy-mineral techniques and their application to Venezuelan stratigraphy. *American Association Petroleum Geologists Bulletin*, vol. 40, pp. 984-1000. (16)

Femor, L. L. 1918. Preliminary note on the burning of coal seem at the outcrop. *Min. Geol. Inst. India, Transactions*, vol. 12; pp. 50-63. (31)

Fischer, A. G. and Arthur, M. A. 1977. Secular variations in the pelagic realm. *In* H. E Cook and P. Enos (eds.), *Deep Water Carbonate Environments*, pp. 19-50. Soc. Econ. Palaeont. Min. Spl. Pub No. 25. (256)

Fischer, A. G. and Judson, S. 1975 (eds.) *Petroleum and Global Tectonics*, Princeton University Press, New Jersey, 322 pp. (201)

Fisk, H. N. 1944. Geological investigation of the alluvial valley of the lower Mississippi River. Mississippi River Commission, Vicksburg, Mississippi, 78 pp. (4)

Flügel, Erik. 1982. Microfacies Analysis of Limestones (trans. K. Christenson). Springer-Verlag, Berlin, 633 pp. (35)

Folk, R. L. 1959. Practical petrographic classification of limestones. *American Association Petroleum Geologists Bulletin*, vol. 43, pp. 1-38. (37, 38, 39, 42, 43, 282)

Folk, R. L. 1968. *Petrology of Sedimentary Rocks*. Hemphill's Book Store, Austin, Texas, 170 pp. 2nd ed. (1974), 182 pp. (17, 22, 23)

Folk, R. L. and Land, L. S. 1975. Mg/Ca ratio and salinity: two controls over crystallization of dolomite. *American Association Petroleum Geologists Bulletin*, vol. 59, pp. 60-68. (43, 44)

Folk, R. L. and Ward, W. C. 1957. Brazos River bar, a study in significance of grain-size parameters. *Journal Sedimentary Petrology*, vol. 27, pp. 3-27. (68, 85)

Fouch, T. D. and Dean, W. E. 1982. Lacustrine and associated clastic depositional environments. *In* P. A. Scholle and D. Spearing (eds.) *Sandstone Depositional Environments*, pp. 87-114. American Association Petroleum Geologists, Tulsa, OK, 410 pp. (175)

Fox, W. T. 1967. Fortran IV program for vector trend analysis of directional data. *Kansas Geological Survey, Computer Contributions*, vol. 11, 36 pp. (285)

Francis, J. R. D. 1973. Experiments on the motion of solitary grains along the bed of a water-stream. *Proc. Royal Society London*. Ser. A. vol. 332, pp. 443-471. (95)

Frey, R. W. 1973. Concepts in the study of biogenic sedimentary structures. *Journal Sedimentary Petrology*, vol.43; pp. 6-19. (133)

Frey, R. W. and Seilacher, A. 1980. Uniformity in marine invertebrate ichnology. *Lethaia*, vol. 13, pp. 183-207. (141)

Friedman, G. M. 1958. Determination of sieve-size distribution from thin section data for sedimentary petrological studies. *Journal Geology*, vol. 66, 394-416. (66, 71)

Friedman, G. M. 1959. Identification of carbonate minerals by staining methods. *Journal Sedimentary Petrology*, vol. 29, pp. 87-97. (282)

Friedman, G. M. 1961. Distinction between dune, beach, and river sands from their textural characteristics. *Journal Sedimentary Petrology*, vol. 31, pp. 514-529. (68, 73)

Friedman, G. M. 1962. Comparison of moment measures for sieving and thin-section data in sedimentary petrological studies. *Journal Sedimentary Petrology*, vol. 32, pp. 15-25. (66)

Friedman, G. M. 1965. Terminology of crystallization textures and fabrics in sedimentary rocks. *Journal Sedimentary Petrology*, vol. 35, pp. 643-655. (45)

Friedman, G. M. 1967. Dynamic processes and statistical parameters compared for size frequency distribution of beach and river sands. *Journal Sedimentary Petrology*, vol. 37, pp. 327-354. (68)

Friedman, G. M. 1979. Differences in size distributions of populations of particles among sands of various origins. Pres. Address IAS. *Sedimentology*, vol. 26, pp. 3-32. (68)

Friedman, G. M. and Sanders, J. E. 1974. Positive relief bedforms on modern tidal flats that resemble molds of flutes and grooves: implications for geopetal criteria and for classification of bedforms. *Journal Sedimentary petrology*, vol. 44, pp. 181-189. (282)

Friedman, G. M. and Sanders, J. E. 1978. *Principles of Sedimentology*. John Wiley & Sons, Inc., New York, 792 pp. (6, 27, 33, 45, 278)

Friedman, G .M. and Sanders, J. .E. 2000. Comments about the relationship between new ideas and geologic terms in stratigraphy and sequence stratigraphy with suggested modifications. *American Association Petroleum Geologists Bulletin*, vol. 84, pp.1274-1280. (278)

Friend, P. F., and Sinha, R. 1993. Braiding and meandering parameters. *In* J.L. Best and C.S. Bristow (eds.) *Braided Rivers. Geological Society London*. Spl. Pub. No.75. pp. 105-111. (179)

Fritz, W. J. and Moore, J. N. 1988. *Basics of Physical Stratigraphy and Sedimentology*. John Wiley & Sons, Inc., New York, 371 pp. (27, 78, 96, 293)

Füchtbauer, Hans. 1974. Some problems of diagenesis in sandstones. *Bulletin Centre Rech*. Pan-SNPA, vol. 8, pp. 391-403. (56)

Gaal, Gabor. 1964. Pre-Cambrian flysch and molasse—tectonics and sedimentation around Rakha mines and Jaikan in Singhbhum District, Bihar, India. Report of XXII International Geological Congress, New Delhi, pt. IV, pp. 331-356. (209)

Galloway, W. E. 1975. Process framework for describing the morphologic and stratigraphic evolution of deltaic depositional systems. *In* M. L. Broussard (ed.), Deltas, pp. 87-98, Houston Geol. Society, Texas, 555 pp. (202, 203, 206)

Galloway, W. E., and Hobday, D. K. 1983. *Terrigenous Clastic Depositional Systems*. Springer-Verlag, New York, 423 pp. (6, 160, 161, 175, 180, 203, 249)

Garde, R. J. and Ranga Raju, K. G. 1977. *Mechanics of Sediment Transportation and Alluvial Stream Problems*. (A Halsted Press Book). John Wiley & Sons, Inc., New York, 483 pp. (97)

Gardner, G. H. F., Gardner, L. W. and Gregory, A. R. 1974. Formation velocity and density, the diagnostic basis for stratigraphic traps. *Geophysics*, vol. 39, pp. 770-780. (267)

Gardner, T. W. 1983. Paleohydrology and Paleomorphology of a Carboniferous, meandering fluvial sandstone. *Journal Sedimentary Petrology*, vol. 53, pp. 991-1005. (295)

Garrels, R. M., and Christ, C. L. 1965. *Solutions, Minerals and Equilibria*. Harper & Row, New York, 450 pp. (11)

Garrels, R. M., and Mackenzie, F. T. 1971. *Evolution of Sedimentary Rocks*. W. W. Norton & Company, Inc. New York, 397 pp. (4, 6, 18)

Geological Survey of India (GSI). 1986. *Analysis of Rocks, Ores and Minerals by X-ray Spectrometry*. Bulletin Series C. No. 1. Airborne Mineral Surveys & Exploration Wing. Geological Survey of India, Bangalore, 486 pp. (31, 34)

Ghosh, J. K., Mazumdar, B. S., Saha, M. R. and Sengupta, Supriya. 1986. Deposition of sand by suspension currents: experimental and theoretical studies. *Journal Sedimentary Petrology*, vol. 56, pp. 57-66. (28)

Ghosh, J. K., Mazumdar, B. S. and Sengupta, S. 1981. Methods of computation of suspended load from bed materials and flow parameters. *Sedimentology*, vol. 28, pp. 781-791. (99)

Ghosh, Parthasarathi. 2000. Estimation of channel sinuosity from paleocurrent data: a method using fractal geometry. *Journal Sedimentary Research*, vol.70, pp. 449-455. (182)

Ghosh, Santosh Kumar, 1972. Comparative study of the Lower Vindhyan Limestones around Churhat, Sidhi District, Madhya Pradesh, *Quart. Journ. Geological, Mining & Metallurgical Society India*, vol. 44, pp. 41-53. (53)

Ghosh, S. K. and Lahiri, Samhita. 1983. Morphology of penecontemporaneous interpenetrative contortions and their modifications by diastrophic movements in the Ghatsila-Galudih area, Singbhum, Eastern India. *In* S. Sinha-Roy (ed.), *Structure and Tectonics of the Precambrian Rocks of India*, pp. 144-157. Hindustan Publishing Corporation, New Delhi, 252 pp. (126)

Ghosh, S. K. and Mukhopadhyay, A. 1985. Tectonic history of the Jharia Basin—an intracratonic Gondwana basin in eastern India. *Quart. Jour. Geological, Mining & Metallurgical Society India*, vol. 57, pp. 33-58. (231, 232)

Gilbert, G. K. 1914. Transportation of debris by running water. U.*S. Geological Survey*, Professional paper, No. 86, 263 pp. (4, 99)

Gilbert, G. K. 1954. *In* H. Williams, F. J. Turner and G. K. Gilbert. *Petrography: An Introduction to the Study of Rocks in Thin Sections*. W. H. Freeman & Co., San Francisco, 406 pp. (21)

Gilluly, J., Waters, A. C. and Woodford, A. O. 1975. *Principles of Geology*. W. H. Freeman & Co., San Francisco, 2nd ed., 527 pp. (5)

Gingerich, P. D. 1969. Markov analysis of cyclic alluvial sediments. *Journal Sedimentary Petrology*, vol. 39, pp. 330-332. (259)

Ginsburg, R. N. (ed.) 1975. *Tidal Deposits*, Springer, New York, 428 pp. (4)

Glennie, K. W. and Evans, G. 1976. A reconnaissance of the Recent sediments of the Rann of Kutch, India. *Sedimentology*, vol. 23, pp. 625-647. (200)

Goldhaber, Martin. 1978. Euxinic facies. *In* R. W. Fairbridge and Joanne Bourgeois (eds.), *Encyclopedia of Sedimentology*, pp. 296-300. Dowden, Hutchinson & Ross, Inc., Stroudsburg, 901 pp. (176)

Goldich, S. S. 1938. A study in rock weathering. *Journal Geology*, vol. 46, pp. 17-58. (12)

Goodwin, A. M. 1982. Distribution and origin of Precambrian banded iron formation (Abs.), *Intl. Symposium on Archean and Early Proterozoic Geological Evolution and Metallogenesis, Salvador, Brazil*, pp. 15-16. (51)

Goodwin, P. W. and Anderson, E. J. 1985. Punctuated aggradational cycles: a general hypothesis of episodic stratigraphic accumulation. *Journal Geology*, vol. 93, pp. 515-533. (257, 258)

Goswami, D. C. 1988. Estimation of bedload transport in the Brahmaputra River, Assam. *Indian Journal Earth Sciences*, vol. 15, pp. 14-26. (97)

Grabau, A. W. 1904. On the classification of sedimentary rocks. *Amer. Geol.* vol. 33, pp. 228-247. (Cited by Pettijohn, 1975). (37, 38, 42, 43)

Graton, L. C. and Fraser, H. J. 1935. Systematic packing of spheres—with particular reference to porosity and permeability. *Jour. Geol.* vol. 43, pp. 785-909. (80)

Griffiths, J. C., 1967. *Scientific Methods in Analysis of Sediments*. McGraw-Hill, New York, 508 pp. (6)

Hand, B. M. and Bartberger, C. E., 1988. Leeside sediment fallout patterns and the stability of angular bedforms, *Journal Sedimentary Petrology*, vol. 58, pp. 33-43. (104)

Hansen, E. C., Newton, R. C. and Janardhan, A. S. 1984. Pressures, temperatures and metamorphic fluids across an unbroken amphibolite facies to granulite facies transition in southern Karnataka, India. *In* A. Kroner et al. (eds.), *Archean Geochemistry*, pp. 161-181. Springer-Verlag, Berlin-Heidelberg, 286 pp. (212)

Haq, B. U., Hardenbol, J. and Vail, P. R. 1987. Chronology of fluctuating sea levels since the Triassic (250 m.y. to present). *Science*, vol. 235, pp. 1156-1167. (269)

Harms, J. C. and Fahnestock, R. K. 1965. *Stratification, bedforms and flow phenomena (with example from the Rio Grande)*. Society of Economic Paleontologists and Mineralogists Spl. Pub. No. 12, pp. 84-115. (113)

Harms, J. C., Southard, J. B., Spearing, D. R. and Walker, R. G. 1975. *Depositional Environments as Interpreted from Primary Sedimentary Structures and Stratification Sequences, Short course* No. 2, Society of Economic Paleontologists and Mineralogists, Tulsa, OK, 161 pp. (111, 117, 293)

Harms, J. C. Southard, J. B. and Walker, R. G. 1982. *Structures and Sequences in Clastic Rocks.* Lecture Notes for Short Course No. 9, Society of Economic Paleontologists and Mineralogists, Tulsa, OK, 8-49 pp. (115, 116)

Harper, C. W. Jr. 1984. Improved methods of facies sequence analysis. *In* R. G. Walker (ed.), *Faciels Models,* 2nd ed. pp. 11-13, Geological Association Canada, Toronto, Ont. 317 pp. (259)

Hatch, F. H. and Rastall, R. H. 1913. *The Petrology of the Sedimentary Rocks.* George Allen, London, 425 pp. (6)

High, L. R. Jr. and Picard, M. D. 1974. Reliability of cross-stratification types as paleocurrent indicator. *Journal Sedimentary Petrology,* vol. 44, pp. 158-168. (284)

Hjulström, F. 1935. Studies of the morphological activity of rivers as illustrated by the river Fyris. *Bulletin of Geological Institution University of Uppasala,* vol. 25, pp. 221-346. (4, 92)

Hoffman, Paul. 1976. Stromatolite morphogenesis in Shark Bay, Western Australia, pp. 261-271. *In* M. R. Walter (ed.), *Stromatolites. Developments in Sedimentology,* vol.20. Elsevier Scientific Publishing Co., Amsterdam, 790 pp. (137, 139, 140)

Hoffman, Paul, Dewey, J. F. and Burke, Kevin. 1974. Aulacogens and their genetic relation to geosynclines, with a Proterozoic example from Great Slave Lake, Canada. *In* R. H. Dott Jr. and R. H. Shaver (eds.), *Geosynclinal Sedimentation,* pp. 38-55. Society of Economic Paleontologists and Mineralogists, Spl. Pub. No. 19, Tulsa, OK, 380 pp. (217)

Holmes, Arthur and Holmes, Doris L. 1978. Holmes Principles of Physical Geology (3rd edition). Thomas Nelson and Sons Ltd. Middlesex 730 pp. (9)

Hough, J. L. 1958. Freshwater environment of deposition of Precambrian banded iron formations. *Journal Sedimentary Petrology,* vol. 28, pp. 414-430. (51)

Howell, D. G. and McLean, Hugh. 1976. Middle Miocene paleogeography, Santa Cruz and Santa Rosa Island. *In* D. G. Howell (ed.), *Aspects of the Geologic History of California Continental Borderland,* pp. 266-293. AAPG Pacific Section Misc. Pub. 24. (155)

Howell, D. G. and Normark, W. R. 1982. Sedimentology of submarine fans. *In* P. A. Scholle and D. Spering (eds.), *Sandstone Depositional Environments,* pp. 365-404, American Association of Petroleum Geologists, Tulsa, OK, 410 pp. (155, 157, 158)

Hsü, K. J. and Siegenthaler, C. 1969. Preliminary experiments on hydrodynamic movement induced by evaporation and their bearing on the dolomite problem. *Sedimentology,* vol. 12, pp.11-25. (45)

Hubbert, M., King, 1967. Critique of the principle of uniformity, *In* C. C. Albritton (ed.), *Uniformity and simplicity,* pp. 3-33, Geological Society of America Spl. paper 89. (5)

Hunter, R. E. 1977. Basic types of stratification in eolian dunes, *Sedimentology,* vol. 24, pp. 363-387. (113, 173)

Hunter, R. E. 1981. Stratification styles in eolian sandstones: some Pennysylvanian to Jurassic examples from the Western Interior USA. *In* F. G. Ethridge and R. M. Floves (eds.), *Recent and Ancient Nonmarine Depositional Environments,* pp. 315-329. Society of Economic Paleontologists and Mineralogists, Spl. Pub. No. 31. (173)

Hutchinson, R. W., and Engels, G. G. 1970. Tectonic significance of regional geology and evaporite lithofacies in northeastern Ethiopia. *Philosophical Transactions Royal Society,* Ser. A. vol. 267, pp. 313-329. (229)

Imam, M. B. and Shaw, H. F. 1984. The diagenesis of Neogene clastic sediments from the Bengal basin, Bangladesh, *Journal Sedimentary Petrology,* vol. 55, pp. 665-671. (55)

James, H. L. 1966. Chemistry of the iron-rich sedimentary rocks. *U.S. Geological Survey Professional Paper 440-W,* 61 pp. (49)

James, N. P. 1984. Shallowing upward sequences in carbonates. *In* R. G. Walker (ed.), *Facies Models,* 2nd ed. pp. 213-228. The Geological Association of Canada, Toronto, Ont. 317 pp. (197, 199)

Jarvis, G. T. 1984. An extensional model of graben subsidence—the first stage of basin evolution. *Sedimentary Geology,* vol. 40, pp. 13-31. (217)

Jarvis, G. T. and McKenzie, D. P. 1980. Sedimentary basin formation with finite extension rates. *Earth & Planetary Science Letters,* vol. 48, pp. 42-52. (215)

Johansson, C. E. 1976. Structural studies of frictional sediments. *Geografiska Annaler,* 58A, pp. 201-300. (121, 285)

Jopling, A. V. 1961. Origin of regressive ripples explained in terms of fluid mechanic processes. *U.S. Geol. Survey Prof. Paper* 424-D, 408 p. (112)

Jopling, A. V. 1965. Hydraulic factors controlling the shape of laminae in laboratory deltas. *Journal Sedimentary Petrology*, vol. 35, 777-791. (112, 118)

Jopling, A. V. 1980. Palaeohydraulics of a tabular, cross-stratified sand in the Brampton esker, Ontario. *Sedimentary Geology*, vol. 25, pp. 169-188. (295)

Jopling, A. V. and Walker, R. G. 1968. Morphology and origin of ripple drift cross-lamination, with examples from the Pleistocene of Massachusetts. *Journal Sedimentary Petrology*, vol. 38, pp. 971-984. (106, 107)

Kaila, K. L., Reddy, P. R., Mall, D. M., Venkateswarlu, N., Krishna, V. G., Prasad, A. S. S. S. R. S., 1992. Crustal structure of the West Bengal basin, India from deep seismic sounding investigations. *Geophysical Journal International*, vol. 111, pp. 45-66. (241, 298)

Kay, Marshall, 1951. North American Geosynclines. *Geological Society America Memoir*, vol. 48, 143 pp. (4, 208, 212)

Kendall, A. C. 1984. Evaporites. *In* R. G. Walker (ed.), *Facies Models*. 2nd ed., pp. 259-296. Geol. Assoc. of Canada, Toronto, Ont., 317 pp. (47)

Kennett, J. P. 1982. *Marine Geology*, Prentice-Hall, Inc., N.J., 813 pp. (160)

King, B. C. 1970. Vulcanicity and rift tectonics in East Africa. *In* T. N. Clifford and I. G. Glass (eds.), *African Magmatism and Tectonics*, pp. 263-283, Oliver & Boyd, Edinburg. (230)

Kolb, C. R. and Van Lopik, J. R. 1966. Depositional environments of the Mississippi River delta plain, southeastern Louisiana, *In* M. L. Shirley and J. A. Regsdale (eds.), *Deltas in Their Geological Framework*, pp. 19-61. Houston Geological Society, Houston, Texas 251 pp. (203, 205, 277)

Komar, P. D. 1986. 'Breaks' in grain-size distributions and application of suspension criterion to turbidites-Reply to a discussion by J. V. Tassell. *Sedimentology*, vol. 33, pp. 437-440. (70)

Kraig, D. E., Lawrence, M. B., Moore, G. F. and Curray, J. R. 1980. Structural framework of the fore-arc basin, NW Sumatra. *Journal Geological Society London*, vol. 137, pp. 77-91. (237)

Krinsley, D. H. and Donahue, J. 1968. Environmental interpretation of sand grain-size textures by electron microscopy. *Geological Society America Bulletin*, vol. 79, pp. 743-748. (168)

Krinsley, D. H. and Doornkamp, J. C. 1973. *Atlas of Quartz Sand Grain Surface Textures*. Cambridge Earth Science Series, Cambridge Univ. Press, New York, 91 pp. (172)

Krishnan, M. S. 1964. Iron-ores in India. Intl. Geological Congress, XXII Session, New Delhi 21 pp. (52)

Krumbein, W. C. 1934. Size frequency distribution of sediments. *Journal Sedimentary Petrology*, vol. 4, pp. 65-77. (62, 63)

Krumbein, W. C. 1959. Trend surface analysis of contour-type maps with irregular control point spacing. *Journal Geophysical Research*, vol. 64, pp. 823-834. (285)

Krumbein, W. C. and Pettijohn, F. J. 1938. Manual of Sedimentary Petrography. Appleton-Century-Crofts, New York, 549 pp. (63, 66)

Krumbein, W. C. and Garrels, R. M. 1952. Origin and classification of chemical sediments in terms of pH and oxidation-reduction potentials. *Journal Geology*, vol. 60, pp. 1-33. (4, 50)

Krumbein, W. C. and Sloss, L. L. 1951. *Stratigraphy and Sedimentation*. W. H. Freeman & Co., San Francisco. 497 pp; 2nd ed. (1963), 660 pp. (6, 176, 250)

Krynine, P. D. 1948. The megascopic study and field classification of sedimentary rocks. *Journal Geology*, vol. 56, pp. 130-165. (4, 27)

Kuenen. Ph. H. 1950. *Marine Geology*. John Wiley, New York, 568 pp. (6)

Kuenen. Ph. H. 1953. Significant features of graded bedding. *American Assoc. Petrol. Geol. Bulletin*, vol. 37, pp. 1044-66. (119)

Kuenen. Ph. H. 1958. Experiments in geology. *Transactions Geological Society Glasgow*, vol. 23, pp. 1-28. (125)

Kuenen. Ph. H. 1960a. Experimental abrasion of sand grains. *International Geological Congress, 21st Session*. Norden, pt. 10, pp. 50-53. (75)

Kuenen. Ph. H. 1960b. Experimental abrasion. 4. Eolian action. *Journal Geology*, vol. 68, pp. 427-449. (75)

Kuenen. Ph. H. 1964. Experimental abrasion. 6. Surf action. *Sedimentology*, vol. 3, pp. 29-43. (75)

Kuenen. Ph. H. 1966. Matrix of turbidites: experimental approach. *Sedimentology*, vol. 7, pp. 267-297. (28)

Kuenen. Ph. H. and Migliorini, C. I. 1950. Turbidity currents as a cause of graded bedding. *Journal Geology*, vol. 58, pp. 91-127. (4, 118)

Kuenen. Ph. H. and Sengupta. S, 1970. Experimental marine suspension currents, competency and capacity. *Geologie en Mijnbouw*, vol. 49 (2), pp. 89-118. (28, 99)

Kukal, Z. 1970. *Geology of Recent Sediments*. Czech. Academy of Science, Prague, 490 pp. (6)

Kumar, R. K. and Nanda, A. C. 1989. Sedimentology of the Middle Siwalik sub-group of Mohand area, Dehra Dun Valley, India. *Jour. Geol. Soc. India*, vol. 34, pp. 597-616. (211)

Kumar, N. and Sanders, J. E. 1974. Inlet sequence: a vertical succession of sedimentary structures and textures created by the lateral migration of tidal inlets. *Sedimentology*, vol. 21, pp. 491-532. (193)

Kumar, S., and Singh, I. B. 1978. Sedimentological study of Gomti River sediments, Uttar Pradesh, India. Example of a river in alluvial plain. *Senckenbergiana Marit*. vol.10, pp.145-211. (190)

Kutty, P. S. N. and Agarwal, V. K. 1993. Application of seismic stratigraphy in oil exploration with examples from western offshore. *Journal Geological Society India*. vol. 41, pp.319-320. (278)

Lambert, A. and Hsü, K. J. 1979. Non-annual cycles of varve-like sedimentation in Walensee, Switzerland. *Sedimentology*, vol. 26, pp. 453-461. (166)

La Pichon, X., Francheteau, J. and Bonnin, J. 1973. *Plate* Tectonics. Elsevier, Amsterdam, 300 pp. (213)

Larsonneur, C. 1975. Tidal deposits, Mont Saint-Michel Bay, France. *In* R. N. Ginsburg (ed.), *Tidal Deposits*, p. 21-30, Springer, New York. (196)

Laursen, E. M. 1958. The total sediment load of streams. *Journal Hydraulics Division*. American Society of Civil Engineers (ASCE), vol. 54, (HY 1). Proceedings. Paper 1530, pp. 1-36. (97)

Lee, Y. I. and Klein. G. de V. 1986. Diagenesis of sandstones in the bac-arc basins of the Western Pacific Ocean. *Sedimentology*, vol. 33, pp. 651-675. (297)

Leeder, Mike 1999. *Sedimentology and Sedimentary Basins*. Blackwell Science, Oxford. 592 pp. (6, 21, 91, 96, 104)

Leggett, J.K. 1985. Deep-sea pelagic sediments and palao-oceanograpy: a review of recent progress. *In* Sedimentology: Recent Developments and Applied Aspects (eds. P.J. Brenchley and B.PJ. williams). Geological Society London Special Publication, 18, pp. 95-121. (159)

Leopold, L.B. and Langbein, W.B. 1966. River meanders. *Scientific American*. vol. 214, pp. 60-70. (180)

Leopold, L. B. and Wolman, M. G. 1957. River channel patterns: braided, meandering and straight. *United States Geological Survey Professional Paper*. No. 282 B, pp. 39-85. (184)

Leopold, L. B., Wolman, M. G. and Miller, J. P. 1964. *Fluvial Processes in Geomorphology*. Freeman, San Francisco, 522 pp. (4, 179, 295, 297)

Levorsen, A. I. 1958. *Geology of Petroleum*. W. H. Freeman & Co., San Francisco. 703 pp. (45, 80)

Lewis, K. B. 1980. Quaternary sedimentation on the Hikurangi oblique-subduction transform margin, New Zealand. *In* P. F. Ballance and R. G. Reading (ed.), *Sedimentation in Oblique-Slip Mobile Zones*, pp. 171-189. International Assoc. Sedimentologists, Spl. Pub. No, 4. (218)

Le Roux J.P. 1992. Determining the channel sinuosity of ancient fluvial systems from paleocurrent data. *Journal Sedimentary Petrology*, vol. 62, pp. 283-291. (182)

Lindholm, Roy, C. 1987. *A Practical Approach to Sedimentology*, Allen & Unwin, Winchester, Mass. (First Indian reprint by CBS Publishers and Distributors, New Delhi, 1991). 276 pp. (6, 141, 143)

Lindsay, J. F., Holliday, D. W. and Hulbert, A. G. 1991. Sequence stratigraphy and the evolution of the Ganges-Brahmaputra delta complex. *American Association Petroleum Geologists Bulletin*, vol.75, pp. 1235-1254. (278)

Ljunggren, Pontus and Sundborg, Åke. 1968. Some aspects of fluvial sediments and fluvial morphology. II. A study of some heavy mineral deposits in the valley of the river Alv. *Geografiska Annaler*, vol. 50, pp. 121-135. (97, 98)

Logan, B. W., Rezak, R. and Ginsburg, R.N. 1964. Classification and environmental significance of algal stromatolites. *Journal of Geology*, vol. 72, pp. 68-83. (135-136, 139)

Longman, M. W. 1981. Carbonate diagenesis as a control on stratigraphic traps (with examples from Williston basin). *A.A.P.G. Education Course Note Series no.* 21, 159 pp. (45)

Lowe, D. R. 1975. Water escape structures in coarse-grained sediments. *Sedimentology*, vol. 22, pp. 157-204. (127)

Mader, Detlef. 1985. Aspects of Fluvial Sedimentation in the Lower Triassic Buntsandstein of Europe. *Lecture Notes in Earth Sciences*, vol. 4, Springer, Berlin. (188)

Mahoney, J. J. MacDougall, J. D., Lugmair, G. W. and Gopalan, K. 1983. Kerguelen hotspot source for Rajmahal Traps and Ninetyeast Ridge? *Nature*, vol. 303, pp. 385-389. (244)

Malur, M.N. and Nagendra, R. 1988. Microstylolites in Late Precambrian carbonate rocks, Karnataka, South India. *Journal Geological Society of India*, vol.32, pp. 430-432. (131)

Martinsson, A. 1970. Toponomy of trace fossils. *In* T. P. Crimes, and J.G. Harper (eds.) Trace Fossils. *Geological Journal spl. issue* 3, pp. 323-330. (141)

Mather, K. F. and Mason, S. L. 1939. *A Source Book in Geology 1400-1900*. Harvard University Press. Cambridge, Mass., 2nd printing (1967), 702 pp. (5)

Maulik, P. K., and Chaudhuri, A. K. 1983. Trace fossils from continental Triassic red beds of the Gondwana sequence, Pranhita-Godavari Valley, South India. *Palaeogeography, Palaeoclimatology, Palaeoecology*, vol.41, 17-34. (145)

Mazumder, Rajat. 2000. Turbulence-particle interactions and their implications for sediment transport and bedform mechanics under unidirectional current: some recent developments. *Earth Science Reviews*, vol. 50, pp. 113-124. (91)

McBride, E. F. 1974. Significance of colour in red, green, purple, olive, brown and gray beds of DiFunta Group, North-eastern Mexico. *Journal Sedimentary Petrology*, vol. 44, pp. 760-773. (31)

McCave, I. N. and Syvitski, J. P. M. 1991. Principles and methods of geological particle size analysis. *In* J. P. M. Syvitski (ed.), *Principles, Methods, and Application of Particle Size Analysis*, pp. 3-21, Cambridge University Press, Cambridge, 368 pp. (64)

McIntyre, D. B. 1963. James Hutton and the philosophy of geology, pp. 1-23. *In* C. C. Albritton (ed.) *The Fabric of Geology*. Addison-Wesley Publishing Co. Reading. 372 pp. (5)

McKee, E. D. 1945. Cambrian history of the Grand Canyon region. Part 1. *Stratigraphy and Ecology of the Grand Canyon Cambrian*. Carnegie Institute Washington Publication 563, pp. 3-168. (251)

McKee, E. D. 1949. Facies changes in the Colorado Plateau. *Geological Society America Memoir*, vol. 39, pp. 35-48. (251)

McKee, E. D. 1954. Stratigraphy and history of the Moenkopi Formation of Triassic age. *Geological Society America Memoir*, vol. 61, 133 pp. (188)

McKee, E. D. 1979. Introduction to a study of global sand seas (pp. 1-19), and sedimentary structures in dunes (pp. 83-134). In E. D. McKee (ed.), *A study of Global Sand Seas. U. S. Geological Survey Professional Paper* 1052, pp. 83-134. (4, 169)

McKee, E. D. and Weir, G. W. 1953. Terminology for stratification and cross-stratification in sedimentary rocks. *Geological Society America Bulletin*, vol. 64, pp. 381-390. (109, 117)

McKee, E. D. and Goldberg, M. 1969. Experiments on formation of contorted structures in mud. *Geological Society America Bulletin*, vol. 80, pp. 231-244. (126, 127)

McKenzie, D. P. 1978. Some remarks on the development of sedimentary basins. *Earth and Planetary Science Letters*, vol. 40, pp. 25-32. (215)

McKenzie, D. P. and Parker, R. L. 1967. The North Pacific: an example of tectonics on a sphere. *Nature*, vol. 216, pp. 1276-1280. (213)

McLane, Michael 1995. *Sedimentology*. Oxford University Press, Oxford. 423 pp. (6, 33, 35, 37, 45)

Mehrotra, M. N. Srivastava, R. A. K. and Sinhamahaptra, P. K. 1975. Physico-chemical behaviour of the carbonate rocks of western Son Valley region, India. *Journal Thermal Analysis*, vol. 7, pp. 667-674. (34)

Miall, A. D. 1973. Markov chain analysis applied to ancient alluvial plain succession. *Sedimentology*, vol. 20, pp. 347-364. (259)

Miall, A. D. 1977. A review of the braided river depositional environment. *Earth Science Reviews*, vol. 13, pp. 1-62. (184)

Miall, A. D. (ed.) 1978. *Fluvial Sedimentology*. Canadian Society of Petroleum Geologists. Calgary, 859 pp. (5)

Miall, A. D. 1978a. Lithofacies types and vertical profile models in braided river deposits : a summary. *In* A. D. Miall (ed.) *Fluvial Sedimentology*, pp. 597-604. Canadian Society of Petroleum Geologists. Calgary, 859 pp. (184)

Miall, A. D. 1984. *Principles of Sedimentary Basin Analysis.* 2nd ed. (1989) Springer, New York, 490 pp. (6, 223)

Miall, A. D. 1984a. Deltas. *In* R. G. Walker (ed.), *Facies Models,* 2nd ed., pp. 105-118. The Geological Society of Canada, Torqnto, Ont., 317 pp. (202)

Middleton, G. V. 1973. Johannes Walther's Law of Correlation of Facies. *Geological Society America Bulletin,* vol. 84, pp. 979-988. (5)

Middleton, G. V. 1978. Sedimentology—History. *In* F. W. Fairbridge and Joanne Bourgeois (eds.), *The Encyclopedia of Sedimentology,* pp. 707-712. Dowden, Hutchinson & Ross. Stroudsburg, 901 pp. (5)

Middleton, G. V. 1978a. Facies. *In* F. W. Fairbridge and Joanne Bourgeois (eds.), *The Encyclopedia of Sedimentology,* pp. 323-325. Dowden, Hutchinson & Ross. Stroudsburg, 901 pp. (147)

Middleton, G. V. and Hampton, M. A. 1976. Subaqueous sediment transport and deposition by sediment gravity flows. *In* D. J. Stanley and D. J. P. Swift (eds.), *Sediment Transport and Environmental Management,* pp. 197- 218. John Wiley, New York. (156)

Miller, M. C., McCave, I. N. and Komar, P. D. 1977. Threshold of sediment motion under unidirectional currents. *Sedimentology,* vol. 24, pp. 507-527. (93)

Miller, R. L. 1956. Trend surfaces: their application to analysis and description of environments of sedimentation. *Journal of Geology,* vol. 64, pp. 425-446. (285)

Milliman, J. D. 1974. *Marine carbonates.* Springer-Verlag, Berlin. 375 pp. (6)

Milner, H. B. 1922. *An Introduction to Sedimentary Petrography,* Murby, London, 125 pp. (6)

Milner, H. B. 1962. *Sedimentary Petrography* (in 2 vols.). Macmillan, New York, 643 pp. and 715 pp. (6)

Mitchum, R. M. 1977. Glossary of terms used in seismic stratigraphy. *In* C. E. Payton (ed.), *Seismic Stratigraphy,* pp. 205-212. AAPG Memoir 26, 516 pp. (271)

Mitchum, R. M. Jr., Vail, P. R. and Thompson, III, S. 1977a. Seismic stratigraphy and global changes in sea level Part 2: The depositional sequence as a basic unit for stratigraphic analysis. *In* C. E. Payton (ed.), *Seismic Stratigraphy,* pp. 53-62. AAPG *Memoir.* 26, 516 pp. (271)

Moiola, R. J. and Weiser, D. 1968. Textural parameters: an evaluation. *Journal Sedimentary Petrology,* vol. 38, pp. 45-53. (68)

Moore, D. G. Curray, J. R., Raitt, R. W. and Emmel, F. J. 1974. Stratigraphic seismic section correlations and implications to Bengal Fan history. *In* C. G. Von der Borch and J. G. Sclater (eds.), *Initial Reports of the Deep Sea Drilling Project,* vol. 22, U.S. Govt. Printing Office. (236)

Mukherjee, K. K., Das, S. and Chakrabarti, A. 1987. Common physical sedimentary structures in a beach-related open-sea siliciclastic tropical tidal flat at Chandipur, Orissa. India, and evaluation of the weather conditions through discriminant analysis. Senckenbergiana Marit, vol. 19, pp. 261-293. (200)

Mukhopadhyay, A. and Chanda, S. K. 1972. Silica diagenesis in the banded hematite jasper and bedded chert associated with the Iron Ore Group of Jamad-Koira Valley, Orissa, India, *Sedimentary Geology,* vol. 8, pp. 113-135. (58)

Mutti, E. and Ricci-Lucchi, F. 1975. Examples of turbidite facies associations from selected formations of the northern Apennines. 9th International Sedimentological Congress, Nice, France, Guidebook to Field Trip A-11, 120 pp. (158)

Naha, K. 1961. Precambrian sedimentation around Ghatsila in east Singhbhum, eastern India. *Proceedings National Institute Sciences India,* vol. 27, pp. 361-372. (209)

Naidu, A. S. 1966. Lithological and chemical facies changes in the Recent deltaic sediments of the Godavari River, India. *In* M. L. Shirley and J. A. Ragsdale (eds.), *Deltas in Their Geological Framework,* pp. 126-157. Houston Geological Society, Houston, Texas, 251 pp. (193)

Narain, Hari and Kaila, K. L. 1982. Inferences about the Vindhyan basin from geophysical data. *In* K. S. Valdiya et al. (eds.), *Geology of Vindyanchal,* pp. 179-192, Hindustan Publishing Corporation (India), Delhi, 231 pp. (227, 228)

National Geographic Society. 1967. Map of the Indian Ocean Floor. *National Geographic Magazine,* October, 1967. (162)

Nelson, C. M. 1985. Facies in stratigraphy: from 'terrains' to 'terranes'. Journal *Geological Education,* vol. 33, pp. 175-187. (5)

Nemec, W. 1988. The shape of the rose. *Sedimentary Geology*, vol. 59, pp. 149-152. (284)

Newton, R. C. 1988. Nature and origin of fluids in granulite facies metamorphism. *Journal Geological Society India*, vol. 31, pp. 103-105. (212)

Newton, R. S. 1968. Internal structure of wave-formed ripple marks in the near shore zone. *Sedimentology*, vol. 11, pp. 275-292. (115)

Nichols, Gary. 1999. *Sedimentology and Stratigraphy*. Blackwell Science Ltd., Oxford. 355 pp. (6, 188, 259, 270, 274)

Normark, W. R. 1978. Fan valleys, channels and depositional lobes on modern submarine fans: characters for recognition of sandy turbidite environments. *American Association Petroleum Geologists Bulletin*, vol. 62, pp. 912-931. (158)

North American Commission on Stratigraphic Nomenclature (NACSN). 1983. North American Stratigraphic Code. *American Association Petroleum Geologists Bulletin*, vol. 67, pp. 841-875. (246, 247, 253)

Novak, I. D. 1973. Predicting coarse sediment transport : the Hjulström curve revisited. *In* M. Morisawa (ed.). *Fluvial Geomorphology*, pp. 13-25. Allen and Unwin, London. 164 pp. (92)

Okada, H. 1966. Non-Greywacke turbidite sandstones in the Welsh geosyncline. *Sedimentology*, vol. 7, pp. 211-232. (27)

Okada, H. 1971. Classification of sandstone : analysis and proposal. *Journal Geology*, vol. 79, pp. 509-525. (21)

Parkash, B., Awasthi, A. K. and Gohain, K. 1983. Lithofacies of the Markanda terminal fan, Kurukshetra District, Haryana, India. *International Association of Sedimentologists, Spl. Pub*. No. 6, pp. 337-344. (185, 186)

Passega, R. 1957. Texture as a characteristic of clastic deposition. *American Association Petroleum Geologists Bulletin*, vol. 41, pp. 1952-1984. (68)

Passega, R. 1964. Grain-size representation by CM patterns as a geological tool. *Journal Sedimentary Petrology*, vol. 34, pp. 830-847. (68)

Paul, D. D. and Lian, H. M. 1975. Offshore Tertiary basins of south-east Asia, Bay of Bengal to South China Sea. *Proceedings IX World Petroleum Congress*, vol. 3, pp. 107-121. (242)

Petters, S. W. 1978. Stratigraphic evolution of the Benue Trough and its implications for the Upper Cretaceous Paleogeography of West Africa. *Journal Geology*, vol. 86, pp. 311-322. (233)

Pettijohn, F. J. 1949. *Sedimentary Rocks*. Harper & Row, New York, 526 pp.; 2nd ed. (1957), 718 pp.; 3rd ed. (1975), 628 pp. (6, 13, 15, 16, 18, 20, 21, 31, 34, 39, 42, 43, 74, 77, 80, 130, 131, 144, 163, 212, 249, 260)

Pettijohn, F. J. 1962. Paleocurrent and paleogeography. *American Association Petroleum Geologists Bulletin*, vol. 46, pp. 1468-1493. (4, 6, 283, 285, 288)

Pettijohn, F. J., Potter, P. E. and Siever, R. 1973. *Sand and Sandstone*. Springer-Verlag, New York, 618 pp.; 2nd ed. (1987), 553 pp. (6)

Picard, M. D. and High, L. R. Jr. 1972. Criteria for recognizing lacustrine rocks. *In* J. K. Rigby and W. K. Hamblin (eds.), *Recognition of Ancient Sedimentary Environment*, pp. 108-145. Society of Economic Paleontologists and Mineralogists, Spl. Pub. No. 16, 340 pp. (178)

Pitman III, W. C. 1978. Relationship between eustacy and stratigraphic sequences of passive margins. *Geological Society America Bulletin*, vol. 89, pp. 1389-1403. (255)

Plummer, P.S. and Gostin, V.A. 1981. Shrinkage cracks: Desiccation or synaeresis? *Journal Sedimentary Petrology*, vol.51, pp.1147-1156. (119)

Posamentier, H. W., Jervey, M. T. and Vail, P. R. 1988. Eustatic controls on clastic deposition. In C. K. Wilgans, B. S. Hastings and G. G. St. C. Kendall (eds.), *Sea Level Research—an Integrated Approach*, pp. 109-154. Society of Economic Paleontologists and Mineralogists, Spl. Pub. No. 42. (275)

Posamentier, H. W., Allen, G. P. and James, D. P. 1992. High resolution sequence stratigraphy—the east Coulee delta, Alberta. *Journal Sedimentary Petrology*, vol. 62, pp.310-317. (278)

Potter, P. E. 1978. Significance and origin of big rivers, *Journal Geology*, vol. 86, pp. 13-33. (15)

Potter, P. E. 1978a. Petrology and chemistry of modern big river sands. *Journal Geology*, vol. 86, pp. 423-449. (15)

Potter, P. E. and Pettijohn, F. J. 1963. *Paleocurrent and Basin Analysis*. Springer-Verlag, New York, 296 pp., 2nd ed. (1977), 425 pp. (6, 284, 287)

Potter, P. E., Maynard, J. B. and Prior, W. A. 1980. *Sedimentology of Shale*. Springer-Verlag, New York, 306 pp. (30, 117)

Powers, M. C. 1953. A new roundness scale for sedimentary particles. *Journal Sedimentary Petrology*, vol. 23, pp. 117-119. (74, 75)

Purkait, Barendra, 2000. Morphology and growth patterns of the Usri River point bars, Bihar, India. *International Journal Sediment Research*, vol.15, pp. 445-457. (181)

Purkait, Barendra. 2002. Patterns of grain-size distribution in some point bars of the Usri River, India. *Journal Sedimentary Research*, vol.72, pp. 367-375. (182)

Raaben, M. E. 1969. Columnar stromatolites and late Precmbrian stratigraphy. *American Journal of Science*, vol.267, pp.1-18. (136, 138)

Raha, P. K. 1972. Note on the new firid of Riphean stromatolites from the Jammu Limestone north of Raisi, Udhampur District, J&K State, India. *Indian Minerals*, Vol. 26(3), pp. 68-69. (139)

Raha, P. K., and Sastry, M. V. A. 1973. Stromatolites from the Jammu Limestone, Udhampur District, J & K State and their stratigraphic and palaeogeographic significance. *Himalayan Geology*, vol.3, pp.135-147. (139)

Raha, P. K., and Sastry, M. V. A. 1976. Stratigraphy of the Lesser Himalayan carbonate formations with special reference to stromatolites. *Geological Survey of India Miscellaneous Publications*, vol.41 (1), pp. 123-142. (139)

Raha, P. K., Chandy, K. C., and Balasubrahmanyan, M. N. 1978. Geochronology of the Jammu Limestone, Udhampur District, Jammu, India. *Journal Geological Society India*, vol.19, pp. 221-223. (139)

Rai, K. L., Sarkar, S. N. and Paul, P. R. 1980. Primary depositional and diagenetic features in the Banded Iron Formation and associated iron-ore deposits of Noamundi, Singhbhum district, Bihar, India. *Mineral. Depostia* (Berl.), vol. 15, pp. 189-200. (51)

Ramsay, A. T. S. 1977. Sedimentological clues, to palaeo-oceanography. *In* Oceanic Micropalacontology (*ed.* A.T.S. Ramsay) pp. 1371-1453. Academic Press, London. (159)

Rao, J. S. and Sengupta, Supriya, 1970. An optimum hierarchical samplig procedure for cross-bedding data. *Journal Geology*, vol. 78, pp. 533-544. (284, 303, 304, 305)

Rao, J. S. and Sengupta, Supriya, 1972. Mathematical techniques for paleocurrent analysis: treatment of directional data. *Mathematical Geology*, vol. 4, pp. 235-248. (284, 303)

Rao, M. B. Ramchandra and Sengupta, S. N. 1964. Hidden ridges in the Indo-Gangetic Plains. *Advancing Frontiers in Geology & Geophysics* (Krishnan Volume). Indian Geophysical Union, Hyderabad, pp. 147-158. (241)

Reading, H. G. (ed.), 1978. *Sedimentary Environments and Facies*. Elsevier, New York, 569 pp.; 2nd ed. (1986), 615 pp. (6, 227, 231, 236, 248, 256, 259)

Reading, H. G. 1982. Sedimentary basins and global tectonics. *Proceedings Geological Association*, 93 (4), pp. 321-350. (215, 218, 219, 227, 229, 230, 233, 236, 237)

Reading, H.G. (ed.). 1996. *Sedimentary Environments: Processes, Facies and Stratigraphy*. 3rd ed. Blackwell Science, Oxford, 688 pp. (6)

Reddy, P. R., Koteswara Rao, P., Venkateswarlu, N. and Prasad, A. S. S. R. S., 1995. Delineation of faults, using deep seismic sounding refraction data, West Bengal Basin. *Journal Geological Society India*, vol. 45, pp. 97-106. (241)

Reineck, H. E. and Singh, I. B. 1973. *Depositional Sedimentary Environments*. Springer-Verlag, New York, 439 pp. 2nd ed. (1980), 549 pp. (6, 118, 124)

Reineck, H. E. and Wunderlich, F. 1968. Classification and origin of flaser and lenticular bedding. *Sedimentology*, vol. 11, pp. 99-104. (108, 113)

Reinson, G. E. 1984. Barrier island and associated strand-plain systems. *In* R. G. Walker (ed.), *Facies Models*, 2nd ed., 119-140. The Geological Society of Canada, Toronto, Ont., 317 pp. (190, 191, 192, 195)

Reyer, T. A. 1983. Transgressive-regressive cycles and the occurrence of coal in some Upper Cretaceous strata of Utah. *Geology*, vol. 11, pp. 207-210. (257)

Ricci-Lucchi, F. 1975. Depositional cycles in two turbidite formations of Northern Apennines (Italy), *Journal Sedimentary Petrology*, vol. 45, pp. 3-43. (158)

Robinson, P. L., 1964. Climates ancient and modern. *In* C. R. Rao and P. Pant (eds.) *Contributions to Statistics*, Pergamon Press, Oxford, and Statistical Publishing Society, Calcutta, pp. 391-410. (132)

Rodgers, John 1959. The meaning of correlation. *American Journal of Science*, vol. 257, pp. 684-691. (259)

Rodolfo, K. S. 1969. Bathymmetry and marine geology of the Andaman Basins and tectonic implications for southeast Asia. *Geological Society America Bulletin*, vol. 80, pp. 1203-30. (236)

Rouse, Hunter, 1938. *Fluid Mechanics for Hydraulic Engineers*. McGraw Hill Book Co. (Dover edition printed in 1961). Dover Publications, Inc., New York, 422 pp. (94, 97)

Roybarman, A. 1983. Geology and hydrocarbon prospects of West Bengal. *In* L. L. Bhandari et al. (eds.), Petroliferous Basins of India, Part I, pp. 51-56, *Petroleum Asia Journal*, V-VI (4), 189 pp. (244)

Roychoudhury, S. C. and Deshpande, S. V. 1982. Regional distribution of carbonate facies, Bombay offshore region, India. *American Association Petroleum Geologists Bulletin*, vol. 66, pp. 1483-1496. (58)

Rubin, D. M. and McCulloch, D. S. 1980. Single and superimposed bedforms: a synthesis of San Francisco Bay and flume observations. *Sedimentary Geology*, vol. 26, pp. 207-231. (101)

Rudra, D. K. and Maulik, P. K. 1987. Stromatolites from Jurassic freshwater limestone, India. *Mesozoic Research*, vol. 1(3), pp. 135-146. (140)

Rust, B. R. 1978. A classification of alluvial channel system. *In* A. D. Miall (ed.), *Fluvial Sedimentology*, pp. 187-198. *Canadian Society of Petroleum Geologists Memoir 5*, Calgary, Alberta, 859 pp. (178)

Rust, B. R. 1978a. Depositional models for braided alluvium. *In* A. D. Miall (ed.), *Fluvial Sedimentology*, pp. 605-625. *Canadian Society of Petroleum Geologists*, Calgary, 859 pp. (184)

Rust, B. R. and Koster, E. M. 1984. Coarse alluvial deposits. *In* R. G. Walker (ed.), *Facies Models*, 2nd ed., pp. 53-70. Geological Association of Canada, Toronto, Ont., 317 pp. (184, 185)

Saha, Dilip and Chakraborty, Tapan, 1990. TILTVEC: A Fortran-77 program for the tilt correction of paleocurrent data with resolution of incongruities. *Computers and Geosciences*, 16, 8, pp. 1193-1207. (284)

Saha, Dilip and Ghosh, Gautam, 1987. Tectonic setting of Proterozoic sediments around Somanpalli, Godavari Valley. *Journal Indian Association Sedimentologists*, vol. 7, pp. 29-45. (53)

Sahu, B. K. 1964. Transformation of weight frequency and number frequency data in size distribution studies of clastic sediments. *Journal Sedimentary Petrology*, vol. 34, pp. 768-773. (66)

Sahu, B. K. 1964a. Depositional mechanisms from the size analysis of clastic sediments. *Journal Sedimentary Petrology*, vol. 34, pp. 73-83. (68)

Sarkar, Bhagabati, 1984. Microfossils from the Banded Iron Formation from the Noamundi basin, Eastern India. *Quarterly Journal Geological, Mining & Metallurgical Society India*, vol. 56, pp. 41-46. (51)

Sarkar, Soumen, 1988. Petrology of caliche-derived peloidal calcirudite/ calcarenite in the Late Triassic Maleri Formation of the Pranhita-Godavari Valley, South India. *Sedimentary Geology*, vol. 55, pp. 263-282. (132)

Sarkar, Soumen and Chaudhuri, A. K. 1992. Trace fossils in Middle to Late Triassic fluvial redbeds, Pranhita-Godavari Valley, South India. *Ichnos*, vol.2, pp. 7-19. (145)

Savrda, C. E. and Bottjer, D. J. 1986. Trace-fossil model for reconstruction of paleo-oxygenation in bottom waters. *Geology*, vol.14, pp. 3-6. (144)

Scheidegger, A. E. and Potter, P. E. 1968. Textural studies of graded bedding. *Sedimentology*, vol. 11, pp. 163-170. (53)

Schenck, H. G. and Muller, S. W. 1941. Stratigraphic terminology. *Geological Society America Bulletin*, vol. 52, pp. 1419-1426. (246)

Schlumberger Ltd. 1970. *Fundamentals of Dipmeter Interpretation*. Schulmberger, Houston, Texas, 145 pp. (293)

Schreiber, B. C., Friedman, G. M., Decima, A. and Schreiber, E. 1976. Depositional environments of Upper Miocene (Messinian) evaporite deposits of the Sicilian basin. *Sedimentology*, vol. 23, pp. 729-760. (47)

Schumm, S. A. 1972. Fluvial Paleochannels. *In* J. K. Rigby and W. K. Hamblin (eds.), *Recognition of Ancient Sedimentary Environments*, pp. 98-107, Society of Economic Paleontologists and Mineralogists, Spl. Pub. No. 16, 340 pp. (295)

Schumm, S. A. 1973. Geomorphic thresholds and complex response of drainage systems. *In* M. Morisawa (ed.), *Fluvial Geomorphology*, State University of New York, Binghampton, Publications in Geomorphology, vol. 4, pp. 299-310. (184)

Schumm, S. A. 1977. *The Fluvial System*. Wiley Interscience, New York, 335 pp. (178, 184)

Seilacher, A. 1962. Paleontological studies on turbidite sedimentation and erosion. *Journal of Geology*, vol.70, pp. 227-234. (144)

Seilacher, A. 1964. Sedimentological classification and nomenclature of trace fossils. *Sedimentology*, vol.3, pp. 253-256. (141, 142)

Seilacher, A. 1967. Bathymetry of trace fossils. *Marine Geology*, vol.5, pp. 413-428. (142, 143)

Seilacher, A., Bose, P.K., and Pflüger, F. 1998. Triploblastic animals more than 1 billion years ago: trace fossil evidence from India. *Science*, vol.282, pp. 80-83. (146)

Selley, R. C. 1968. A classification of paleocurrent models. *Journal Geology*, vol. 76, pp. 99-110. (291, 292)

Selley, R. C. 1976. *An Introduction to Sedimentolgy*. Academic Press London, 408 pp. (6)

Sellwood, B. W. 1986. Shallow-marine carbonate environments. *In* R. G. Reading (ed.), *Sedimentary Environments and Facies*, 2nd ed. pp. 283-342. Blackwell Scientific Publications, Oxford, 615 pp. (57)

Simpson, S. 1975. Classification of trace fossils, pp. 39-54. *In* R. W. Frey,. (ed.), *The Study of Trace Fossils*, Springer-Verlag, New York, 562 pp. (141)

Srivastava, B. N., Rana, M. S. and Verma, N. K. 1983. Geology and hydrocarbon prospects of the Vindhyan basin. *In* L. L. Bhandari et al. (eds.) Petroliferous Basins of India, vol. I, pp. 179-189. *Petroleum Asia Journal*, V-VI, 189 pp. (227)

Sen, D. P. 1991. Sedimentation patterns of the Talchir Group in the Giridih Gondwana basin, India: A case of multiple glacial advance and retreat. *Palaeogeography, Palaeoclimatology, Palaeoecology*, vol. 86, pp. 339-352. (19)

Sen, D. P., Bhattacharyya, S. C. and Ray, S. K. 1969. Turbidite deposits around the Simla hills. A study from the Simla series, Himachal Pradesh (India). *Sedimentary Geology*, vol. 3, pp. 317-329. (209)

Sengupta, Supriya. 1957. Petrology of para-lavas of eastern part of Jharia coalfield. *Quarterly Journal Geological, Mining & Metallurgical Society India*, vol. 29, pp. 79-101. (31)

Sengupta, Supriya. 1966. Studies on orientation and imbrication of pebbles with respect to cross-stratification. *Journal Sedimentary Petrology*, vol. 36, pp. 363-369. (121, 122, 188, 285)

Sengupta, Supriya. 1966a. Geological and geophysical studies in western part of the Bengal basin, India. *American Association Petroleum Geologists Bulletin*, vol. 50, pp. 1001-1017. (241, 243, 244, 268, 298)

Sengupta, Supriya. 1970. Gondwana sedimentation around Bheemaram (Bhimaram), Pranhita-Godavari Valley, India. *Journal Sedimentary Petrology*, vol. 40, pp. 140-170. (188, 189, 251, 289, 290)

Sengupta, Supriya. 1974. Fluviatile cross-stratifications: ancient, recent and experimental. Q. J. Geological, Mining & Metallurgical Society of India, Golden Jubilee Volume (vol. 46), pp. 87-99. (113, 114)

Sengupta, Supriya. 1975. Size-sorting during suspension transportation - log normality and other characteristics. *Sedimentology*, vol. 22, pp. 257-273. (70, 99)

Sengupta, Supriya. 1979. Grain-size distribution of suspended load in relation to bed materials and flow velocity. *Sedimentology*, vol. 26, pp. 63-82. (70, 99)

Sengupta, S. 2003. Gondwana sedimentation in the Pranhita-Godavari Valley: a review. *Journal Asian Earth Sciences*, vol.21, pp. 633-642. (189, 190, 251)

Sengupta, S., Bose, D., Siva Prasad, K. and Das, S. S. 1988. Karharbari and Barakar Sedimentation in the Giridih basin. *Indian Journal Geology*, vol. 60, pp. 35-56. (286, 296)

Sengupta, Supriya, Ghosh, J. K. and Mazumdar, B. S. 1991. Experimental-theoretical approach to interpretation of grain-size frequency distributions. *In* J. P. M. Syvitski (ed.), *Principles, Methods and Application of Particle Size Analysis*, pp. 264-280. Cambridge University Press, Cambridge, 368 pp. (71, 182)

Sengupta, S., Das, S. S., and Gupta, A. S. 2005. Current crescent as indicator of flow velocity. pp. 76-77 *In* Bora, S. N. (ed) *Some Aspects of Environmental Fluid Mechanics*. Proceedings ICEFM '05, Guwahati. (121)

Sengupta, S., Das, S. S., and Maji, A. K. 1999. Sediment transportation and sorting processes in streams. *Proceedings Indian National Science Academy (PINSA)*, vol. 65-A, pp. 167-206. (71, 99, 182)

Sengupta, Supriya and Rao, J. S. 1966. Statistical analysis of cross-bedding azimuths from the Kamathi Formation around Bheemaram, Pranhita-Godavari Valley, Sankhyā, *Indian Journal Statistics*, vol. 28-B, pp. 165-174. (189)

Sengupta, Supriya and Veenstra, H. J. 1968. On sieving and settling techniques for sand analysis. *Sedimentology*, vol. 11, pp. 83-98. (67)

Shannon, P. M. and Naylor, D. 1989. *Petroleum Basin Studies*. Graham & Trotman, London, 206 pp. (6, 216, 224, 225, 238, 239, 241)

Shea, J. H. 1982. Uniformitarianism and sedimentology, *Journal Sedimentary Petrology*, vol. 52, pp. 701-702 and Reply to discussions in v. 53, (1983), p. 680. (5)

Shen, H. W. 1978. Sediment transport models. *In* A. D. Miall (ed.), *Fluvial Sedimentology*, pp. 49-60. Canadian Society of Petroleum Geologists, Calgary, 859 pp. (97)

Shrock, R. R. 1948. *Sequence in Layered Rocks*. McGraw-Hill, New York, 507 pp. (6, 41)

Siever, Raymond. 1983. The dynamic earth. *Scientific American*, vol. 249, pp. 30-39. (2)

Sindowski, F. K. H. 1949. Results and problems of heavy-mineral analysis in Germany; a review of sedimentary-petrological papers, 1936-1948. *Journal Sedimentary Petrology*, vol. 19, pp. 3-25. (13)

Simons, D. B., Richardson, E. V. and Nordin, C. F. Jr. 1965. Sedimentary structures generated by flow in alluvial channels. *In* G. V. Middleton (ed.), *Primary Sedimentary Structures and Their Hydrodynamic Interpretations*, pp. 34-52. Society of Economic Paleontologists and Mineralogists, Tulsa, OK., Spl. Pub. No. 12, 265 pp. (99, 100)

Simpson, S. 1975. Classification of trace fossils, pp. 39-54 *In* Frey, R.W. (ed.), *The study of Trace Fossils*. Springer-Verlag, New York, Heidelberg, Berlin, 526 p. (141)

Singh, A., and Bhardwaj, B. D. 1991. Fluvial facies model of the Ganga River sediments, India. *Sedimentary Geology*, vol.72, pp. 135-146. (185)

Singh, I. B. 1980. The Bijāigarh Shale, Vindhyan System (Precambrian), India–an example of a lagoonal deposit. *Sedimentary Geology*, vol. 25, pp. 83-103. (31, 194)

Singh, I. B. 1980a. Precambrian sedimentary sequences of India: their peculiarities and comparison with modern sediments. *Precambrian Research*, vol. 12, pp. 411-436. (194)

Singh, I.B. 1996. Geological evolution of Ganga plain. *Journal Palaeontological Society of India*, vol.41, pp. 99-137. (185)

Sinha, R. C. and Singh, R. M. 1964. Distribution and behaviour of trace elements during the diagenesis of Vindhyan sediments of the Son Valley near Robertsganj, U.P., India. *Report of the XXII International Geological Congress*, New Delhi, Part XV, pp. 205-225. (177)

Sippel, R. F. 1968. Sandstone petrology: evidence from luminescence petrography. *Journal Sedimentary Petrology*, vol. 38, pp. 530-554. (55)

Slingerland, R. L. and Williams, E. G. 1979. Paleocurrent analysis in light of trough cross-stratification geometry. *Journal Geology*, vol. 87, pp. 724-732. (284)

Sloss, L. L. 1988. Forty years of sequence stratigraphy. *Geological Society America Bulletin*, vol. 100, pp. 1661-1665. (267, 269)

Smith, A. E. Jr. 1966. Modern Deltas: Comparison Maps, pp. 233-251. *In* M.L. Shirley and A. Regsdale (eds.), *Deltas* (Appendix). Houston Geological Society, Houston, Texas, 251 pp. (204)

Smith, A. J. 1963. Evidence for a Talchir (Lower Gondwana) glaciation: striated pavement and boulder bed at Irai, Central India. *Journal Sedimentary Petrology*, vol. 33, pp. 739-750. (168)

Smith, D. G. and Smith, N. D. 1980. Sedimentology in anastomosing river systems: examples from alluvial valleys near Banff, Alberta, *Journal Sedimentary Petrology*, vol. 50, pp. 157-164. (184, 187)

Sneed, E. D. and Folk, R. L. 1958. Pebbles in the lower Colorado River, Texas - a study in particle morphogenesis. *Journal Geology*, vol. 66, pp. 114-150. (77, 78)

Southard, J. B. and Boguchwal, L. A. 1990. Bed configuration in steady unidirectional water flows. Part 2. Synthesis of flume data. *Journal Sedimentary Petrology*, vol. 60, pp. 658-679. (100, 101)

Srinivasan, R., Naqvi, S. M., Uday Raj, B., Subba Rao, D. V. and Balaram, V. 1989. Geochemistry of the Archaean graywackes from the northwestern part of the Chitradurga Schist Belt, Dharwar Craton, South India—evidence for granitoid upper crust in the Archaean. *Journal Geological Society India*, vol. 34, pp. 505-516. (222, 223, 224)

Srinivasan, R., and Ojakangas, R. W. 1986. Sedimentology of quartz-pebble conglomerates and quartzites of the Archean Bababudan Group, Dharwar Craton, South India: evidence for early crustal stability. *Journal Geology*, vol. 94, pp. 199-214. (19)

Srivastava, B. N., Rana, M. S. and Verma, N. K. 1983. Geology and hydrocarbon prospects of the Vindhayan basin. *In* L. L. Bhandari et. al. (eds.) Petroliferous Basins of India, vol. I, pp. 179-189. *Petroleum Asia Journal*, V-VI, 189 pp. (227)

Stanton, R. L. 1972. Ore *Petrology*. McGraw-Hill Book Co., New York, 713 pp. (51)

Stolum, Hans-Henrik. 1996. River meandering as a self organisation process. *Nature*, vol. 271, pp.1710-1713. (182)

Strahler, Alan, and Strahler, Arthur, 2002. *Physical Geography*. John Wiley & Sons (Asia) Pvt. Ltd., Singapore, 748 pp. (12)

Sundborg, Åke. 1956. The River Klarälven. A Study of Fluvial Process. *Geografiska Annaler*, 38, 2 : 127-316. (4, 90, 94, 97, 181)

Suttner, L. J., Basu, Abhijit and Mack, G. H. 1981. Climate and the origin of quartz arenites. *Journal Sedimentary Petrology*, vol. 51, pp. 1235-1246. (14)

Suttner, L. J. and Dutta, P. K. 1986. Alluvial sandstone composition and paleoclimate, I. Framework mineralogy. *Journal Sedimentary Petrology*, vol. 56, pp. 329-345. (29, 30)

Swift, D. J. P., Figueiredo, A. G. (Jr.), Freeland, G. L. and Oertel, G. F. 1983. Hummocky, cross-stratification and megaripples: a geological double standard? *Journal Sedimentary Petrology*, vol. 53, pp. 1295-1317. (117)

Syvitski, J. P. M. (ed.) 1991. *Principles, Methods and Application of Particle Size Analysis*. Cambridge University Press, Cambridge, 368 pp. (6)

Tada, Ryuji and Siever, Raymond. 1989. Pressure solution during diagenesis. *Annual Review Earth and Planetary Sciences*, vol. 17, pp. 89-118. (55)

Tandon, S. K. 1976. Siwalik sedimentation in a part of Kumaun Himalaya, India. *Sedimentary Geology*, vol. 16, pp. 131-154. (209)

Tandon, S. K. 1990. The Himalayan foreland: focus on Siwalik basin. *In* S. K. Tandon et al. (eds.), *Sedimentary Basins of India: Tectonic Context*, pp. 171-201, Gyanodaya Prakashan, Nainital. (209, 210)

Tandon, S. K., Andrews, J. E., Sood, A., and Mittal, S., 1998 Shrinkage and sediment supply control on multiple calcrete profile development: a case study from the Maastrichtian of Central India. *Sedimentary Geology*, vol. 119, pp. 25-45. (132)

Tandon, S. K., and Friend, P. F., 1989. Near-surface shrinkage and carbonate replacement processes, Arran Cornstone Formation, Scotland. *Sedimentology*, vol. 36, pp.1113-1126. (131)

Tandon, S. K., and Gibling, M. R., 1994. Calcrete and coal in Late Carboniferous cyclothems of Nova Scotia, Canada: climate and sea-level changes linked. *Geology*, vol. 22, pp. 755-758. (132)

Tandon, S. K., and Gibling, M. R., 1997. Calcretes at sequence boundaries in Upper Carboniferous cyclothems of the Sydney basin, Atlantic Canada. *Sedimentary Geology*, vol.112, pp. 43-67. (132)

Tandon, S. K., Kumar R. and Singh, P. 1985. Syntectonic controls of palaeoflow reversals and variability: sediment-vector sequences in the late orogenic fluvial Siwalik basin, Punjab, sub-Himalaya, India, *Sedimentary Geology*, vol. 41, pp. 97-112. (209)

Tandon, S. K. and Narayan, D., 1981. Calcrete conglomerate, case-hardened conglomerate and cornstone—a comparative account of pedogenic and non-pedogenic carbonates from the continental Siwalik Group, Punjab, India. *Sedimentology*, vol. 28, pp. 353-367. (131)

Terry, R. D. and Chilingar, G. V. 1955. Summary of 'Concerning some additional aids in studying sedimentary formations' by M. S. Shvetsov (in Russian). *Journal Sedimentary Petrology*, vol. 25, pp. 229-234. (59)

Thakur, V. C. and Virdi, N. S. 1979. Lithostratigraphy, structural framework, deformation and metamorphism of the SE region of Ladakh, Kashmir Himalaya, India. *Himalayan Geology*, vol. 9, pp. 63-78. (164)

Thiel, G. A. 1945. Mechanical effects of stream transportation on mineral grains of sand size (Abs.), *Geological Society America Bulletin*, vol. 56, pp. 1207. (13)

Trevena, A. S. and Nash, W. P. 1981. An electron microprobe study of detrital feldspar. *Journal Sedimentary Petrology*, vol. 51, pp. 137-150. (14)

Tripathi, S. C. and Lahiri, T. C. 2000. Marine oscillation event stratification: an example from Late Cretaceous Bagh carbonate sequence of Narmada Valley, India. *Memoir Geological Society India*, vol. 46. pp.15-24. (278)

Tucker, M. E., Wright, V. P. and Dickson, J. A. D. 1990. *Carbonate Sedimentology*. Balckwell Science, Oxford, 482 pp. (6, 35, 38, 43, 129, 159, 160)

Twenhofel, W. H. 1928. *Treatise on Sedimentation*. Williams and Wilkens, Baltimore, 2nd ed. (1932), 926 pp. (4, 6)

Twenhofel, W. H. 1950. *Principles of Sedimentation*. McGraw Hill, New York, 673 pp. (6)

Udden, J. A. 1914. The mechanical composition of clastic sediments. *Geological Society of America Bulletin*, vol. 25, pp. 655-744. (4, 62)

Vail, P. R. 1987. Seismic stratigraphy interpretation using sequence stratigraphy. Part I. Seismic stratigraphy interpretation. *In* A. W. Bally (ed.), *Atlas of Seismic Stratigraphy*, vol. I, *AAPG Studies in Geology*, vol. 27, pp. 1-10. (269)

Vail, P. R., Mitchum, R. M. and Thompson, III S. 1977. Seismic stratigraphy and global changes of sea level. *In* C. E. Payton (ed.), S*eismic Stratigraphy—Applications to Hydrocarbon Exploration*, pp. 83-97. American Association Petroleum Geologists, Memoir no. 26, Tulsa, OK. 516 pp. (256, 257)

Vail, P. R., Todd, R. G. and Sangree, J. B. 1977. Chronostratigraphic significance of seismic reflections. *In* C. E. Payton (ed.), *Seismic Stratigraphy—Applications to Hydrocarbon Exploration*, pp. 99-116. American Association of Petroleum Geologists, Memoir no. 26, Tulsa, OK. 516 pp. (256, 257)

Valdiya, K.S. 1969. Stromatolites of the Lesser Himalayan carbonate formations and the Vindhyans. *Journal Geological Society India*, vol.10, pp. 1-25. (139)

Valdiya, K. S. 1970. Simla Slates. The Precambrian flysch of the Lesser Himalaya, its turbidites, sedimentary structures and paleocurrents. *Geological Society America Bulletin*, vol. 81, pp. 451-468. (209)

van Houten, F. B. 1974. Northern Alpine molasse and similar Cenozoic sequences of Southern Europe. *In* R. H. Dott and R. H. Shaver (eds.), *Modern and Ancient Geosynclinal Sedimentation*, pp. 260-273. Society of Economic Paleontologists and Mineralogists, Spl. Pub. No. 19, Tulsa, OK, 380 pp. (209)

Vanoni, V. A., Brooks, N. A. and Kennedy, J. F. 1961. Lecture notes on sediment transportation and channel stability. Report No. KH-R-1, W. M. Keck Laboratory of Hydraulics and Water Resources, California Institute of Technology, Pasadena, California. (69, 82)

van Straaten, L. M. J. U. 1954. Composition and structure of Recent marine sediments in the Netherlands. *Leidse Geol. Medel.*, vol. 19, pp. 1-110. (4)

van Straaten, L. M. J. U. 1954a. Sedimentology of Recent tidal flat deposits and the psammites du Condroz (Devonian). *Geologie Mijnbouw*, vol. 16, pp. 25-47. (4)

Van Wagoner, J. C., Mitchum, R. M., Jr., Campion, K. M. and Rahmanian, V. D. 1990. Siliciclastic sequence stratigraphy in well logs, cones and concepts for high resolution correlation of time and facies. *AAPG Methods in Exploration* Series No. 7, 55 pp. (271, 272, 276)

Vine, J. D. and Tourtelot, E. B. 1970. Geochemistry of black shale deposits—A summary report. *Economic Geology*, vol. 65, pp. 253-272. (177)

Virdi, N. S. 1986. Indus-Tsangpo suture in the Himalaya: Crustal expression of a palaeo-subduction zone. *Annales Soc. Geol. Poloniae*, vol. 56, pp. 3-31. (164)

Visher, G. S. 1969. Grain-size distributions and depositional processes. *Journal of Sedimentary Petrology*, vol. 39, pp. 1074-1106. (68)

von Engelhardt, Wolf, 1977. *The Origin of Sediments and Sedimentary Rocks*. (trans. W. D. Johns). A Halsted Press Book. John Wiley & Sons, New York, 2nd rev. ed., 359 pp. (56)

Wadell, H. 1932. Volume, shape and roundness of rock particles. *Journal Geology*, vol. 40, pp. 443-451. (4, 61)

Wadell, H. 1935. Volume, shape and roundness of quartz particles. *Journal Geology*, vol. 43, pp. 250-280. (63)

Walcott, R. I. 1978. Geodetic strains and large earthquakes in the axial tectonic belt of North Island, New Zealand. *Journal Geophysical Research*, vol. 83, pp. 4419-4429. (218)

Walker, R. G. 1984a. General introduction: facies, facies sequences and facies models. In R. G. Walker (ed.). *Facies Models* (2nd ed.), pp. 1-10. Geological Association of Canada, Toronto, Ont., 317 pp. (147, 248, 259)

Walker, R. G. 1984b. Shelf and shallow marine sands. In R. G. Walker (ed.) *Facies Models* (2nd ed.), pp. 141-170. Geological Association of Canada, Toronto, Ont., 317 pp. (116, 163)

Walker, R. G. and Cant, D. J. 1984. Sandy fluvial systems. In R. G. Walker (ed.), *Facies Models*, 2nd ed., pp. 71-89. Geological Association of Canada, Toronto, Ont. 317 pp. (183, 185)

Walter, M.R. (ed.) 1976. *Stromatolites, Developments in Sedimentology*, vol.20, Elsevier Scientific Publishing Co., Amsterdam, 790 pp. (134)

Wasson, R. J. and Hyde, R. 1983. Factors determining desert dune type. Nature, vol. 304, pp. 337-339. (169, 172)

Watts, A. B., Karner, G. D. and Steckler, M. S. 1982. Lithospheric flexure and the evolution of sedimentary basins. *Philosophical Transactions Royal Society, London*, Ser. A. vol. 305, pp. 249-281. (256)

Wilcox, R. E., Harding, T. P. and Seely, D. R. 1973. Basic wrench tectonics. *American Association Petroleum Geologists Bulletin*, vol. 57, pp. 74-96. (219)

Weber, J. N., Williams, E. G., and Keith, M. L., 1964. Paleoenvironmental significance of carbon isotopic composition of siderite nodules in some shales of Pennsylvanian age. *Journal Sedimentary Petrology*, vol. 34, pp. 814-881. (131)

Weeks, L: G. 1952. Factors of sedimentary basin development that control oil occurrence. *American Association Petroleum Geologists Bulletin*, vol. 36, pp. 2071-2124. (298)

Weimer, R. J. 1976. Deltaic and shallow marine sandstones: Sedimentation, tectonics and petroleum occurrences. *AAPG Education Course Note Series* no. 2, 167 pp. (255)

Wentworth, C. K. 1922. A scale of grade and class terms for clastic sediments. *Journal Geology*, vol. 30, pp. 377-392. (4, 62)

Wernicke, B. P. 1981. Low angle normal faults in the Basin and Range Province; nappe tectoncis in an extending orogen, *Nature*, vol. 291, pp. 645-648. (215)

Wilson, I. G. 1972. Aeolian bedforms–their development and origins. *Sedimentology*, vol. 19, pp. 193-210. (169)

Wilson, J. L. 1975. *Carbonate Facies in Geologic History*. Spinger-Verlag, New York, 471 pp. (6, 149, 151, 154)

Winkelmolen, A. M. 1971. Rollability as a function of shape property of sand grains. *Journal Sedimentary Petrology*, vol. 41, pp. 703-714. (78)

Winkelmolen, A. M. 1972. Shape sorting in Lower Oligocene, North Belgium. *Sedimentary Geology*, vol. 7, pp. 183-227. (78)

Wolf, K. H. 1978. Limestones, pp. 434-446. In R. W. Fairbridge and J. Bourgeois (eds.) *The Encyclopedia of Sedimentology*. Dowden, Hutchison & Ross, Inc., Stroudsburg, 901 pp. (36)

Wright, V. P. and Marriott, S. B. 1993. Sequence stratigraphy of fluvial depositional systems—the role of floodplain sedimentary storage. *Sedimentary Geology*, vol. 86, pp. 203-210. (278)

Zenger, D. H. 1986. Lyell and Episodicity, *Journal Geological Education*, vol. 34, pp. 10-13. (5)

Zingg, Th. 1935. Beitrage zur Schotteranalyse. Min. Petrog. Mitt. Schweiz., vol. 15, pp. 39-140 (cited by Pettijohn, 1975). (4, 76, 77)

SUBJECT INDEX